Lecture Notes in Computer Science 13236

Services Science

Subline of Lectures Notes in Computer Science

More information about this series at https://link.springer.com/bookseries/558

Hakim Hacid · Monther Aldwairi ·
Mohamed Reda Bouadjenek ·
Marinella Petrocchi · Noura Faci ·
Fatma Outay · Amin Beheshti ·
Lauritz Thamsen · Hai Dong (Eds.)

Service-Oriented Computing – ICSOC 2021 Workshops

AIOps, STRAPS, AI-PA and Satellite Events
Dubai, United Arab Emirates, November 22–25, 2021
Proceedings

 Springer

Editors
Hakim Hacid ⓘ
Zayed University
Dubai, United Arab Emirates

Mohamed Reda Bouadjenek ⓘ
Deakin University
Waurn Ponds, VIC, Australia

Noura Faci ⓘ
University of Lyon
Villeurbanne, France

Amin Beheshti ⓘ
Macquarie University
Sydney, NSW, Australia

Hai Dong ⓘ
Royal Melbourne Institute of Technology
Melbourne, VIC, Australia

Monther Aldwairi ⓘ
Zayed University
Abu Dhabi, United Arab Emirates

Marinella Petrocchi ⓘ
National Research Council C.N.R.
Pisa, Italy

Fatma Outay ⓘ
Zayed University
Dubai, United Arab Emirates

Lauritz Thamsen ⓘ
Technical University of Berlin
Berlin, Germany

ISSN 0302-9743   ISSN 1611-3349 (electronic)
Lecture Notes in Computer Science
ISBN 978-3-031-14134-8   ISBN 978-3-031-14135-5 (eBook)
https://doi.org/10.1007/978-3-031-14135-5

This Springer imprint is published by the registered company Springer Nature Switzerland AG
The registered company address is: Gewerbestrasse 11, 6330 Cham, Switzerland

Preface

This volume presents the proceedings of the scientific satellite events that were held in conjunction with the 19th International Conference on Service-Oriented Computing (ICSOC 2021), held virtually during November 22–25, 2021. The satellite events provide an engaging space for specialist groups to meet, generating focused discussions on specific sub-areas within service-oriented computing, which contributes to ICSOC community building. These events significantly enrich the main conference by both expanding the scope of research topics and attracting participants from a wider community.

As is customary to ICSOC, this year, these satellite events were organized around three main tracks, a workshop track, a demonstration track, and a tutorials track, along with a PhD symposium.

The ICSOC 2021 workshop track consisted of the following three workshops covering a wide range of topics that fall into the general area of service computing:

- 2nd International Workshop on Artificial Intelligence for IT Operations (AIOps 2021)
- 3rd Workshop on Smart Data Integration and Processing on Service-based Environments (STRAPS 2021)
- 2nd International Workshop on AI-enabled Process Automation (AI-PA 2021)

This year in the workshop track, the theme of artificial intelligence and its applications in service computing was particularly noticeable. Workshops were selected based on the submission of a detailed description to the Conference Workshops Co-chairs. After a review of all the submissions, three workshops were selected by the Conference Workshops Co-chairs, in consultation with the General Co-chairs. All submitted papers to the workshops went through a heavy review process where each paper was reviewed by at least three members of the Program Committee of the workshop to which it was submitted. The Conference Workshops Co-chairs checked the assignments and the reviews before the final decisions were made. A total of 47 papers were submitted to the workshops and 20 were accepted, giving an acceptance rate of around 42%. The workshops were held on November 22, 2021, and included keynote talks from prominent speakers from industry and academia.

The PhD symposium is an international forum for PhD students to present, share, and discuss their research in a constructive and critical atmosphere. It also provides students with fruitful feedback and advice on their research approach and thesis. The PhD symposium was held over a half-day session and included four accepted papers. This year, and due to COVID-19, the conference supported all PhD students and their participation was fully free of charge.

The demonstration track offers an exciting and highly interactive way to show research prototypes/work in service-oriented computing and related areas. The demonstration track was held over a two-hour session for the presentations and then dedicated sessions for real-time demonstrations, running in parallel. Four demonstrations were accepted and presented during the conference.

ICSOC 2021 also featured four tutorials, held on the same day as the workshops, which offered comprehensive overviews and deeper insights on subjects ranging from AI-enabled processes and service-oriented architectures for blockchain to distributed IoT systems and service robots.

We would like to thank all the authors for submitting their work to the satellite events, as well as the various committee members, who together contributed to these important events of the conference. We hope that these proceedings will serve as a valuable reference for researchers and practitioners working in the service-oriented computing domain and its emerging applications.

July 2022

Hakim Hacid
Monther Aldwairi
Mohamed Reda Bouadjenek
Marinella Petrocchi
Noura Faci
Fatma Outay
Amin Beheshti
Lauritz Thamsen
Hai Dong

Organization

Organizing Committee

General Co-chairs

Hakim Hacid — Zayed University, UAE
Odej Kao — TU Berlin, Germany

Hakim Hacid	Zayed University, UAE
Odej Kao	TU Berlin, Germany

Program Co-chairs

Massimo Mecella	Sapienza Università di Roma, Italy
Naouel Moha	École de technologie supérieure, Canada
Helen Paik	University of New South Wales, Australia

Industrial Track Co-chairs

Jorge Cardoso	Huawei and University of Coimbra, Portugal
Anup K. Kalia	IBM T. J. Watson Research Center, USA

Workshop Co-chairs

Monther Aldwairi	Zayed University, UAE
Reda Bouadjenek	Deakin University, Australia
Marinella Petrocchi	Institute of Informatics and Telematics, CNR, Italy

Special Sessions and Tutorials Co-chairs

Amin Beheshti	Macquarie University, Australia
Zhihui Lv	Fudan University, China
Lauritz Thamsen	TU Berlin, Germany

Demonstrations Co-chairs

Hai Dong	RMIT, Australia
Yucong (Henry) Duan	Hainan University, China
Imran Junejo	Zayed University, UAE

PhD Symposium Co-chairs

Noura Faci	Claude Bernard Lyon 1 University, France
Honghao Gao	University of Shanghai, China
Fatma Outay	Zayed University, UAE

Sponsorship Co-chair

Haseena Al Katheeri	Zayed University, UAE

Finance Chair

Hakim Hacid   Zayed University, UAE

Local Arrangement Co-chairs

Ons Al-Shamaileh   Zayed University, UAE
Eleana Kafeza   Zayed University, UAE
Andrew Leonce   Zayed University, UAE

Publication Chair

Francis Palma   Linnaeus University, Sweden

Publicity, Website, and Social Media Co-chairs

Manel Abdellatif   Polytechnique Montréal, Canada
Andrew Leonce   Zayed University, UAE
Xiao Xue   Tianjin University, China
Sami Yangui   IRIT, France

ICSOC Steering Committee Representative

Michael Papazoglou   University of Tilburg, The Netherlands

PhD Symposium Program Committee

Angelo Spognardi   Sapienza University of Rome, Rome, Italy
Michela Fazzolari   IIT-CNR, Italy
Ilaria Matteucci   IIT-CNR, Italy
Manuel Pratelli   IMT Lucca School for Advanced Studies, Italy
Michele Starnini   ISI Foundation, Italy
Francesco Pierri   Politecnico di Milano, Italy
Vincenzo Ciancia   ISTI-CNR, Italy
Simone Raponi   CMRE, Italy
Abderrahmane Maaradji   ECE Paris, France
Amin Beheshti   Macquarie University, Australia
Helen Paik   University of New South Wales, Australia
Fatma Outay   Zayed University, UAE
Tetsuya Yoshida   Nara Women's University, Japan

Second International Workshop on Artificial Intelligence for IT Operations (AIOPS 2021)

Workshop Co-chairs

Roberto Natella   University of Naples, Italy
Jasmin Bogatinovski   Technical University of Berlin, Germany

Workshop Program Committee

Florian Schmidt	Technische Universität Berlin
Dušan Okanović	Novatec Consulting GmbH, Germany
Vladimir Podolskiy	Technical University of Munich, Germany
Philippe Fournier-Viger	Harbin Institute of Technology, China
Tse-Hsun (Peter) Chen	Concordia University, Canada
Dan Pei	Tsinghua University, China
Weiyi Shang	Concordia University, Canada
Ahmed E. Hassan	Queen's University, Canada
Steffen Becker	University of Stuttgart, Germany
Michael R. Lyu	Chinese University of Hong Kong, Hong Kong
Filipe Araujo	University of Coimbra, Portugal
Domenico Cotroneo	University of Naples Federico II, Italy
Anshul Jindal	Technical University of Munich, Germany
André van Hoorn	University of Hamburg, Germany
Gjorgji Madjarov	Ss. Cyril and Methodius University of Skopje, North Macedonia
Pietro Liguori	University of Naples Federico II, Italy
Sasho Nedelkoski	Technische Universität Berlin, Germany
Alexander Acker	Technische Universität Berlin, Germany

Third Workshop on Smart daTa integRation And Processing on Service-based environments (STRAPS 2021)

Workshop Co-chairs

Chirine Ghedira	IAE, Jean Moulin Lyon 3 University and LIRIS, France
Genoveva Vargas-Solar	CNRS, LIRIS, France
Nadia Bennani	INSA Lyon, LIRIS, France

Workshop Program Committee

Ali Akouglu	Arizona University, USA
Joao Batista Souza	IFRN, Brazil
Khalid Belhajjame	LAMSADE, Université Paris-Dauphine_PSL, France
Umberto Costa	UFRN, Brazil
Javier A. Espinosa Oviedo	ERIC, University of Lyon, France
Carmem Hara	Universidade Federal do Parana, Brazil
Michael Mrissa	University of Primorska, Slovenia
Alex Palesandro	D2SI, France
Pierluigi Plebani	Politecnico di Milano, Italy
Placido Antonio Souza Neto	IFRN, Brazil
Nicolas Travers	De Vinci Research Centre, France
José Luis Zechinelli Martini	Universidad de las Américas Puebla, México

Second International Workshop on AI-enabled Process Automation (AI-PA 2021)

Workshop Co-chairs

Amin Beheshti	Macquarie University, Australia
Boualem Benatallah	UNSW, Australia
Hamid Motahari	UpBrains AI, Inc., USA

Workshop Program Committee

Schahram Dustdar	TU Wien, Austria
Michael Sheng	Macquarie University, Australia
Fabio Casati	ServiceNow, USA
Aditya Ghose	University of Wollongong, Australia
Anup Kalia	IBM Research, USA
Jian Yang	Macquarie University, Australia
Mehdi Elahi	University of Bergen, Norway
Enayat Rajabi	Dalhousie University, Canada
Fabrizio Messina	University of Catania, Italy
Azadeh Ghari Neiat	Deakin University, Australia
Rama Akkiraju	IBM Watson, USA
Marcos Baez	University of Trento, Italy
Adrian Mos	NAVER LABS Europe, France

Contents

PhD Symposium

Demonstrations

Tutorials

AIPA

A Study of Artificial Intelligence Frameworks and Their Capability to Diagnose Major Depressive Disorder

Oluwafeyisayo Oyeniyi[1], Shreyansh Sandip Dhandhukia[2], Amartya Sen[1(✉)], and Kenneth K. Fletcher[2]

[1] Oakland University, Rochester, MI 48309, USA
{ooyeniyi,sen}@oakland.edu
[2] University of Massachusetts Boston, Boston, MA 02125, USA
{S.Dhandhukia001,kenneth.fletcher}@umb.edu

Abstract. The accurate diagnosis of mental illness is challenging because mental illness does not result in evident physical symptoms as compared to physical illness like the common cold. As a result, no definitive medical tests exist for mental illnesses. This situation is further aggravated by the fact that many of the symptoms of various mental illnesses overlap. Further, traditional means of mental care and therapy are not easily accessible to a majority of the population in developed and developing countries alike. In addition, openly discussing mental illness in major parts of society is still considered taboo. Therefore, a plausible way to improve mental illness diagnosis and address the aforementioned challenges is by using Artificial Intelligence (AI). This paper presents a comprehensive survey of AI-enabled approaches to Major Depressive Disorder (MDD) diagnosis and outlines some future research directions and challenges in this field. The paper also presents a preliminary system architecture of an AI-enabled approach to diagnose mental health with the objectives of making the underlying system more user-centric, scalable and accessible.

Keywords: Artificial intelligence · Explainable AI · Machine learning · Mental health

1 Introduction

A human being's good mental health is vital to ensure emotional, psychological, and social well-being, and any condition that affects this well-being is known as a mental disorder. One such disorder is Major Depressive Disorder (MDD), or simply depression. According to the World Health Organization (WHO)[1], symptoms of depression are characterized by an individual's lack of interest in previously enjoyable activities and persistent sadness.

Motivation: Depression is a leading cause of disability worldwide, with almost 75% of individuals suffering from MDD in developing countries who remain

---

[1] https://www.who.int/health-topics/depression.

© Springer Nature Switzerland AG 2022
H. Hacid et al. (Eds.): ICSOC 2021 Workshops, LNCS 13236, pp. 3–17, 2022.
https://doi.org/10.1007/978-3-031-14135-5_1

untreated and approximately 1 million susceptible cases lead to suicide [3]. In the United States (U.S.) alone, MDD affects more than 16.1 million American adults each year, which constitutes about 6.7% of the U.S. population aged 18 and older [3]. Additionally, due to the global pandemic from SARS-CoV-2 (COVID-19) in the year 2020 itself, depression diagnoses were up by 873% [17].

The demand for mental health services is rapidly increasing, as shown in a 2018 survey by National Council on Behavioral Health (NCHB)[2]. The survey concluded that at least 56% of people want access to mental health services but face many barriers. These barriers can be attributed to a lack of resources and/or trained healthcare providers, the social stigma associated with mental disorders, and inaccurate assessment[3]. Another survey in 2018 [14], showed that there is a shortage of mental health care professionals in every state across the U.S. This situation is further aggravated due to the high cost and insufficient insurance coverage for mental health conditions, leading to the difficulty of accessing mental health services by economically challenged population.

Nonetheless, most people suffering from mental illnesses have a prevalent behavior of wanting to be alone. Due to this isolation, mental disorder patients seek online venues like Twitter, Facebook, or Reddit to openly or anonymously share about discomforts and anxieties. This gives rise to data repositories comprising a variety of user-curated contents on social media platforms such as personal status, user's pictures, and geo-location, which can be mined for information. While humans have limited capacity to learn, an AI actuated system can easily access thousands of medical information sources and help in the early detection of chronic mental health diseases in patients.

Background: The Need for AI. Traditional means of providing screening (diagnosis) and monitoring, although good, are not sufficient today due to two different reasons. First is the Big Data Connection. The information generated from pervasive devices with Internet connectivity and social media platforms can aid in detecting and monitoring depressive behavior. However, analyzing Big Data generated from online platforms is not supported by traditional care and therapy. Second is the ease of accessible care and constant monitoring of MDD patients. Traditional means cannot address the accessibility issue and provide constant monitoring in contrast to AI-enabled frameworks to address mental well-being. So it is necessary to support and extend the methods of traditional care by using AI-based frameworks to leverage the advancement in computer technology for social good.

A diverse technical solution has been adopted and implemented to help solve the limited access to a medical professional. The solutions range from chatbots, IoT/Wearable devices, mobile and web applications to behavioral technologies [1,7,8,11,13,18]. These solutions have helped to significantly increase the access to the professional help needed for individuals suffering from MDD. The nature of these solutions makes them easily accessible, thereby reducing the stigma that is associated with MDD. Further, The type of study conducted in this paper

---

[2] https://rb.gy/8hpftt/.
[3] https://rb.gy/zgznyj.

will also benefit other domains like Analysis of Behavioral Disorders in business processes such as banking recommender systems and cognitive recommender systems [4,5].

Objectives and Contributions: Given the importance of AI-enabled frameworks towards diagnosing and monitoring MDD, this work surveys and discusses the different AI-enabled approaches to MDD that have been proposed in the recent literature. The contributions of this work are as follows:

- Review existing literature on AI-enabled frameworks for diagnosing mental health illnesses and summarize existing research challenges and some future directions.
- Propose a decentralized system design of an accessible and scalable AI-enabled approach for diagnosing mental health illnesses.

2 AI-Enabled Approaches

The **systematic search strategy** along with some exclusion criteria were as follows. The paper identified related works based on their solutions that addressed features like reliability, accuracy, accessibility, and explainability. The works prior to 2015 was excluded along with any work that had no testing or implementation, or if the data was not gathered from a valid source like health agencies or social media. For recent literature surveyed belonging to each domain, the paper presents a discussion on their algorithmic description, and the input and output type.

2.1 Expert System and Fuzzy Logic

In [2], the authors proposed an expert system that can be used to diagnose depression in an individual. Due to the nature of depression diagnosis, the expert system would help the psychologist to appropriately diagnose an individual. The expert system was created using Simpler Level 5 (SL5) object language, with its engine implemented in Delphi Embarcadero RAD Studio XE6. The proposed expert system interacted with the human subjects by asking them a set of Boolean questions. Based on the answers chosen, the expert system outlined a diagnosis and recommendation to the user. The knowledge base for the expert system was created with the information documented from experienced psychologists and specialized websites for depression. The proposed expert system thereafter was evaluated by an experienced psychologist, who verified the correctness of the system output. The benefits of the expert system can be categorized by its ease of use. Further, the system can be deployed as a standalone system, without the need for an intervention from a medical professional.

The authors in [18] designed a web-based expert system using fuzzy logic to diagnose individuals suffering from depression and determine the level of severity. The system was designed to be user-centric; it could be implemented in specialized settings such as in a psychiatric office as well as being used by the user solely.

Fuzzy logic was used to address the uncertainty or ambiguity present in human knowledge and the decision-making process. Knowledge was acquired from several resources and professionals, with the Diagnostic and Statistical Manual of Mental Disorders (DSM-V) being the main source. The scale for the severity of depression disorder used was "very low, low, medium, high and very high". The expert system was implemented in Jess. The system's predictive results were evaluated by psychological consultants who verified the accuracy of the result. As reported by the authors, their proposed system also achieved a 79% and 98% outcome for the metrics of sensitivity and specificity, respectively.

Using Fuzzy logic gives allowance for uncertainty in decision making which makes it flexible and easily adaptable to different scenarios. The level of uncertainty or vagueness present in the decision-making process makes the preciseness and accuracy of the results obtained questionable.

2.2 Machine Learning Approaches

The current trends in digitization have penetrated many industries including healthcare. The concurrent growth in medical demands and improved efficacy of machine/deep learning algorithms can help medical institutions and their professionals to detect early signs of chronic diseases, which helps to reduce the time and lower the costs of treatment. This section summarizes the various Machine Learning (ML) based frameworks that have been proposed. The literature within this category is further analyzed based on sub-categories like the type of ML algorithms ((un)supervised, semi-supervised, and reinforcement learning), along with the proposed model's respective algorithmic input and output. Although, a net total of about 15 works have been reviewed in this category (Sect. 4), owning to page limitations, in this section, the description of only a select few notable papers have been presented. For a complete description of the reviews, readers are kindly referred to the following document - https://rb.gy/ajeq5q.

Supervised Learning. Machine learning approaches that belong to this algorithmic category utilizes historical and classified input and output to train themselves and facilitate the intended functionalities.

Summary: To begin with, one such literature that belongs to this algorithmic category of supervised learning is [19]. It discusses the prevalence of Major Depressive Disorder (MDD) and Generalized Anxiety Disorder (GAD) in youth attending universities. The authors state that if such issues are not diagnosed at an early stage then it leads to drinking-related harm and alcohol abuse. The paper [19] proposes a machine learning model that predicts if an individual suffers from MDD and GAD by creating a so-called Electronic Health Records (EHR) data set. The EHR comprised of an undergraduate student's biometric and demographic data with the exclusion of all psychiatric features to preserve privacy. An ensemble algorithm consisting of Support Vector Machine (SVM), XG Boost, K-Nearest Neighbor (KNN), Random Forest, Logistic Regression, and Neural Network was deployed using Bayesian hyperparameter optimization.

Algorithmic Input: The input to the machine learning model consisted of an individual's physiological data like blood pressure, Body Mass Index (BMI), heart rate, and demographic data like housing status, health insurance, and age. The authors used non-psychological factors such as age, housing zip code, and health insurance from Electronic Health Record (EHR) data set to predict if a college student is suffering from Major Depressive Disorder (MDD) or General Anxiety Disorder (GAD), and explain the model's result.

Algorithmic Output: As per the author's reporting, the ensemble model achieved 70% accuracy and some of the top features were satisfaction with living conditions, public health insurance, and parental home. XG Boost classifier was used to predict an individual status if they are suffering from depression or not. Overall, the Area Under Curve (AUC) scores obtained from each model in the ensemble model indicated the model accurately predicted an individual's state, and also Shapely Additive Explanations (SHAP) was used to explain how the importance of each feature affects the overall result of a model's performance. The obtained results allowed the authors to validate that depression can be diagnosed using non-psychiatric features.

Summary: In [12], the authors have proposed the incorporation of digital intervention in predicting depression and other forms of anxiety symptoms. The authors aimed to evaluate the improvement experienced by a patient who received care via digital intervention after hospital discharge from illness around work-related stress. This research was conducted in a randomized controlled trial of 632 people who received the digital intervention. This group of users was split into two groups - a group using the digital interventions for their follow-up and the other group receiving information from a professional. This study was done for 9 months after the hospital's discharge.

Algorithmic Input: The authors implemented an ensemble model which had 2 layers to analyze the data. The first layer, known as the base model consisted of ridge regression, random forests, general linear models, Gaussian process, support vector machines, K-nearest neighbors. The second layer was the averaging layer, which took as input the mean prediction for a given subject across the different models and validation folds.

Algorithmic Output: Based on the output of the results, the authors stated that their ML model predicted a change in the depression of a person. The authors also stated that the ML model could be capable of guiding a clinician to know the best way to allocate resources in caring for a user, either using a higher level of traditional care or a low-level resource consisting of digital intervention.

Summary: The authors in [8] aimed to evaluate chatbots utilization by users. Tess is the chatbot that was used as a case study in this paper. Tess deciphers a user's emotional needs through the content of their conversation. Tess has 12 modules and its questionnaires or content spans across different areas of interest such as self-compassion, cognitive distortion, coping statements, and so on. These areas make Tess a robust chatbot but also present a herculean task for the users to go through the modules in good time.

Algorithmic Input: Tess was deployed on Facebook messenger with anonymous data collection. The authors aimed to analyze the usage of Tess, understand user's flows between the various modules and the usage of each module.

Algorithmic Output: From the analysis carried out, the authors showed that a chatbot is an effective tool for mental health, just the major modules required to assist a user should be implemented in a chatbot, and the overall usage across the different modules was based on some factors such as the questions asked and the time required to complete a module.

Summary: In this work [1], the authors developed a depression diagnosis algorithm using an individual's sequence of responses to a virtual agent and determined their depression severity.

Algorithmic Input: This research involved diagnosing depression through sequencing of responses obtained from a user using audio and text features. The responses were categorized as responses to specific questions asked from the Patient Health Questionnaire (PHQ) and responses that are not dependent on the questions asked. The authors carried out this experiment using three different scenarios to evaluate the effectiveness of their model. In the first scenario called the context-free modeling, a regularized logistic regression was deployed and the time a question was asked was the key factor and not the question type. In the second scenario called weighted modeling, a regularized logistic regression was deployed and the question type, not the timing of the question was the key factor. In the third scenario called sequence modeling, a bi-directional Long Short-Term Memory (LSTM) was implemented.

Algorithmic Output: The authors discovered that in the cases of context-free modeling and sequence modeling, the text features gave a better result. Whereas, in the case of weighted modeling, the audio features performed better. Overall, the best result was obtained from the weighted modeling for both the text and audio data.

Unsupervised Learning is a type of machine learning technique where the label is not pre-defined but determined by the clustering of a set of data points. The labels are defined by the patterns discovered in the data set. In this subsection, the paper presents a discussion on existing works that fall in the category of unsupervised learning to address Major Depressive Disorder.

Summary: In [21], the authors implemented an unsupervised machine learning algorithm using K-Means Clustering to predict if an individual is suffering from depression along with predicting their depression severity. K-Means clustering uses the position of each data point instead of the summary of data points, thereby making its prediction output more holistic in nature. Moreover, the model's results were compared with the results from a traditional norm-based classification to evaluate the model's performance. Middle and High school Chinese students were the focus of this study.

Algorithmic Input: The K-Means clustering was implemented in a 13-dimensional space with 13 features obtained from Beck Depression Inventory

(BDI) questionnaire. The clusters centers were determined using the maximum and minimum points. The authors classified the severity of depression as none, mild, moderate, and severe.

Algorithmic Output: The model was able to correctly predict if an individual was suffering from depression, and when compared to the traditional norm-based classification models, the model had higher accuracy and AUC score.

Semi-supervised Learning is a combination of supervised and unsupervised learning. In semi-supervised learning, a small amount of labeled data is used along with a large amount of unlabeled data, which can be helpful when there is an unavailability of a large amount of labeled data.

The research work presented in [22] is an example of semi-supervised learning wherein the research objective was to monitor the clinical depressive symptoms from Twitter data that imitates the PHQ-9 survey used by clinicians. The authors proposed two approaches to detect the possibility of users suffering from depression. The former was a bottom-up approach where authors used distributional semantics to unwrap the symptoms of depression. The first approach is based on Latent Dirichlet Allocation (LDA), which is specifically unsupervised learning, views tweets as a mixture of latent topics, where a topic is a distribution of co-occurring words. However, the topics learned by LDA are not specific enough to correspond to depressive symptoms. The latter approach was based on a top-down approach where the authors added supervision to LDA by using a probabilistic topic modeling approach, named semi-supervised topic modeling over time (ssToT).

Algorithmic Input: The authors collected data of 45000 Twitter users with self-declared depression and another 2000 tweets of undeclared users. After the examination, it was seen that most users talked about their family and companion issues and the requirement for their help. The authors compared the ssToT learned topics with existing semi-supervised and unsupervised learning approaches such as k-means clustering, LSA, LDA, BTM, Partially Labeled LDA. These experiments suggested that ssToT outperformed all the state-of-art models paying little heed to the corpus that probabilities are acquired from. ssToT model was also tested as a multi-label classifier, using 10400 tweet dataset in 192 buckets. Each bucket contains tweets that are posted by the client inside a range of 14 days. The ssToT model was trained on an unlabeled data set and the performance was estimated by utilizing the labeled data-set.

Algorithmic Output: Accuracy and precision were measured for the ability to predict the presence of 9 depressive symptoms (in compliance with PHQ9) and they were found to be 0.68 and 0.72 on average. The obtained results suggest that semi-supervised topic modeling overtime was successfully able to capture depression symptoms from the Twitter data-set which was competitive with a fully supervised approach.

Reinforcement Learning enables agents to learn by their interaction with the environment by trial and error using feedback from past actions and experiences.

This technique is based on rewards and punishment based on the agent's set of actions to perform a task.

Summary: In [6], the authors aim to understand the relations between model-derived reinforcement learning parameters with the depression symptoms and symptom change after the treatment. Studies were conducted on 101 adults of which 68.3% were females. The authors also assessed the changes in model-learning parameters and symptoms after some of the participants received Cognitive Behavioral Therapy (CBT).

Algorithmic Input: The participants responded to the Mood and Anxiety Symptom Questionnaire (MASQ), a validated self-report measure of the symptom of anhedonia, negative affect, and arousal, as well as the Beck Depression Inventory to assess overall depression severity, the Wechsler Test of Adult Reading to estimate verbal IQ, and a demographic questionnaire. Out of 101 participants in the study, a total of 69 participants suffered from MDD and 32 participants had no history of MDD. The computational model analysis of behavior choices and neural data identified contemporary learning with symptoms during reward and loss learning.

Algorithmic Output: The results showed that during reward learning, the reward values increased slowly with increased anhedonia. The anhedonia was associated with model-derived parameters such as learning rate, outcome sensitivity, and neural signals. The results during the loss function manifested that learning parameter was associated with negative affect were found to be outcome shift, and the model-learning parameter associated with disrupted neural encoding of learning signals. After mapping the reinforcement learning model parameters with the symptoms of MDD it was observed that after CBT the features associated with the symptoms had shown possible learning-based therapeutic processes.

3   Explainable AI (XAI)

The recent advances in the utilization of AI for the medical field have been restricted due to the absence of interpretation for complicated models. This decreases the trust of clients when it comes to the output generated by AI-enabled models. The AI models that explain its outcomes are better known as explainable AI (XAI) and can be classified into two major categories [15]. First, as Ante-hoc, where the framework incorporates logic in the model. Second, as Post-hoc, where the frameworks utilize a more simple model to clarify a black-box model. In the following section, the paper discusses some of the notable literature from the mental health domain that proposes the usage of XAI.

Summary: In [24], authors propose to detect the early signs of depression from users' social media posts. They combined XAI with natural language processing (NLP). The XAI model used Local Interpretable Model-Agnostic Explanations (LIME) [9] for interpretation. LIME is a local model interpretation method utilizing local surrogate models to surmise forecasts of black-box models. The authors observed that LIME interpretation was sensitive to pre-processing of the data set, especially the choice of whether to remove stop-words from the data set. The authors removed stop-words from the data set to study the behavior of occurrence of personal pronouns in the depression data set.

Algorithm Input: consisted of Urdu and English text data from the social media platform Reddit and applied NLP algorithms to predict if a user is suffering from depression. The author used bag-of-Words (BoW) and term-frequency times inverse document-frequency (TF-IDF) features and used them in machine learning classifiers like Logistic Regression and Random Forest classifiers.

Algorithm Output: for the author's experimental setup, the Logistic regression classifiers performed better with an F1 score of 0.89 for TF-IDF features as compared to the Random Forest classifier (F1 score of 0.84).

Summary: In [23], the framework proposed by the authors incorporated the SHapely Additive exPlanations (SHAP) [16] for the purpose of model interpretation. SHAP uses values that are an average of marginal contributions of a single feature value across all possible partnerships of the features. It helps to represent the impact certain features have on the model outcome with which these shapely values are associated. The authors in [23], used these shapely values to assess and foster a novel AI approach for predicting mental health risk in individuals with diabetes mellitus.

Algorithm Input: Data was collected from 142,432 people suffering from diabetes, and their mental health status was verified using two sources - claims data and medication prescription data. The participant's behavior was classified into four categories including demographics and glucometer, coaching, and event data. Data sets were then collected to make member period occurrences, and descriptive analyses were performed to comprehend the connection between psychological well-being status and passive sensing signals. The model used for training the data set was an ensemble of ten LightGBM models and it was evaluated using sensitivity, specificity, precision, the area under the curve, F1 score, accuracy, and confusion matrix.

Algorithm Output: The outcome as reported by the authors showed that the proposed model performed with a score of more than 0.5 for sensitivity, specificity, AUC, and accuracy. The SHAP values determined the elements that offer more towards the classification output, such as demographics (race and gender), participant's emotional state during blood glucose checks, time of day of blood glucose checks, and blood glucose values. The authors were able to effectively anticipate the mental health risk in individuals experiencing diabetes and showed the component significance of elements that contributed most towards the classification output of the model.

Summary: The authors in [20] introduced a framework (What-If tool or WIT) that can visually analyze the ML systems. The WIT tool supports both local and global interpretation, that is it can analyze a local data point as well as analyze model behavior across the whole data set. This framework allows the users to find the nearest counterfactual to better understand the model's behavior. The tool also enables users to analyze the relationship between the features and the predicted output using partial dependence plots, which helps to better understand what features contribute more to the prediction outcome.

Algorithm Input: The authors gave three contextual analyses to show the utilization of WIT to analyze the model performance. The first study, directed by ML specialists, utilized the WIT to analyze regression models. A sample of 2000 data-set was stacked in the regression model. In the second study, a programmer of a large technology company used WIT to predict the health metric for clinical patients. WIT coupled with two regression models tracked down that the inference results and the data-point visualization results were bigger for the first regression model than the second. More interestingly, it was observed that there was a bug in the calculation that caused the first and the last value of the input feature to trade. The error matrix was sensible for the model, but the model was tackling an alternate issue because of this bug. Finally, the third study was led by computer science students from M.I.T examining the legitimacy of the stop-and-frisk practices of the Boston police department. They used a data set of 150,000 records and prepared a linear classifier model. The data-set had features like age, race, sex, and criminal record of an individual being halted by police alongside the date and location of the offender.

Algorithm Output: The first study observed that the partial dependence plot was flat for some features of every data point. In the second case study, the WIT tool found a bug in the software that prevented a certain feature to be fed to the model, which resulted in a wrong decision. The output of the classifier in the third study was either positive or negative i.e. if the offender was searched or not searched during the confrontation. This outcome was coupled with WIT that identified features like age, gender, and race played a key role in determining whether the offender was frisked or not.

4   Discussions

In this section, Fig. 1 and 2 summarizes the literature classified using different categories like machine learning algorithm types, input data type, and so forth. The numerical values in these illustrations represent the number of reviewed papers that belong to a respective category. Further discussions presented in this section illustrate some of the design challenges identified in this field, the scope of future directions, and the tentative design of an AI system to address mental health illnesses keeping in mind computational design parameters like accessibility, efficiency, and effectiveness.

4.1   Lessons Learned: Current Challenges and Future Directions

An important finding of this survey is that majority of the AI-enabled approaches belong to the category of Machine or Deep Learning frameworks (ML). This is

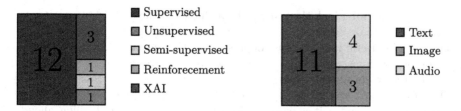

Fig. 1. Literature classification - machine learning algorithm types

Fig. 2. Literature classification - algorithm input types

owing to the dynamic nature of the information associated with depression and the growing number of patients, which makes fuzzy and expert systems an ill-fit in terms of computational design criteria. The process of developing expert systems or fuzzy logic-based systems requires extensive information gathering, which can inherently be tedious. Thereafter, these systems require a knowledge base, to be developed from the gathered information. These knowledge bases can quickly become outdated and in addition, are challenged with issues of scalability and accuracy. Therefore, the majority of the existing works use ML frameworks since a ML model can be built using a lot of data making it robust and can be retrained with new data, making it accurate, scalable, and up-to-date. A model can be easily trained and deployed quickly as the professional input needed can be obtained from a variety of sources in a short time.

The second finding from the survey is that, within the domain of ML frameworks, the majority of the existing works incorporate the supervised learning methodology compared to unsupervised, semi-supervised, or reinforcement learning methods. The reason behind the popularity of supervised learning can be attributed to a few cases. First, in the mental health research domain, the majority of the AI-enabled frameworks consider the problem (e.g. MDD prediction) as binary classification - depressed or not depressed. This makes labeling easier when compared to say unsupervised learning, which tries to infer its clusters based on the data set which can be challenging at times. Hence, there is a preference for supervised learning frameworks, where the classes are already defined and the models know what they are looking for. Secondly, in these domains, the availability of a reliable, accurate, and a reasonably substantial amount of data is a bottleneck. This bottleneck proportionately impacts the performance of any ML-based AI framework. Supervised learning when compared to unsupervised learning does not require too much data. Third, given the current state of the art of AI systems, in the medical field, it is still favorable to have a human expert intervention. Supervised learning allows this process at an early stage i.e. when the models are being trained and thereby allowing for more accurate models when compared to unsupervised or semi-supervised learning methods.

Additionally, the reinforcement learning framework is not best suited for this nature of problem i.e. depression diagnosis. In reinforcement learning, the agent goes through different states, and depending on the outcome, they either get rewarded or punished. Due to the dynamic nature of how depression can be diagnosed in an individual, there might be a chance of the agent not prioritizing a particular state or discounting a state because it does not seem like a 'norm' when it can be useful in diagnosing depression in the individual. Also, there is a question of how long would the agent run before it arrives at a decision and can prove the result obtained is valid.

The functional benefits of supervised learning algorithms over their counterparts currently make it a more preferable choice. However, looking ahead, researchers in this domain need to assess an important question - if there is a paradigm shift towards unsupervised (or semi-supervised) learning algorithms, what kind of benefits and challenges will be encountered? To begin with, the benefits of such a paradigm shift will be in the ability to identify patterns that exist in diagnosing depressive disorders that are currently unknown even within medical professionals. With this in mind, the outcome of the AI models no longer needs to be a simple yes or no classification. The ranking of depression severity would become more flexible as these algorithms will be able to analyze patterns, relationships, and causality that exist in the data points. This will help to identify features that are easily susceptible to a depression level in an individual.

However, the primary challenge will be in terms of data gathering. The unsupervised and semi-supervised models require a lot of data. Thereafter, functionally designing these frameworks will also be challenging. An imperative question in this regard is - as the patterns are being established across the data points, if the first data set classifies individuals suffering from depression, would the second iteration from the result obtained from the first iteration be used to evaluate depression severity? Researchers will also need to address the issues in a model's bias. In case a data set is obtained from a set of the population (say, individuals attending college), it is possible the features used to predict depression would be different from another set of the population (say, middle-aged career individuals). As such, one will not be able to design a one-size-fits-all framework.

In terms of data collection, currently, the common means through which most AI-enabled technologies gather data is through text. This text could be through social media sites or users answering questionnaires delivered through chatbots. According to the investigations done in this field, a person's social media activity via content posted such as the kind of words used and the frequency of the words used can serve as an indicator for recognizing if a person is suffering from depression. Additionally, the use of biological data is an effective way of diagnosing individuals who are suffering from depression due to the information that can be easily seen when the data is being read. Moreover, using non-psychiatric features to predict depression is also a helpful way of predicting if a user is suffering as it helps to minimize the stigma that is attached to a depression diagnosis. It also helps in estimating or predicting the likelihood of a certain class of individuals suffering from depression if certain features are identified.

Finally, the researchers in this domain need to keep in mind that the acceptance of their proposed framework by the end-users can only come through model outcomes that are trustworthy and reliable. This can be achieved by making the proposed frameworks more transparent to end-users. These requirements necessitate the integration of explainable AI (XAI) frameworks to some degree. The incorporation of XAI frameworks will allow end-users with or without prior experience in the domain to be able to accept (trust) or reject a prediction if they understand the reasoning behind it. For example, if a system predicts if a person suffers from a disease, the expert (doctor) can see the symptoms that contribute to the prediction to either accept or reject a prediction.

4.2 Proposed AI System Design

In this section, based on the challenges discussed in the previous section, the paper presents a system architecture of an AI-enabled approach for the diagnosis of mental health disorders like MDD. The goal of this work in progress is to design a decentralized system that will be capable of leveraging different AI-enabled approaches and work with different input data types with an explanation of the generated output. The motivation behind such a computational design is driven by the need for increased scalability and accessibility by improving framework flexibility, which has not been addressed in the existing frameworks. The system design for the proposed AI framework is illustrated in Fig. 3.

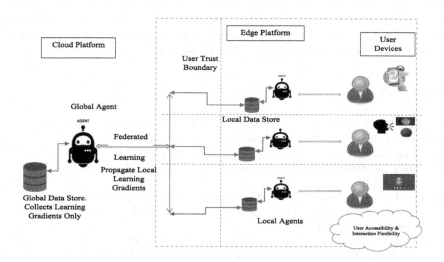

Fig. 3. Proposed system architecture for AI framework

As shown in Fig. 3, the system is divided into three distinct platforms. First, the user-centric platform consisting of various user devices like cellphones, tablets, or desktops. Second, the Edge platform will host a local agent (or LAI), which will improve the quality of service and preserve the privacy of user data. These local agents will learn from the data collected through the interactions with the users. A user will have the flexibility of interacting in any way that they want - either through text chats, audio interactions, or audio-video interactions.

Depending on the user's choice, a suitable local agent will be deployed ad-hoc to their Edge platform. The Cloud platform will host the global agent (or GAI), whose objective will be to synchronize the activities between different local AI agents hosted on various Edge Platforms. The learning on the Cloud platform will be done by receiving the transmitted learning gradients from the local agents instead of actual user data thereby improving user privacy and incorporating the concepts of federated machine learning. Given the decentralized and distributed nature of the proposed framework, it can then be dispensed as-a-service on an ad-hoc basis with applications similar to IBM Watson's Personality Insights [10] service but more geared toward diagnosing depression.

5   Conclusion

In this paper, a comprehensive survey was presented for AI-enabled approaches proposed in the literature for diagnosing mental illnesses like depression. The survey was carried out by categorizing the various studies by their incorporated methodology like Expert Systems, Fuzzy Logic, or Machine Learning ((un-)supervised, semi-supervised, reinforcement). The paper also summarized the survey by presenting some of the existing challenges in this research domain, the scope of future work, and a design schema of an AI system that can address some of the required computational design requirements for this research area along the lines of accessibility, scalability, privacy, and quality of service.

References

1. Alhanai, T., Ghassemi, M., Glass, J.: Detecting depression with audio/text sequence modeling of interviews. Interspeech, September 2018
2. Alshawwa, I.A., Elkahlout, M., El-Mashharawi, H.Q., Abu-Naser, S.S.: An expert system for depression diagnosis. Int. J. Acad. Health Med. Res. (IJAHMR) **3**, 20–27 (2019)
3. Anxiety and Depression Association of America: Understand anxiety and depression, facts and statistics. https://adaa.org/understanding-anxiety/facts-statistics. Accessed 10 Sept 2021
4. Beheshti, A., Moraveji-Hashemi, V., Yakhchi, S., Motahari-Nezhad, H.R., Ghafari, S.M., Yang, J.: personality2vec: enabling the analysis of behavioral disorders in social networks. In: The Thirteenth ACM International Conference on Web Search and Data Mining (WSDM 2020) (2020). https://doi.org/10.1145/3336191.3371865
5. Beheshti, A., Yakhchi, S., Mousaeirad, S., Ghafari, S.M., Goluguri, S.R., Edrisi, M.A.: Towards cognitive recommender systems. Algorithms **13**(8), 176 (2020). https://doi.org/10.3390/a13080176. http://www.mdpi.com/journal/algorithms
6. Brown, V.M., et al.: Reinforcement learning disruptions in individuals with depression and sensitivity to symptom change following cognitive behavioral therapy. JAMA Psychiatry **78**(10), 1113–1122 (2021)
7. Cupkova, D., Kajati, E., Mocnej, J., Papcun, P., Koziorek, J., Zolotova, I.: Intelligent human-centric lighting for mental wellbeing improvement. Int. J. Distrib. Sens. Netw. **15**(9), 1550147719875878 (2019)

8. Dosovitsky, G., Pineda, B.S., Jacobson, N.C., Chang, C., Escoredo, M., Bunge, E.L.: Artificial intelligence chatbot for depression: descriptive study of usage. JMIR Form. Res. **4**(11), e17065 (2020). https://doi.org/10.2196/17065
9. Guestrin, C., Singh, S., Ribeiro, M.T.: Why should i trust you? Explaining the predictions of any classifier. ACM, August 2016
10. IBM Watson Labs. https://cloud.ibm.com/docs/personality-insights?topic=personality-insights-about. Accessed 26 Aug 2021
11. Inkster, B., Sarda, S., Subramanian, V.: An empathy-driven, conversational artificial intelligence agent(Wysa) for digital mental well-being: real-world data evaluation mixed-methods study. JMIR Mhealth Uhealth **6**(11), e12106 (2018)
12. Jacobson, N.C., Nemesure, M.D.: Using artificial intelligence to predict change in depression and anxiety symptoms in a digital intervention: evidence from a transdiagnostic randomized controlled trial. Psychiatry Res. **295**, 113618 (2021). https://doi.org/10.1016/j.psychres.2020.113618
13. Johansson, R., Andersson, G.: Internet-based psychological treatments for depression. Expert Rev. Neurother. **12**(7), 861–870 (2012). https://doi.org/10.1586/ern.12.63
14. Kaiser Family Foundation: Mental health care health professional shortage areas (HPSAs). https://www.kff.org/other/state-indicator/mental-health-care-health-professional-shortage-areas. Accessed 10 Sept 2021
15. Khedkar, S., Subramanian, V., Shinde, G., Gandhi, P.: Explainable AI in healthcare. In: ICAST (2019)
16. Lundberg, S.M., Fischer, A., Holt-Gosselin, B., and L.W.: A unified approach to interpreting model predictions. In: NIPS, November 2017
17. Mental Health America: 2021: Covid-19 and mental health: A growing crisis. https://mhanational.org/sites/default/files/Spotlight2021-COVID-19andMentalHealth.pdf. Accessed 16 Feb 2021
18. Mohammadi Motlagh, H.A., Minaei Bidgoli, B., Parvizi Fard, A.A.: Design and implementation of a web-based fuzzy expert system for diagnosing depressive disorder. Appl. Intell. **48**(5), 1302–1313 (2017). https://doi.org/10.1007/s10489-017-1068-z
19. Nemesure, M.D., Heinz, M.V., Huang, R., Jacobson, N.C.: Predictive modeling of depression and anxiety using electronic health records and a novel machine learning approach with artificial intelligence. Sci. Rep. **11**, 1980 (2021). https://doi.org/10.1038/s41598-021-81368-4
20. Wexler, J., Pushkarna, M., Bolukbasi, T., Wattenberg, M., Viegas, F., Wilson, J.: The what-if tool: interactive probing of machine learning models. IEEE Trans. Vis. Comput. Graph. **26**(1), 56–65 (2019)
21. Yang, Z., Chen, C., Li, H., Yao, L., Zhao, X.: Unsupervised classifications of depression levels based on machine learning algorithms perform well as compared to traditional norm-based classifications. Front. Psychiatry **11**, 45 (2020). https://doi.org/10.3389/fpsyt.2020.00045
22. Yazdavar, A.H., et al.: Semi-supervised approach to monitoring clinical depressive symptoms in social media. In: IEEE/ACM Advances in Social Networks Analysis and Mining, July 2017
23. Yu, J., Chiu, C., Wang, Y., Dzubur, E., Lu, W., Hoffman, J.: A machine learning approach to passively informed prediction of mental health risk in people with diabetes: retrospective case-control analysis. JIMIR (2021)
24. Zainab, R., Chandramouli, R.: Detecting and explaining depression in social media text with machine learning. In: KDD 2020, August 2020

Say No2Ads: Automatic Advertisement and Music Filtering from Broadcast News Content

Shayan Zamanirad[(✉)] and Koen Douterloigne

Isentia Media Intelligence, Sydney 2012, Australia
{shayan.zamanirad,koen.douterloigne}@isentia.com

Abstract. The incredible growth in available news content has been met with steeply increasing demand for news amongst the general population. The 24/7 news cycle gives people an awareness of events, activities and decisions that may have an impact on them (e.g. the latest updates on the COVID-19 outbreak). Despite the flourish of social networks, recent research suggests radio and especially TV are still the main sources of news for many people. However, unlike in social media, the content aired on radio and TV requires people to listen to every single advertisement and music (for radio) before consuming the next item. For this reason, media monitoring companies have to dedicate considerable amount of resources on processing or manually filtering the advertising content (which is blended with the actual news). Often their clients still receive ads. To mitigate this problem, in this paper, we propose *No2Ads*, an autoregressive deep convolutional neural network (CNN) model that is trained on over 500 h of human annotated training samples to remove ads and music from broadcast content. *No2Ads* reached very high performance results in our tests, achieving 97% and 95% in precision and recall on detecting ads/music for radio channels; 95% precision and 98% recall for TV channels. Between March to September 2021, across 261 radio and TV channels in Australia and New Zealand, *No2Ads* has detected and filtered out 22,161 h of all captured broadcast content as either *advertisements* or *music*.

1 Introduction

Local news is critical for people in a locality to learn about activities that may have an impact on them. Having national focused news allows people to gain a national perspective, especially for countries that have many different population centres. News from foreign countries also plays an important role in today's global economy, by giving people a glimpse of each other's ways of life and cultural differences. News can even shape us in surprising ways, from what we see in our dreams, to the possibility of having a heart attack [23]. Unusual or impactful events such as a pandemic, protests or an economic recession can make people pay more attention to the news than ever before [9,44].

While there is a growing trend towards using social media to receive the latest news, the threat of misinformation on social media, such as fake news [6], means

© Springer Nature Switzerland AG 2022
H. Hacid et al. (Eds.): ICSOC 2021 Workshops, LNCS 13236, pp. 18–31, 2022.
https://doi.org/10.1007/978-3-031-14135-5_2

that TV and radio remain the main source of news for many people [1,8,10,34]. Thus, broadcast news remains critical for media monitoring companies and their customers who want to monitor and make sense of the world's conversations in real-time. The media monitoring process starts with data ingestion where news is captured 24/7. This captured data is then transcribed using an automatic speech to text model. The textual information, attached as meta-data to the captured content, then goes through an enrichment pipeline where each item is complemented with extra attributes such as topics (e.g. *finance, sport, traffic*), entities (e.g. *person, political group, city*) and sentiment (e.g., *strongly negative, weakly positive*). Such attributes are then processed by decision-making algorithms to deliver relevant news items to customers.

Aired news, however, often comes with a high number of advertisements and music (especially in the case of radio) [28]. This mixture of content degrades the quality of the data that is presented to customers. Trimming unwanted advertising content improves the quality of the service for clients and lowers the processing cost for media monitoring companies. To remove non-news content, these companies often employ workers to monitor captured items one by one, non-stop, and around the clock. Although this approach offers very accurate results, it suffers from significant cost as well as processing bottlenecks and long delays for content delivery. Unexpected incidents, like the COVID-19 outbreak [15,39], can worsen the situation even further, because of increases in news content when news coverage is extended. Thus, an automated approach such as exploiting a deep learning model, is a more efficient solution.

In this paper, we propose *No2Ads*, an autoregressive convolutional neural network (CNN) that converts captured broadcast content into 2D images by extracting *Mel-frequency cepstral coefficients (MFCCs)* features. The CNN then classifies each image into one of the categories *ads/music* or *news*. Based on our analysis, broadcast content always appears in a *chain pattern*. In other words, the likelihood of an ad appearing after another ad is higher than the likelihood of an ad appearing after the news. Thus, we designed our model in autoregressive [24] form to observe such patterns and find correlations between preceding and succeeding content while making predictions. *No2Ads* was trained on over 500 h of annotated news items. We achieve 97% precision and 95% recall in filtering ads and music from radio, and 95% precision and 98% recall for TV. Our model reduces the extensive effort and time that is required to manually remove ads and music from aired news content with low latency and negligible deployment cost. Between March and September 2021, across 261 radio and TV channels in Australia and New Zealand, *No2Ads* has detected and filtered out 22,161 h of all captured broadcast content as either *advertisements* or *music*.

2   Related Work

2.1   News Classification

Grouping news into different categories using deep learning models has received much attention in recent years, with huge leaps in reported results. Muhammad

Ali et al. [40] proposed a text-based CNN model to classify Indonesian news, while [37] presented a large scale topic classifier based on BERT and SparkNLP. By leveraging headlines, [19] identified political biases, [14] detected clickbait, and [13,21,22,42] categorised news based on their sentiment tones which later showed to be a strong feature in predicting finance news [32]. A two step BERT-based system proposed by Ciara et al. [11] first verified news as fact-based or opinion-based, and then as real or fake. Related to this work, [27] used unigram features and a Linear SVM classifier to detect fake news. An ensemble approach by combining Decision Tree and Random Forests was introduced by [2]. Saeed Amer et al. [4] suggested that sequence models can be better solutions when news content is short, while [36] introduced a hybrid approach by merging CNN and RNN models to utilise local features and long-term dependencies to classify news.

In contrast with the above mentioned works, our approach requires no semantic understanding of news and non-news items for classification. This makes our approach more robust to changes in content of news and ads.

2.2 Audio Classification

Leveraging neural models for audio recognition and classification has drawn a lot of attention recently. Some of the most popular application domains include speech recognition, music classification and environmental sound recognition. For instance, Aclnet [26] is an environment sound classification network with two separate models, a 1D CNN model to produce low-level features (spectral-like) from raw inputs, followed by the high-level classifier model similar to the VGG architecture. Along with this work, [41] and [33] demonstrated that single layer CNN models can achieve high accuracy by augmenting environmental sounds. Similarly, Loris et al. [35] proposed ensemble CNN models that utilise augmentation techniques for audio classification. Their results demonstrate that the performance of these models increases if they leverage fine-tuned CNNs with different architectures. Using an SVM model to extract acoustical features such as sub-band power and pitch information to classify audio samples was the approach demonstrated by Lin et al. [29]. Similarly, Lu et al. [30] showed that if the audio features are known beforehand, having a hierarchical SVM-based system that classifies sub-segment of audio clips into classes can achieve promising results [30]. Sander et al. [17] showed that, when given raw audio inputs, CNN models are able to learn useful features and discover frequency decompositions for tagging music tasks.

Compared to the mentioned approaches, our model does not require any manual feature extraction. Furthermore, it attends to recent inputs and their predictions by leveraging temporal representations of sound waves. This enables our model to unlock hidden patterns and relations that exist in sequences of broadcast news items.

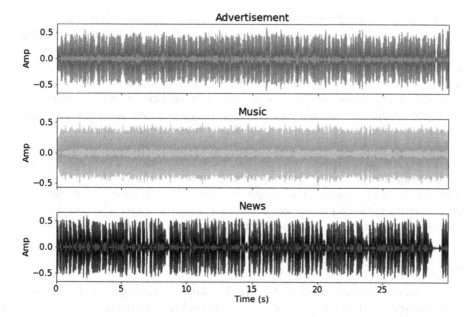

Fig. 1. Waveforms of 30 s audio samples: *advertisement, music* and *news*. Broadcasted in Sydney 2 GB radio channel on 03/07/2020. For demonstration purpose waveforms are downsampled to 1600 Hz.

3   Filtering Approach

3.1   Concept

When you listen to a radio station in an unknown foreign language, you can usually distinguish between the parts that are normal spoken text (e.g. news stories, people having an interview), and the parts that are advertisements or music, even when you do not understand what is being said. Semantic understanding of the content is not required to identify the content class. This implies that speech transcription is not essential when designing a model to classify ads and news.

To illustrate this concept, Fig. 1 shows the waveform of 30 s of radio audio for advertisement, music and news clips. The spoken news is marked by frequent short pauses, whereas the music clip looks like a continuous stream of audio. Advertisements are often a combination of background music and speech, or have the natural speech pauses artificially removed to decrease the advertisement's length and cost. Note that Fig. 1 only shows the amplitude of the audio signal over time. The frequency over time is not visible, even though the frequency is also a source of information that can be used to classify the type. For example, musical instruments often produce tones that are higher or lower in pitch than what can be produced by the human voice.

One way to visually represent both the amplitude and the frequency in a single image is by creating a spectrogram. In order to create a spectrogram, discrete Fourier transforms (DFT) are applied to the audio signal over a sliding window of a certain sample-size. Each Fourier transform results in a set of values that represent the signal strength per frequency. The signal strength values are then color-mapped and displayed as a vertical band. The combination of vertical bands for all sliding windows gives an image of the frequency strength over time. This spectrogram image has been successfully used for audio classification tasks in [25].

A downside of the spectrogram is that it does not take the physical properties of the human vocal and auditory system into account. In No2Ads, we want to distinguish normal speech from abnormal speech or non-speech, and thus it is important that the input to our classifier contains features that are good at identifying this difference. One such representation is offered by *MFCC* features.

3.2 MFCC Features

Mel-frequency cepstral coefficients (MFCC) were originally created for speech representation and recognition, and to aid in speaker identification [16,31]. They have since been used in numerous areas related to speech processing. The MFCC features are computed from a modified version of the cepstrum [12], which identifies repeating patterns in the frequency spectrum. The MFCC features are designed to mimic the human hearing system, by applying a filter bank to the audio signal, where the filters are distributed linearly for low frequencies and logarithmically for high frequencies.

MFCCs are computed in the following steps[1]:

1. Compute the Fast Fourier Transform (FFT) over a sliding window on the audio signal.
2. For each window frame, apply a filter bank with overlapping bandpass filters, either triangular or cosine in shape. The widths of the filters are according to the mel scale, which represent equidistant pitches under human perception. The filter bank thus maps the spectrum onto the mel scale.
3. Compute the logarithm at each frequency in the mel domain.
4. Apply a discrete cosine transform (DCT) to the collection of mel logarithms.
5. The resulting amplitudes of the DCT are the mel-frequency cepstral coefficients.

In Fig. 2, we show an example of MFCC outputs for 6 samples of 30 s of audio, for different modalities. In each image, a single vertical band represents the color-mapped values of 100 MFCCs computed on a sliding window. It is clear that advertisements and music have a very different character to normal speech, making this a good input to our model.

---

[1] For further details, we refer the interested reader to https://www.intechopen.com/chapters/63970.

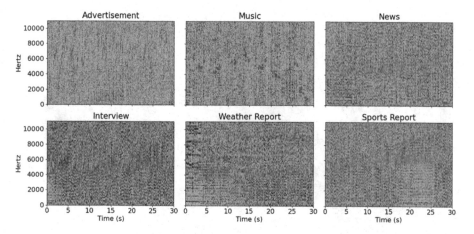

Fig. 2. MFCCs extracted from six types of audio samples, which were broadcast on the Sydney 2 GB radio channel on 2020-07-03 - sample rate = 16000 Hz, features = 100, duration = 30 s. Points with *warmer* color depict higher MFCC values.

3.3   Model Architecture

No2Ads is a convolutional neural network (CNN) model that uses two core ideas. First, images that are generated from ads and music have areas with higher values for the MFCCs, while images from news content generally have lower MFCCs. This corresponds to music and ads simply being louder on average than spoken content. Figure 2 shows this clearly. Second, images of the same content class tend to appear in sequence, thus the CNN model can use the type of the n previously classified images to aid in the current decision (autoregressive approach). In our experiments, we found that $n = 5$ achieves the highest accuracy.

Our model has four 2D convolution layers that sit next to each other with four different filter sizes (6×100, 7×100, 9×100, 10×100). Each filter focuses on detecting patterns/features at a specific scale in the input image. The filter weights are randomly initialized in the beginning of the training process, but they get updated as the model learns to recognize features of the input (through backpropagation). Filters scan the input image from left to right to create a convoluted image a.k.a *feature map* [5]. Extracted features in feature maps are then passed into a Rectified Linear Unit (ReLU) to increase the non-linearity by ignoring the non-positive features. This normalization step helps to prevent the exponential growth in computation and therefore accelerate the learning process. The rectified feature maps are then used by a pooling layer to select the maximum value (*max pooling* [5]) in each region within a feature map. The results are down-sampled feature maps that highlight the most dominant features in regions. Reducing the dimensions of feature maps results in decreasing the number of parameters to learn by the model while it becomes more robust to variations in the position of features in the input image. Pooled features from all four layers are then concatenated and flattened (unroll all values into a 1D

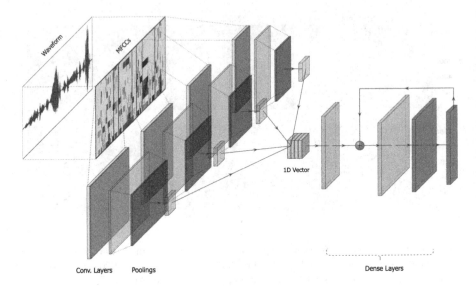

Fig. 3. *No2Ads* model architecture - MFCC generated images are fed through convolution and pooling layers to extract features and predict target labels.

vector). This flattened vector is then passed onto a fully connected layer to perform the actual classification task.

The prediction result is then concatenated with the last $n = 5$ results. The concatenated result is used by two stacked fully connected layers followed by a sigmoid function to make the final prediction (between 0 for *ads/music* and 1 for *news*). The architecture of No2Ads model, specific to radio, is shown in Fig. 3. For TV broadcast we use a slightly modified (simplified) version of our model with only two convolution layers, of filter sizes (5×100, 6×100).

3.4 Dataset

A high quality training dataset is essential to create a well performing AI model. Creating such a dataset is costly as it involves a lot of manpower to annotate high-fidelity labels [45]. Unlike in the natural language processing (NLP) field where cleaned training data is readily available [3,43], or where it can be generated automatically [47], in the field of digital audio processing finding a domain-specific dataset among general-usage datasets [7,20,38] can be an unsuccessful effort. Thus, we decided to create an annotated dataset specifically for the Australia/New Zealand market.

Our dataset consists of 1,800 h of audio, with the audio sampled from 70 different radio and TV stations, both public and commercial. Our sampling focuses on news, interviews and talkshows, interspersed with ads and music. The available classes for labeling are *advertisement, music, interview, sports, weather, finance,* and *general news*. To the best of our knowledge, this is the first dataset with these particular classes.

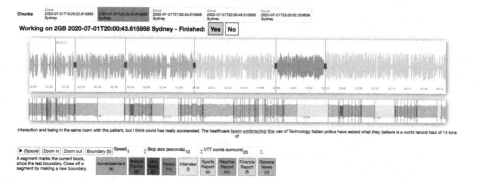

Fig. 4. Annotation tool to create labeled data from broadcast news

A common problem when creating training data are human mistakes that lead to incorrect annotations [46], which impacts the performance of trained models. To mitigate such errors we employed two experienced media monitors[2] for three months, who are very familiar with listening to lengthy audio streams and interpreting the content, to create high quality labelled data. The use of experienced annotators increased annotation speed up to 3x (i.e. annotating 1 h of audio took on average 20 min). It also reduced the need for extensive reviews on the labeled data.

We implemented a custom web interface to allow quick annotation of audio streams. The interface was built leveraging the peaks.js library [18]. Figure 4 shows our annotation tool, highlighting the display of blocks of content in an audiostream, represented as a waveform. To speed up loading times in the browser, mp3 files of 1 h long were generated together with data representing their waveform. Machine-generated transcriptions (from a third party speech-to-text provider) were available for all audio and were shown on the page synchronized to the audio. However, the annotators did not find the transcript useful and gave feedback that they ignored this extra information while creating annotations. Hotkeys were added to the interface to speed up the work, namely hotkeys for easy *speedup* and *slowdown* of the audio playback, hotkeys for quick *jumping forward* and *backward* of a variable amount of seconds (15 s by default), and hotkeys for annotation labels.

4 Experimental Setup

In this section, we first describe our experimental setup and then present our main results.

---

[2] A media monitor is a person who summarises media content such as news, interviews, etc.

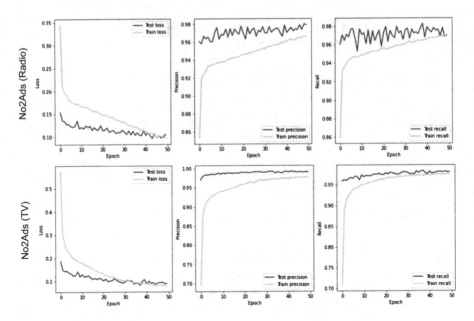

Fig. 5. Loss vs Precision and Recall during training the models

Dataset. Our human annotated dataset consists of audio samples with varying lengths. In order to train and test our model, we need to normalize these samples to a fixed length of n seconds. As we capture broadcast content in chunks of 10 s (a setting outside of our control), we decided to choose $n = 10$. This means that our classification works on a granularity of 10 s, i.e. each chunk of 10s is classified as being either spoken text or an advertisement/music. For audio samples shorter than 10 s we added silence to the end of the sample. Eighty percent of our dataset was used as a training set while the remaining 20% was a test set. Table 1 shows the number of hours for each set after normalization.

Table 1. Number of hours of *training* and *testing* samples used to build and validate No2Ads model

	Radio			TV		
	Total	Train	Test	Total	Train	Test
Ads/Music	161	129	32	57	45	12
News	264	211	53	159	127	32
	425	**340**	**85**	**216**	**172**	**43**

Metrics. We report *Accuracy, Precision* and *Recall* in Figs. 5 and 6 as measurement criteria for the performance of our models.

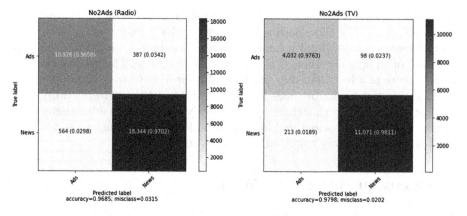

Fig. 6. Confusion matrices to report precision and recall for both models

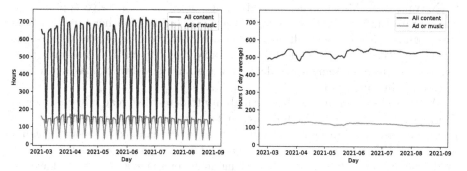

Fig. 7. The total number of hours of data analyzed per day, and the number of hours of ads or music identified. Less media content is captured during weekends, resulting in large fluctuations. The right graph shows a weekly (7 day) average.

4.1 Test Results

After training our radio and TV models for 50 epochs, we achieved an accuracy of 96.85% for the radio model and an accuracy of 97.98% for the TV model. Figure 5 shows the training and test loss, precision and recall for both models per epochs. As one can see, the TV model reached a steady state at around epoch 37 whereas it took all 50 epochs for the radio model to converge. Figure 6 summarizes the precision and recall for each model per sample type.

4.2 Live Results

The No2Ads model ran on production volumes between March and September 2021. Over these 6 months a total of 97,212 h of mixed broadcast content were analyzed. Of those, about 22,161 h (22.8%) were identified as *advertisements* or *music* and were filtered out. Figure 7 compares the daily number of hours of analyzed radio and TV content with the number of hours of detected advertisements and music. We show both a daily count and a weekly average. Despite a gradual increase in the total volume of analyzed content, a very slow decline in the count of ad hours is visible. This decline can be explained by gradual changes in the nature of advertisements, and thus of their MFCC, over time. The data used to train our model becomes less representative of the current reality, and thus our model has a higher chance of incorrectly classifying an advertisement as normal spoken text. For optimal long-term performance we suggest to finetune the No2Ads model every couple of months on recent data.

5   Discussion and Future Work

We have presented an approach to perform a binary classification of audio data into either spoken text or music and advertisements. The combination of a deep CNN trained on MFCC features, with a temporal autoregressive approach to incorporate the likelihood of having consecutive blocks of either type, leads to very high precision and recall values. We achieved the best results when separating the data into two models, one for radio and one for TV. The presented approach requires no semantic understanding of the audio, which makes it robust to changes in content of the ads or the news.

A downside to our approach is its inability to detect live read advertisements. This is where the radio or TV host presents a scripted message from a sponsor or makes an unscripted promotional statement. The appearance of the audio signal and the MFCC is nearly identical to an actual news item or conversation, so our model can not identify this speech as being an advertisement. A potential avenue for future work is to investigate whether the inclusion of semantic understanding, i.e. transcription, can allow for this type of ad to be detected as well. Existing work on classification of spam in tweets could be a good starting point.

Our training set was created in English, and our model is used exclusively on English content. It would be interesting to analyze the performance on non-English languages as well. Since our model does not require semantic understanding, we anticipate good results out-of-the-box, especially for languages and cultures that are related to English.

References

1. Aelst, P.V., et al.: Does a crisis change news habits? A comparative study of the effects of COVID-19 on news media use in 17 European countries. Digit. Journal. **9**, 1–31 (2021). https://doi.org/10.1080/21670811.2021.1943481
2. Ahmad, I., Yousaf, M., Yousaf, S., Ahmad, M.O.: Fake news detection using machine learning ensemble methods. Complexity **2020** (2020)
3. Akbik, A., Bergmann, T., Blythe, D., Rasul, K., Schweter, S., Vollgraf, R.: Flair: an easy-to-use framework for state-of-the-art NLP. In: Proceedings of the 2019 Conference of the North American Chapter of the Association for Computational Linguistics (Demonstrations), pp. 54–59 (2019)

4. Alameri, S.A., Mohd, M.: Comparison of fake news detection using machine learning and deep learning techniques. In: 2021 3rd International Cyber Resilience Conference (CRC), pp. 1–6. IEEE (2021)
5. Albawi, S., Mohammed, T.A., Al-Zawi, S.: Understanding of a convolutional neural network. In: 2017 International Conference on Engineering and Technology (ICET), pp. 1–6. IEEE (2017)
6. Apuke, O.D., Omar, B.: Fake news and COVID-19: modelling the predictors of fake news sharing among social media users. Telematics Inform. **56**, 101475 (2021)
7. Ardila, R., et al.: Common voice: a massively-multilingual speech corpus. arXiv preprint arXiv:1912.06670 (2019)
8. Barthel, M., et al.: Measuring news consumption in a digital era (2020). https://www.pewresearch.org/journalism/2020/12/08/measuring-news-consumption-in-a-digital-era/. Accessed 14 Sept 2021
9. Beheshti, A., Hashemi, V.M., Yakhchi, S., Motahari-Nezhad, H.R., Ghafari, S.M., Yang, J.: personality2vec: enabling the analysis of behavioral disorders in social networks. In: WSDM 2020: The Thirteenth ACM International Conference on Web Search and Data Mining, Houston, TX, USA, 3–7 February 2020, pp. 825–828. ACM (2020)
10. Beheshti, S.-M.-R., Benatallah, B., Motahari-Nezhad, H.R., Sakr, S.: A query language for analyzing business processes execution. In: Rinderle-Ma, S., Toumani, F., Wolf, K. (eds.) BPM 2011. LNCS, vol. 6896, pp. 281–297. Springer, Heidelberg (2011). https://doi.org/10.1007/978-3-642-23059-2_22
11. Blackledge, C., Atapour-Abarghouei, A.: Transforming fake news: Robust generalisable news classification using transformers. arXiv preprint arXiv:2109.09796 (2021)
12. Bogert, B.P.: The quefrency alanysis of time series for echoes; cepstrum, pseudo-autocovariance, cross-cepstrum and saphe cracking. Time Ser. Anal. 209–243 (1963)
13. Bostan, L.A.M., Kim, E., Klinger, R.: GoodNewsEveryone: a corpus of news headlines annotated with emotions, semantic roles, and reader perception. In: Proceedings of the 12th Language Resources and Evaluation Conference, Marseille, France, pp. 1554–1566. European Language Resources Association, May 2020. https://aclanthology.org/2020.lrec-1.194
14. Bourgonje, P., Moreno Schneider, J., Rehm, G.: From clickbait to fake news detection: an approach based on detecting the stance of headlines to articles. In: Proceedings of the 2017 EMNLP Workshop: Natural Language Processing meets Journalism, Copenhagen, Denmark, pp. 84–89. Association for Computational Linguistics, September 2017. https://doi.org/10.18653/v1/W17-4215. https://aclanthology.org/W17-4215
15. Cajner, T., Figura, A., Price, B., Ratner, D., Weingarden, A.: Reconciling unemployment claims with job losses in the first months of the COVID-19 crisis (2020)
16. Davis, S.B., Mermelstein, P.: Comparison of parametric representations for monosyllabic word recognition in continuously spoken sentences. IEEE Trans. Acoust. Speech Signal Process. **28**, 357–366 (1980)
17. Dieleman, S., Schrauwen, B.: End-to-end learning for music audio. In: 2014 IEEE International Conference on Acoustics, Speech and Signal Processing (ICASSP), pp. 6964–6968 (2014). https://doi.org/10.1109/ICASSP.2014.6854950
18. Finch, C., Parisot, T., Needham, C.: bbc/peaks.js: 0.21.0, April 2020. https://github.com/bbc/peaks.js

19. Gangula, R.R.R., Duggenpudi, S.R., Mamidi, R.: Detecting political bias in news articles using headline attention. In: Proceedings of the 2019 ACL Workshop Black-boxNLP: Analyzing and Interpreting Neural Networks for NLP, Florence, Italy, pp. 77–84. Association for Computational Linguistics, August 2019. https://doi.org/10.18653/v1/W19-4809. https://aclanthology.org/W19-4809

20. Gemmeke, J.F., et al.: Audio set: an ontology and human-labeled dataset for audio events. In: Proceedings of IEEE ICASSP 2017, New Orleans, LA (2017)

21. Ghodratnama, S., Beheshti, A., Zakershahrak, M., Sobhanmanesh, F.: Intelligent narrative summaries: from indicative to informative summarization. Big Data Res. **26**, 100257 (2021)

22. Ghodratnama, S., Zakershahrak, M., Beheshti, A.: Summary2vec: learning semantic representation of summaries for healthcare analytics. In: International Joint Conference on Neural Networks, IJCNN 2021, Shenzhen, China, 18–22 July 2021, pp. 1–8. IEEE (2021)

23. Gorvett, Z.: How the news changes the way we think and behave (2020). https://www.bbc.com/future/article/20200512-how-the-news-changes-the-way-we-think-and-behave. Accessed 10 Sept 2021

24. Gregor, K., Danihelka, I., Mnih, A., Blundell, C., Wierstra, D.: Deep autoregressive networks. In: International Conference on Machine Learning, pp. 1242–1250. PMLR (2014)

25. Gwardys, G., Grzywczak, D.M.: Deep image features in music information retrieval. Int. J. Electron. Telecommun. **60**(4), 321–326 (2014)

26. Huang, J.J., Leanos, J. J. A.: AclNet: efficient end-to-end audio classification CNN. arXiv preprint arXiv:1811.06669 (2018)

27. John, A., Meenakowshalya, A.: Fake news detection using n-gram analysis and machine learning algorithms. J. Mob. Comput. Commun. Mob. Netw. **8**(1), 33–43 (2021)

28. Kelly, V.: Quality of radio ads poor because advertisers are greedy (2020). https://mumbrella.com.au/quality-of-radio-ads-poor-because-advertisers-are-greedy-634180. Accessed 22 Aug 2021

29. Lin, C.C., Chen, S.H., Truong, T.K., Chang, Y.: Audio classification and categorization based on wavelets and support vector machine. IEEE Trans. Speech Audio Process. **13**(5), 644–651 (2005). https://doi.org/10.1109/TSA.2005.851880

30. Lu, L., Zhang, H.J., Li, S.Z.: Content-based audio classification and segmentation by using support vector machines. Multimedia Syst. **8**(6), 482–492 (2003)

31. McKinney, M., Breebaart, J.: Features for audio and music classification (2003)

32. Moore, A., Rayson, P.: Lancaster a at SemEval-2017 task 5: evaluation metrics matter: predicting sentiment from financial news headlines. In: Proceedings of the 11th International Workshop on Semantic Evaluation (SemEval-2017), Vancouver, Canada, pp. 581–585. Association for Computational Linguistics, August 2017. https://doi.org/10.18653/v1/S17-2095. https://aclanthology.org/S17-2095

33. Mushtaq, Z., Su, S.F.: Environmental sound classification using a regularized deep convolutional neural network with data augmentation. Appl. Acoust. **167**, 107389 (2020)

34. Naeem, S.B., Bhatti, R., Khan, A.: An exploration of how fake news is taking over social media and putting public health at risk. Health Inf. Libr. J. **38**(2), 143–149 (2021)

35. Nanni, L., Maguolo, G., Brahnam, S., Paci, M.: An ensemble of convolutional neural networks for audio classification. Appl. Sci. **11**(13), 5796 (2021)

36. Nasir, J.A., Khan, O.S., Varlamis, I.: Fake news detection: a hybrid CNN-RNN based deep learning approach. Int. J. Inf. Manag. Data Insights **1**(1), 100007 (2021)

37. Nugroho, K.S., Yudistira, N.: Large-scale news classification using BERT language model: spark NLP approach. arXiv preprint arXiv:2107.06785 (2021)
38. Panayotov, V., Chen, G., Povey, D., Khudanpur, S.: Librispeech: an ASR corpus based on public domain audio books. In: 2015 IEEE International Conference on Acoustics, Speech and Signal Processing (ICASSP), pp. 5206–5210. IEEE (2015)
39. Petrosky-Nadeau, N., Valletta, R.G., et al.: An unemployment crisis after the onset of COVID-19. FRBSF Econ. Lett. **12**, 1–5 (2020)
40. Ramdhani, M.A., Maylawati, D.S., Mantoro, T.: Indonesian news classification using convolutional neural network. Indones. J. Electr. Eng. Comput. Sci. **19**(2), 1000–1009 (2020)
41. Salamon, J., Bello, J.P.: Deep convolutional neural networks and data augmentation for environmental sound classification. IEEE Signal Process. Lett. **24**(3), 279–283 (2017)
42. Wang, S., et al.: Assessment2Vec: learning distributed representations of assessments to reduce marking workload. In: Roll, I., McNamara, D., Sosnovsky, S., Luckin, R., Dimitrova, V. (eds.) AIED 2021. LNCS (LNAI), vol. 12749, pp. 384–389. Springer, Cham (2021). https://doi.org/10.1007/978-3-030-78270-2_68
43. Weischedel, R., et al.: Ontonotes release 4.0. LDC2011T03. Linguistic Data Consortium, Philadelphia (2011)
44. Wu, L., Li, X., Lyu, H.: The relationship between the duration of attention to pandemic news and depression during the outbreak of coronavirus disease 2019: the roles of risk perception and future time perspective. Front. Psychol. **12**, 564284 (2021)
45. Yaghoub-Zadeh-Fard, M.A., Benatallah, B., Casati, F., Barukh, M.C., Zamanirad, S.: User utterance acquisition for training task-oriented bots: a review of challenges, techniques and opportunities. IEEE Internet Comput. **24**(3), 30–38 (2020). https://doi.org/10.1109/MIC.2020.2978157
46. Yaghoub-Zadeh-Fard, M.A., Benatallah, B., Chai Barukh, M., Zamanirad, S.: A study of incorrect paraphrases in crowdsourced user utterances. In: Proceedings of the 2019 Conference of the North American Chapter of the Association for Computational Linguistics: Human Language Technologies, Volume 1 (Long and Short Papers), Minneapolis, Minnesota, pp. 295–306. Association for Computational Linguistics, June 2019. https://doi.org/10.18653/v1/N19-1026. https://aclanthology.org/N19-1026
47. Yaghoub-Zadeh-Fard, M.A., Benatallah, B., Zamanirad, S.: Automatic canonical utterance generation for task-oriented bots from API specifications. In: EDBT, pp. 1–12 (2020)

Hybrid Recommendation of Movies Based on Deep Content Features

Tord Kvifte$^{(\boxtimes)}$, Mehdi Elahi, and Christoph Trattner

MediaFutures: Research Centre for Responsible Media Technology and Innovation, Department of Information Science and Media Studies, University of Bergen, Fosswinckels Gate 6, 5007 Bergen, Norway
{Tord.Kvifte,mehdi.elahi,Christoph.Trattner}@uib.no

Abstract. When a movie is uploaded to a movie Recommender System (e.g., YouTube), the system can exploit various forms of descriptive features (e.g., tags and genre) in order to generate personalized recommendation for users. However, there are situations where the descriptive features are missing or very limited and the system may fail to include such a movie in the recommendation list. This paper investigates hybrid recommendation based on a novel form of content features, extracted from movies, in order to generate recommendation for users. Such features represent the visual aspects of movies, based on Deep Learning models, and hence, do not require any human annotation when extracted. We have evaluated our proposed technique using a large dataset of movies and shown that automatically extracted visual features can mitigate the cold-start problem by generating recommendation with a superior quality compared to different baselines, including recommendation based on human-annotated features.

Keywords: Recommender systems · Visually-aware · New item

1 Introduction

Recommender systems are intelligent tools that can support users in their decision making process by suggesting a shortlisted set of items tailored to their personal needs and constraints [1,21,28,39]. These systems can learn from the particular tastes and interests of the users and generate recommendation that can better match their interests and tastes [13,38].

There exists a wide range of approaches that can be adopted to create personalized recommendations for users. Content-Based Filtering (CBF) is among popular approaches that can exploit the content features associated to videos (e.g., tag, and genre) and recommends to a target user the videos with the content similar to the videos that she liked in the past [5,34,40,45]. Collaborative Filtering (CF), on the other hand, is another popular approach which focuses on exploiting patterns among the user preferences (e.g., ratings or likes) and recommends to a target user those videos that have been highly co-rated by like-minded users similar to her [11,22,23,48].

© Springer Nature Switzerland AG 2022
H. Hacid et al. (Eds.): ICSOC 2021 Workshops, LNCS 13236, pp. 32–45, 2022.
https://doi.org/10.1007/978-3-031-14135-5_3

While either of these approaches can be effective in generating relevant recommendation for users, they may fall short to recommend videos whose descriptive data is missing or very limited and hence the system do not have sufficient information about those videos [14, 16]. This is a common problem in recommender systems called *New Item* as part of a bigger challenge called *Cold Start*. New item problem in video streaming applications happens when a new video has been uploaded to the system where the users have not provided neither rating nor any other form of the data, e.g., tag or comments. In such a case, almost all recommender approaches may fail to include such a video when generating personalized recommendation for users. Apart from the new item problem, the process of collecting quality data to represent the videos is itself another major problem. Some forms of data (e.g., genre), a group of experts are essentially required to manually annotate every, and other forms (e.g., rating and tag) may need a large community of users willing to provide the data. This makes the aforementioned data to be very expensive and extremely sparse to collect [3, 4, 9, 30, 46].

In this paper, we address the above-mentioned problems by proposing a novel recommendation technique that exploits visual features to generate personalized recommendation for users. We have adopted a hybrid Matrix Factorisation (MF) algorithm [24], implementing different optimization methods, i.e., *BPR, WARP,* and *Logistic*. The proposed visual features can be extracted in a completely automatic way, using Deep Learning models and hence they require no (expensive) user annotation. This enables our proposed technique to effectively cope with the cold-start problem, when no or limited human-annotated data is available.

We have extracted a large dataset of visual features from 12,875 of the trailers of the movies that exist in the Movielens dataset. Movie trailers have shown to exhibit high visual similarity compared to their full length movies [8]. In addition to visual features, we have also collected a rich dataset of movie subtitles and generated recommendation based on them and considered it as one of the baselines. We evaluated our proposed recommendation technique using the dataset with hundreds of thousands of ratings. The results show the superior performance of our proposed technique compared with a number of baselines, i.e., recommendation based on tag, genre, and subtitle.

The main contributions of this work can be summarized as follows:

1. The proposal of a novel hybrid recommendation technique based on visual features considering different optimization methods. e.g., *BPR, WARP,* and *Logistic*, and comparing it with different baselines with regards to different evaluation metrics;
2. extracting a large dataset with visual features, using an advanced deep learning model; Dataset will be published publicly upon the acceptance of the paper;
3. collecting a large dataset of subtitles from full length movies and exploiting them in a baseline recommendation technique.

2   Related Work

One of the most popular types of recommender systems are based on the Content-based Filtering (CBF) technique. In this technique, the items are represented by their content and the users by associating their preferences with the item content [18,21,28,29,36]. In movie domain, the item content are described with a set of representative features describing different aspects of the movie content. Traditional examples of content features are genre and tag, representing some form of *semantics* within the movies.

Recent approaches based on content-based filtering have adopted a novel form of movie content based on visual features [8,15,49] illustrating a more *stylistic* representation of the movies. This type of novel features, in contrast to the traditional features, does not need any expensive human-annotation and can be extracted automatically adopting *Computer Vision* methods. Hence, they could be a potential solution for movie recommendation in cold start, i.e., when recommending movies with no descriptive features [12]. Another advantage of the visual features is that they can be more representative of the production style and can enable movie recommender systems to become *style-aware* [2,19,26,50].

Visual features, extracted from movie content, can have different classes, each of which illustrating a different representation of the movies [31]. One class of visual features can describe movies from a *high-level* perspective while another class can describe them from a *low-level* perspective. The former type of features typically provide a more semantic representation of the movies (e.g., sun shining in the a movie scene) while the latter type focus more on low level aspects (e.g., colorfulness and brightness in a movie).

A number of prior work have proposed recommender systems capable of using visual features. As an example, the authors in [49] proposed a recommendation approach by combining semantic and visual content features. Another example is [50] that proposed integration of multiple ranking lists, each of which generated by a set of semantic or visual features. The authors of [7] proposed a recommendation technique based on a selection of handcrafted visual features including shot length, object motion, color, and lighting. [41] is another work where authors explored the different potentials of visual features in movie recommender systems. In [6,42], a set of audio-visual features have been exploited to generate movie recommendation. In [27] and [35], the authors proposed a video recommender system that takes advantage of Deep Learning methods based on Convolutional Neural Networks (CNN). Finally, few prior works attempted to address the research gap between video classification, and search & recommendation by proposing a more unified solutions. An example is [25] where the authors proposed a model based on a deep learning approach (i.e., CNN) utilizing a set of audio-visual features and showed to be effective in the noted tasks.

Our work differs compared to the work mentioned above in the following aspects. First of all, these works adopted a one-size-fits-all approach by considering a single optimization method when building their recommendation model. However, different methods may better suit different type of content data (e.g., visual features, genre and tag). Hence, we adopted different optimization methods, based on different loss functions, for different types of data. We have used a large dataset of movies and compared the performances of different optimization methods for the task of recommendation. To the best of our knowledge, non of the prior works has performed such a comparison. Furthermore, we have considered a novel baseline, i.e., recommendation based on movie subtitle and compared it with our proposed recommendation technique (visual features) as well as more traditional baselines (genre and tags) taking into account different evaluation metrics, i.e., Precision@K, Recall@K, AUC, and Reciprocal Rank.

3 Methodology

We used a large dataset of key-frames from 12875 movie trailers collected from YouTube. According to prior work, there is a high similarity between the visual features extracted from full-length movies and their respective movie trailers [8]. The following list represents the entire methodology: *Extracting Visual Features*: Every key-frame is analyzed using a pre-trained CNN model [44], resulting in feature labels. *Aggregating Features*: Visual features are aggregated using two different methods, resulting in two different sets of feature vectors. *Training and predicting*: The feature vectors are used to train the prediction models.

3.1 Feature Extraction

Our feature extraction can be divided into two parts. First part includes the extraction of visual features from movie trailers, and the second part encompasses the collection of movie subtitles.

Visual Feature Extraction. We extracted visual feature labels by applying the VGG-19 image classification model [44], a 19-layer network trained on ImageNet, to the key-frames of every movie trailer in the key-frame dataset. The model was implemented in Python, using the Keras API, which is built on top of the TensorFlow framework [32]. The output of the model consists of a label, representing the predicted classes of the input image, as well as a confidence value representing the certainty of the prediction being correct. The resulting dataset of labels for 12,875 movies includes 997 unique feature labels in total.

Subtitle Collection and Pre-Processing. Subtitles were collected using a public API [33][1], then parsed and pre-possessed, resulting in a dataset of English subtitles from 1514 different movies. Among the pre-processing steps were removal of timestamps and subtitle-specific data, stop word removal, part-of-speech filtering, and lemmatization. The resulting dataset includes 62664 unique features.

---

[1] http://www.opensubtitles.org.

3.2 Feature Aggregation

To form the final feature embeddings of a movie, we have aggregated the extracted features. Visual features were aggregated using two different methods, producing two separate feature matrices, *Deep Visual-f* and *Deep Visual-c.*

Deep Visual-f. Visual features were weighted using *Term Frequency-Inverse Document Frequency (TF-IDF)* [43]. TF-IDF can recognize the importance of each word in a document in the context of a corpus of documents. If a word has low occurrence across the corpus, while having high frequency in one (or few) document, it likely plays a key role in that specific document. In our case, a movie is considered as a document, and the labels of the movie are considered as words of that document. Furthermore, the collection of all movies and their respective labels corresponds to the corpus of documents.

Deep Visual-c. Important elements in a movie can be assumed to be emphasized visually, and thereby more likely to be predicted with a higher confidence, computed by the image classification model. Based on this assumption, visual features were weighted according to the mean confidence value of each label occurring in a movie.

Subtitles. Subtitle features were weighted using the frequency of the words, occurring in subtitles for different movies, and normalized afterwards by applying *min-max* normalization.

3.3 Recommendation Algorithm

We built a hybrid recommender system that extends the Matrix Factorization model and enables it to exploit different types of data. Hence, the recommender system has become capable of using heterogeneous data including different types of side information (visual features & genre of movies, ratings & tags of users). The implementation of the hybrid recommender algorithm has been done using a popular library, i.e., *LightFM* [24]. The hybrid recommender system can learn the latent embeddings for users and items and encodes the user preferences over items. When these representations are multiplied together, they create scores for every item given a user. Representations of users and items are expressed by representations of their features. Feature representations are derived at by estimating an embedding for every feature and summing the embeddings together to arrive at user and item representations. The embeddings are learned with the use of stochastic gradient descent methods.

We considered different optimization methods with different loss functions: *Weighted Approximate-Rank Pairwise (WARP)* [47], *Bayesian Personalized Ranking (BPR)* [37], and *logistic loss.* The WARP loss function is defined as [20, 47]:

$$Err_{\text{WARP}}(x_i, y_i) = L\left[rank(f(y_i|x_i))\right] \tag{1}$$

where the function $rank(f(y_i|\mathbf{x}_i))$ measures the number of negative labelled instances that are "wrongly" given a higher rank than this positive example $\mathbf{x}_i$:

$$rank(f(y_i|\mathbf{x}_i)) = \sum_{(\mathbf{x}',y')\in C_u^-} \mathbb{I}\left[f(y'|\mathbf{x}') \geq f(y|\mathbf{x}_i)\right] \tag{2}$$

where $\mathbb{I}(\mathbf{x})$ is the indicator function, and $L(\cdot)$ transforms this rank into a loss:

$$L(r) = \sum_{j=1}^{r} \tau_j, \text{with } \tau_1 \geq \tau_2 \geq \cdots \geq 0. \tag{3}$$

This class of functions allows one to define different choices of $L(\cdot)$ with different minimizers. Minimizing L with $\tau_1 = 1$ and $\tau_{i>1} = 0$, the precision at 1 is optimized, $\tau_j = \frac{1}{Y-1}$ would optimize the mean rank, while for $\tau_{i\leq k} = 1$ and $\tau_{i>k} = 0$ the precision at k is optimized. For $\tau_i = 1/i$ a smooth weighing is given, where the top position is given more weight, with rapidly decreasing weight for lower positions. This is useful when opimizing Precision@K for a range of different values at K is desirable.

BPR [37] is one of the state-of-the-art algorithms exploit homogeneous implicit feedbacks. It assumes that a user prefers a consumed item to an unconsumed item, denoted as $(u,i) \succ (u,j)$ or $\hat{r}_{uij} > 0$. Mathematically, BPR solves the following minimization problem [37]:

$$\min_{\Theta} \sum_{(u,i,j):(u,i)\succ(u,j)} f_{uij}(\Theta) + \mathcal{R}_{uij}(\Theta) \tag{4}$$

where the loss function $f_{uij}(\Theta) = -\ln \sigma(\hat{r}_{uij})$ is designed to encourage pairwise competition with $\sigma(\mathbf{x}) = 1/(1 + \exp(-\mathbf{x}))$ and $\hat{r}_{uij} = \hat{r}_{ui} - \hat{r}_{uj}$. Note that $\mathcal{R}_{uij}(\Theta) = \frac{\alpha}{2} \|U_u.\|^2 + \frac{\alpha}{2}(\|V_i.\|^2 + \|V_j.\|^2) + \frac{\alpha}{2}(\|B_i\|^2 + \|B_j\|^2)$ is the regularization term used to prevent overfitting, and $\hat{r}_{ui} = \langle U_u., V_i.\rangle + b_i$ is the prediction rule based on user u's latent feature vector $U_u. \in \mathbb{R}^{1\times d}$, item i's latent feature vector $V_i. \in \mathbb{R}^{1\times d}$ and item bias $B_i \in \mathbb{R}$.

4   Experiments and Results

4.1   Evaluation Methodology

We have evaluated our proposed recommendation technique based on (automatic) visual features considering different optimization methods, i.e., WARP, BPR, and logistic loss functions utilizing both item features and user interactions. Each model was trained on one of two types of automatic features (i.e., item embeddings), namely *Deep Visual-f, Deep Visual-c*. For the baselines we, have considered recommendation based on *subtitles, tags,* or *genre*. While subtitle can be automatically extracted, both genre and tags requires human-annotation. In addition to item features, MovieLens1M dataset [17] has been utilized. In order to simulate the cold-start scenario, we have randomly sampled the dataset. The final result contained 272,515 ratings for 1514 items provided by 6040 users.

The train and test sets were built by following a hold-out methodology, i.e., randomly splitting the dataset into 80% (train) and 20% (test) disjoint subsets. The proposed recommendation models have been trained using the train set and evaluated using the test set. Hyperparameter tuning has been performed using a random search to fit LightFM models with random hyperparameter values and evaluating the model performance on the validation set. Based on the hyperparameter tuning result, models were trained over 25 epochs with AdaGrad [10] as learning rate schedule and learning rate of 0.06 (Table 1).

Table 1. Comparison of the recommendation quality based on automatic features and manual features.

Feature	Type	Precision@K	Recall@K	AUC	Reciprocal rank
Tag	*manual*	0.027	0.080	0.518	0.084
Genre	*manual*	0.040	0.024	0.698	0.118
Subtitle	*automatic*	0.070	0.048	0.849	0.179
Deep Visual-c	*automatic*	0.157	0.103	0.846	**0.342**
Deep Visual-f	*automatic*	**0.166**	**0.109**	**0.860**	0.354

4.2   Experiment A: Recommendation Quality

In the first set of experiments, we have measured the quality of the recommendation based on automatic visual features, extracted by the deep learning model. Figure 1 represents the results obtained in this experiment.

First of all, as it can be seen, both version of our proposed recommendation technique (Deep Visual-f and Deep Visual-c), based on visual features, outperform all the other different baselines. In terms of Precision@K, Deep Visual-f achieves the score of 0.166 and Deep Visual-c achieves score of 0.157. The next best precision score is obtained by recommendation based on movie subtitles with the score of 0.070, where recommendation based on manual features, i.e., genre and tag, received the lowest scores, i.e., 0.040 and 0.027, respectively. In terms of Recall@K, similarly, both Deep Visual-f and Deep Visual-c achieved the best results with the scores of 0.109 and 0.103, respectively. The next best performance has been observed for recommendation based on the subtitle with the score of 0.048. The recommendation based on genre and tag have performed the worst with the scores of 0.24 and 0.080, respectively.

In terms of AUC, recommendation based on subtitle has achieved a great score of 0.849, however, Deep Visual-f still has obtained the best score of 0.860. Recommendation based Deep Visual-c has obtained the next best result with the score of 0.846. Recommendation based on genre and tag have received the lowest scores, i.e., 0.698 and 0.518, respectively. Finally, in terms of Reciprocal

Rank, again, proposed recommendation technique based on either Deep Visual-f and Deep Visual-c has achieved the highest scores. While the observed scores for Deep Visual-f and Deep Visual-c were 0.354 and 0.341, the next best score was almost half of these values, observed for recommendation based on subtitle with a score of 0.179. As expected, both genre and tag have shown the worst performance with the scores of 0.118 and 0.084.

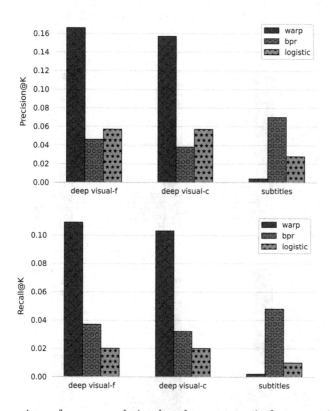

Fig. 1. Comparison of recommendation based on automatic features using different optimization methods in terms of (top) Precision and (bottom) Recall.

4.3 Experiment B: Comparing Loss Functions

In the second set of experiments, we have compared the recommendation based on automatic features when different types of optimization algorithms have used. The results have been illustrated in Fig. 2 and 3.

First of all, as it can be seen, different loss function (hence optimization algorithm) can yield different recommendation quality for each type of automatic features. For the visual features, either deep visual-c or deep visual-f, the best results have been achieved using *warp* loss function, considering all metrics, i.e., Precision@K, Recall@K, AUC, and Reciprocal Rank. Surprisingly, *bpr* loss function does not perform well and in some cases (e.g., Precision) it yields the worst results.

For the subtitle features, on the other hand, the best results have been achieved by bpr loss function for all metrics. In contrary, the worst results are obtained by warp loss function. This is another surprising result as both types of visual and subtitle features are of categorical type and might be expected to share similarities in their nature. However, apparently, they represent different aspects of the videos that are perhaps different and hence shall be handled differently.

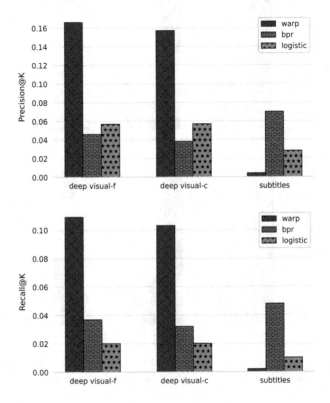

Fig. 2. Comparison of recommendation based on automatic features using different optimization methods in terms of (top) Precision and (bottom) Recall.

Overall, these promising results have shown the excellent performance of hybrid recommendation based on visual features, using different optimization methods. The results have clearly illustrated the substantial potential behind these features that can exploited when no other types of content features are provided to a movie recommender system.

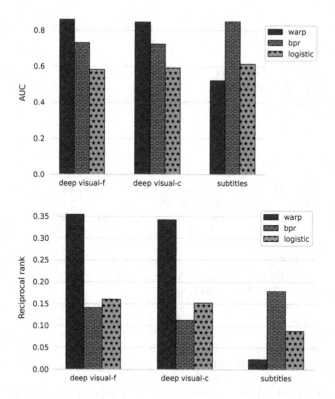

Fig. 3. Comparison of recommendation based on different automatic features using different optimization methods in terms of (top) AUC and (bottom) Reciprocal Rank.

5   Conclusions and Future Work

This paper focuses on the new item problem as part of cold start in recommender systems and proposes a hybrid technique to generate recommendation based on visual features, automatically extracted from movies. The visual features have been extracted using a deep learning network (i.e., CNN) and exploited to generate movie recommendation. The proposed technique can be fully automated and does not require any human involvement and hence can be utilized when recommending movies that have neither any rating nor content features.

The proposed hybrid technique has been evaluated using a large dataset of movie trailers and compared against recommendation based on other features, i.e., subtitle, genre and tags. The results have shown that our proposed recommendation technique can outperform the other techniques with regards to all the evaluation metrics.

In future, we would like to extend these experiments by taking into account the datasets, collected from other social networks (e.g., Instagram). In addition to that we will extend our feature set by considering other types of features that can be extracted automatically. Finally, we will adopt other feature fusions when aggregating the visual features.

Acknowledgements. This work was supported by industry partners and the Research Council of Norway with funding to MediaFutures: Research Centre for Responsible Media Technology and Innovation, through The Centres for Research-based Innovation scheme, project number 309339.

References

1. Aggarwal, C.C., et al.: Recommender Systems, vol. 1. Springer, Cham (2016). https://doi.org/10.1007/978-3-319-29659-3
2. Canini, L., Benini, S., Leonardi, R.: Affective recommendation of movies based on selected connotative features. IEEE Trans. Circuits Syst. Video Technol. **23**(4), 636–647 (2013)
3. Cantador, I., Bellogín, A., Vallet, D.: Content-based recommendation in social tagging systems. In: Proceedings of the Fourth ACM Conference on Recommender Systems, pp. 237–240. ACM (2010)
4. Cantador, I., Konstas, I., Jose, J.M.: Categorising social tags to improve folksonomy-based recommendations. Web Semant. **9**(1), 1–15 (2011)
5. de Gemmis, M., Lops, P., Musto, C., Narducci, F., Semeraro, G.: Semantics-aware content-based recommender systems. In: Ricci, F., Rokach, L., Shapira, B. (eds.) Recommender Systems Handbook, pp. 119–159. Springer, Boston, MA (2015). https://doi.org/10.1007/978-1-4899-7637-6_4
6. Deldjoo, Y., et al.: Audio-visual encoding of multimedia content for enhancing movie recommendations. In: Proceedings of the 12th ACM Conference on Recommender Systems, RecSys 2018, pp. 455–459. Association for Computing Machinery, New York (2018)
7. Deldjoo, Y., Elahi, M., Cremonesi, P., Garzotto, F., Piazzolla, P., Quadrana, M.: Content-based video recommendation system based on stylistic visual features. J. Data Semant. **5**(2), 99–113 (2016). https://doi.org/10.1007/s13740-016-0060-9
8. Deldjoo, Y., Elahi, M., Cremonesi, P., Garzotto, F., Piazzolla, P., Quadrana, M.: Content-based video recommendation system based on stylistic visual features. J. Data Semant. **5**, 1–15 (2016)
9. Di Noia, T., Mirizzi, R., Ostuni, V.C., Romito, D., Zanker, M.: Linked open data to support content-based recommender systems. In: Proceedings of the 8th International Conference on Semantic Systems, pp. 1–8. ACM (2012)
10. Duchi, J., Hazan, E., Singer, Y.: Adaptive subgradient methods for online learning and stochastic optimization. J. Mach. Learn. Res. **12**, 2121–2159 (2011)
11. Elahi, M.: Empirical evaluation of active learning strategies in collaborative filtering. Ph.D. thesis, Ph.D. Dissertation, Free University of Bozen-Bolzano (2014)
12. Elahi, M., et al.: Recommending videos in cold start with automatic visual tags. In: Adjunct Proceedings of the 29th ACM Conference on User Modeling, Adaptation and Personalization, pp. 54–60 (2021)
13. Elahi, M., Beheshti, A., Goluguri, S.R.: Recommender systems: challenges and opportunities in the age of big data and artificial intelligence. In: Data Science and Its Applications, pp. 15–39. Chapman and Hall/CRC (2021)
14. Elahi, M., Braunhofer, M., Gurbanov, T., Ricci, F.: User Preference Elicitation, Rating Sparsity and Cold Start: Algorithms, pp. 253–294 (2018)

15. Elahi, M., Deldjoo, Y., Bakhshandegan Moghaddam, F., Cella, L., Cereda, S., Cremonesi, P.: Exploring the semantic gap for movie recommendations. In: Proceedings of the Eleventh ACM Conference on Recommender Systems, pp. 326–330. ACM (2017)

16. Elahi, M., Ricci, F., Rubens, N.: Active learning strategies for rating elicitation in collaborative filtering: a system-wide perspective. ACM Trans. Intell. Syst. Technol. (TIST) **5**(1), 13 (2013)

17. Harper, F.M., Konstan, J.A.: The movielens datasets: history and context. ACM Trans. Interact. Intell. Syst. (TiiS) **5**(4), 19 (2016)

18. Hawashin, B., Lafi, M., Kanan, T., Mansour, A.: An efficient hybrid similarity measure based on user interests for recommender systems. Expert Syst. **37**, e12471 (2019)

19. Hazrati, N., Elahi, M.: Addressing the new item problem in video recommender systems by incorporation of visual features with restricted boltzmann machines. Expert. Syst. **38**(3), e12645 (2021)

20. Hong, L.J.: Pairwise loss (warp) (2012). Accessed 21 Jan 2021

21. Jannach, D., Zanker, M., Felfernig, A., Friedrich, G.: Recommender Systems: An Introduction. Cambridge University Press, Cambridge (2010)

22. Koren, Y., Bell, R.: Advances in collaborative filtering. In: Ricci, F., Rokach, L., Shapira, B., Kantor, P.B. (eds.) Recommender Systems Handbook, pp. 145–186. Springer, Boston (2011). https://doi.org/10.1007/978-0-387-85820-3_5

23. Koren, Y., Bell, R., Volinsky, C.: Matrix factorization techniques for recommender systems. Computer **42**(8), 30–37 (2009)

24. Kula, M.: Metadata embeddings for user and item cold-start recommendations. In: Bogers, T., Koolen, M. (eds.) Proceedings of the 2nd Workshop on New Trends on Content-Based Recommender Systems co-located with 9th ACM Conference on Recommender Systems (RecSys 2015), Vienna, Austria, 16–20 September 2015. CEUR Workshop Proceedings, vol. 1448, pp. 14–21. CEUR-WS.org (2015)

25. Lee, J., Abu-El-Haija, S., Varadarajan, B., Natsev, A.: Collaborative deep metric learning for video understanding. In: Proceedings of the 24th ACM SIGKDD International Conference on Knowledge Discovery and Data Mining, KDD 2018, pp. 481–490. Association for Computing Machinery, New York (2018)

26. Lehinevych, T., Kokkinis-Ntrenis, N., Siantikos, G., Dogruöz, A.S., Giannakopoulos, T., Konstantopoulos, S.: Discovering similarities for content-based recommendation and browsing in multimedia collections. In: 2014 Tenth International Conference on Signal-Image Technology and Internet-Based Systems (SITIS), pp. 237–243. IEEE (2014)

27. Li, Y., Wang, H., Liu, H., Chen, B.: A study on content-based video recommendation. In: 2017 IEEE International Conference on Image Processing (ICIP), pp. 4581–4585 (2017)

28. Lops, P., de Gemmis, M., Semeraro, G.: Content-based recommender systems: state of the art and trends. In: Ricci, F., Rokach, L., Shapira, B., Kantor, P.B. (eds.) Recommender Systems Handbook, pp. 73–105. Springer, Boston, MA (2011). https://doi.org/10.1007/978-0-387-85820-3_3

29. Martins, E.F., Belém, F.M., Almeida, J.M., Gonçalves, M.A.: On cold start for associative tag recommendation. J. Assoc. Inf. Sci. Technol. **67**(1), 83–105 (2016)

30. Milicevic, A.K., Nanopoulos, A., Ivanovic, M.: Social tagging in recommender systems: a survey of the state-of-the-art and possible extensions. Artif. Intell. Rev. **33**(3), 187–209 (2010)

31. Moghaddam, F.B., Elahi, M., Hosseini, R., Trattner, C., Tkalčič, M.: Predicting movie popularity and ratings with visual features. In: 2019 14th International Workshop on Semantic and Social Media Adaptation and Personalization (SMAP), pp. 1–6. IEEE (2019)
32. Open-source and Google. Keras (2020). Accessed 21 Jan 2021
33. opensubtitles.org. Opensubtitles (2020). Accessed 01 Nov 2020
34. Pazzani, M.J., Billsus, D.: Content-based recommendation systems. In: Brusilovsky, P., Kobsa, A., Nejdl, W. (eds.) The Adaptive Web. LNCS, vol. 4321, pp. 325–341. Springer, Heidelberg (2007). https://doi.org/10.1007/978-3-540-72079-9_10
35. Rassweiler Filho, R.J., Wehrmann, J., Barros, R.C.: Leveraging deep visual features for content-based movie recommender systems. In: 2017 International Joint Conference on Neural Networks (IJCNN), pp. 604–611. IEEE (2017)
36. Renckes, S., Polat, H., Oysal, Y.: A new hybrid recommendation algorithm with privacy. Expert. Syst. **29**(1), 39–55 (2012)
37. Rendle, S., Freudenthaler, C., Gantner, Z., Schmidt-Thieme, L.: BPR: Bayesian personalized ranking from implicit feedback. In: Proceedings of the Twenty-Fifth Conference on Uncertainty in Artificial Intelligence, pp. 452–461. AUAI Press (2009)
38. Resnick, P., Varian, H.R.: Recommender systems. Commun. ACM **40**(3), 56–58 (1997)
39. Ricci, F., Rokach, L., Shapira, B.: Recommender systems: introduction and challenges. In: Ricci, F., Rokach, L., Shapira, B. (eds.) Recommender Systems Handbook, pp. 1–34. Springer, Boston (2015). https://doi.org/10.1007/978-1-4899-7637-6_1
40. Rimaz, M.H., Elahi, M., Bakhshandegan Moghadam, F., Trattner, C., Hosseini, R., Tkalčič, M.: Exploring the power of visual features for the recommendation of movies. In: Proceedings of the 27th ACM Conference on User Modeling, Adaptation and Personalization, pp. 303–308 (2019)
41. Rimaz, M.H., Elahi, M., Bakhshandegan Moghadam, F., Trattner, C., Hosseini, R., Tkalčič, M.: Exploring the power of visual features for the recommendation of movies. In: ACM UMAP 2019 - Proceedings of the 27th ACM Conference on User Modeling, Adaptation and Personalization, pp. 303–308 (2019)
42. Rimaz, M.H., Hosseini, R., Elahi, M., Moghaddam, F.B.: AudioLens: audio-aware video recommendation for mitigating new item problem. In: Hacid, H., et al. (eds.) ICSOC 2020. LNCS, vol. 12632, pp. 365–378. Springer, Cham (2021). https://doi.org/10.1007/978-3-030-76352-7_35
43. Robertson, S.E., Jones, K.S.: Relevance weighting of search terms. J. Am. Soc. Inf. Sci. **27**(3), 129–146 (1976)
44. Simonyan, K., Zisserman, A.: Very deep convolutional networks for large-scale image recognition. In: 3rd International Conference on Learning Representations, ICLR 2015 - Conference Track Proceedings, pp. 1–14 (2015)
45. Soares, M., Viana, P.: Tuning metadata for better movie content-based recommendation systems. Multimedia Tools Appl. **74**(17), 7015–7036 (2014). https://doi.org/10.1007/s11042-014-1950-1
46. Wang, L., Zeng, X., Koehl, L., Chen, Y.: Intelligent fashion recommender system: fuzzy logic in personalized garment design. IEEE Trans. Hum.-Mach. Syst. **45**(1), 95–109 (2015)

47. Weston, J., Bengio, S., Usunier, N.: Wsabie: scaling up to large vocabulary image annotation. In: Proceedings of the Twenty-Second International Joint Conference on Artificial Intelligence - Volume Volume Three, IJCAI 2011, pp. 2764–2770. AAAI Press (2011)
48. Bo Yang, Yu., Lei, J.L., Li, W.: Social collaborative filtering by trust. IEEE Trans. Pattern Anal. Mach. Intell. **39**(8), 1633–1647 (2016)
49. Yang, B., Mei, T., Hua, X.S., Yang, L., Yang, S.Q., Li, M.: Online video recommendation based on multimodal fusion and relevance feedback. In: Proceedings of the 6th ACM International Conference on Image and Video Retrieval, pp. 73–80. ACM (2007)
50. Zhao, X., et al.: Integrating rich information for video recommendation with multi-task rank aggregation. In: Proceedings of the 19th ACM International Conference on Multimedia, pp. 1521–1524. ACM (2011)

Automatic Image Annotation Using Quantization Reweighting Function and Graph Neural Networks

Fariba Lotfi$^{(\boxtimes)}$, Mansour Jamzad, and Hamid Beigy

Sharif University of Technology, Tehran, Iran
flotfi@ce.sharif.edu, {jamzad,beigy}@sharif.edu

Abstract. This paper investigates the issues in image annotation, which automatically assigns appropriate tags to a given image describing its content the best. Due to the introduction of deep learning methods and the use of graph neural networks (GNNs), automatic image annotation has made significant progress in recent years. An image may have multiple tags associated with it, and a tag may appear in several images within the dataset; therefore, it is inefficient to study each tag individually. Some studies have attempted to model the dependencies between tags using vocabulary to improve the performance of automatic image annotation. However, it remains unclear how to create an appropriate vocabulary graph. We propose to construct this graph by modeling the relationship between tags. In the tag graph, edges are reweighted based on cosine similarity and a quantization function. To represent each node in the graph, we use two methods of word embedding. We then use graph neural networks to extract graph features. From the graph and image features, we obtain our output vector (set of class probabilities). The proposed approach is evaluated using *precision, recall, F_1*, and N^+ performance measures on two public benchmark datasets (Corel5k, and ESP Game). Results of experiments show that our method is superior to current state-of-the-art methods. On Corel5k, we achieved the best performance with N^+ and *recall*, the second-best performance with F_1. The second-best performance with N^+ and *precision* and the best F_1 are also achieved on ESP Game.

Keywords: Automatic image annotation · Deep learning · Graph neural networks · Quantization function · Tag graph · Word embedding

1 Introduction

Advances in machine learning, especially in deep learning architectures, have led to an increase in neural network development and deployment. This development has enabled a variety of new applications in the fields of digital image processing, computer vision, and natural language processing. In addition, the popularity of image retrieval (IR) systems [31] has led researchers to focus more

© Springer Nature Switzerland AG 2022
H. Hacid et al. (Eds.): ICSOC 2021 Workshops, LNCS 13236, pp. 46–60, 2022.
https://doi.org/10.1007/978-3-031-14135-5_4

on automatic image annotation tasks, which is a primary step in many computer vision applications. Image annotation aims to identify image content by assigning tags or keywords to them; image classification, retrieval, scene identification, and autonomous driving are some of its applications. Three primary industries can be improved by automatic image annotation, such as agriculture, healthcare, and transportation. Some studies solve the task of image annotation by using the structure of RNNs [12,23,26], while others use the generative adversarial networks (GANs) [4]. Moreover, some other approaches utilise K-nearest neighbors [17,24]. For image feature extraction, convolutional neural networks have proved to perform much better than traditional methods. Although many efforts have been made to correlate image features and tags [3,30], finding an appropriate solution remains a challenge, leaving a gap between these two concepts of data. This gap is known as the semantic gap [32]. As a result, to bridge the semantic gap reasonably, we must model how tagged concepts are related to extracted images. We intend to assign to each image I a few keywords that most accurately describe its contents. The output is a vector $Y \in \{0,1\}^K$, where K is the total number in the tags vocabulary. In general, an image annotation model consists of two components: (i) image feature extraction and (ii) output vector prediction, which is based on combining all the extracted features and visual information from the network. Adding more information, however, can improve the performance of an image annotation model. For example, there are semantically related tags within each dataset, and the possibility of their appearance in a single image is relatively high. Thus, by modeling relationships between tags, automatic image annotation can be made more accurate. Moreover, our datasets are highly imbalanced, and modeling the relationship between tags can help us to improve the annotation performance on images with low-frequency tags. Our dataset imbalance problem has been discussed in detail in Sect. 4.1.

In this paper, we develop a vocabulary graph based on the tags in the dataset. We adjust the weights of edges between each pair of nodes using a new quantization method. The nodes are the tags (word embedding vectors represent each node), and the edges are weighted based on the cosine similarity between the two nodes (images). We then apply our quantization function and a layer of graph neural network. Quantizing tag graph edge based on cosine similarity and applying graph convolution has not been used in image annotation and these two benchmark datasets to the best of our knowledge. The four standard evaluation metrics for automatic image annotation are *precision*, *recall*, F_1, and N^+. Additionally, we present several experiments to validate our method and demonstrate that it is as competitive as or better than the state-of-the-art results on the two datasets usually used as benchmarks in automatic image annotation.

Here is a list of the contributions to this paper: (1) Construct a tag graph, and we introduce a new quantization method to reweight the similarity scores calculated for each pair of nodes in the graph. (2) Propose a method for automatically annotating images. A quantization method and two methods of word embedding are used to construct the tag graph. Then apply a graph convolutional layer to our tag graph. This is multiplied by the image features extracted by a pre-trained network. (3) Our model is evaluated using two benchmark test-standard datasets, and the results indicate that it achieves state-of-the-art (SOTA) level performance, while other scores are highly competitive.

The remainder of this paper is organized as follows. The related work is discussed in Sect. 2. We describe our methodology in detail in Sect. 3. Datasets, evaluation metrics, and experimental results are described in the Sect. 4. We conclude the paper in Sect. 5.

2   Related Work

Automatic image annotation is a challenging problem that has gained a lot of attention from the machine vision research community lately. Different approaches have been proposed to address this problem. After discussing a brief review of the state-of-the-art methods in this field in Sect. 2.1, we introduce three famous architectures in graph neural networks in Sect. 2.2.

2.1   Automatic Image Annotation

Image annotation has long been a challenge addressed in several ways. The MVG-NMF framework is one of its common approaches [5]. This framework factors data into a series of non-negative bases and coefficients corresponding to several views, and each view depends on different dimensions and features. A second approach utilises two different strategies to alleviate the problems caused by imposing a fixed order on the labels [15]. The nodes represent semantic concepts, and the edges represent co-occurrence patterns in [16]. In this work, co-occurrences are arranged hierarchically and based on visual features, concept signatures (semantic descriptions of images) are probabilities of each concept being refined. In 2PKNN [17], image-label and image-image similarities are calculated in two steps. Hence, the comparison measure is important. RIA model [12] solves the task of image annotation as a sequence generation problem so that it can predict the number of output tags according to image contents.

Diverse image annotation (DIA) [22] is a new annotation method that treats DIA as a subset selection problem and uses the conditional detrimental point process (DPP) model. Moreover, it explores semantic hierarchy and synonyms among candidate tags in order to establish weighted semantic paths. $D^2IA - GAN$ [4] creates relevant, yet distinct and diverse annotations for an image. A generative model in $D^2IA - GAN$ is adversarially trained using a generative adversarial network (GAN) model. SSL-AWF [8] obtains a strong classifier that can make full use of unlabeled data. It is a semi-supervised approach based on adaptive weighted fusion for automatic image annotation. Li et al. [23] propose a recurrent highlight network that focuses on the most relevant regions in the image using attention mechanisms. By using multiplicative gates, they develop a gated recurrent relation extractor (GRRE) to model the tag relations. The CCA model [25] uses Convolutional Neural Network features extracted from an image and word embedding vectors to represent the tags in vocabulary. 2PKNN deep [24] analyzes the semantic information extracted by pre-trained deep features and evaluates their performance in k nearest neighbor-based approaches. Results from CNN-RNN [26] are significantly better when regularised semantically. The changes in the image embedding layer and the introduction of semantic regularisation to the CNN-RNN architecture makes the training faster and more stable. MS-CNN architecture [27] extracts and fuses features at different scales corresponding to visual concepts at different levels of abstraction. The model is capable of estimating the optimal label quantity for an image. SEM [28] divides images with similar features into semantic neighbor groups. It then distributes the tags belonging to the semantic neighbor group to the query image.

2.2 Graph Neural Networks

Many approaches have been developed to handle graph-structured data that operate directly on graphs [13, 21, 29]. The Graph Convolutional Network (GCN) approach [13] introduces a simple layer-wise propagation rule for neural network models. It is motivated by first-order approximations of spectral graph convolutions. GCN was designed to be learned with the presence of both training and test data. Then, FASTGCN model [29] is enhanced by importance sampling. The system eliminates the need to rely on test data, but it also allows for controllable per-batch computation costs. GraphSage [21] extends GCNs to inductive unsupervised learning and proposes a generalization framework using trainable aggregation functions. In particular, we follow the idea of the GCN approach [13] as a baseline to apply convolution on our tag graph.

3   Method

In this section, we present our automatic image annotation model. Our approach uses an architecture based on our previous research [3] but uses another famous tags embedding method [19] and two edge calculation methods in the tag graph. First, we explain the construction of an undirected graph based on dataset tags in Sect. 3.1. Then, we explain the details of the base architecture in Sect. 3.2. Finally, the implementation details are discussed in Sect. 3.3, and the block diagram of our model is presented in Fig. 1.

3.1   Tag Graph

Our goal is to model the relationship between tags by constructing an undirected graph. Each graph $G = (V, E)$ is represented with its nodes (V) and the set of edges (E). We define the nodes and edges of this graph in the following sections in detail.

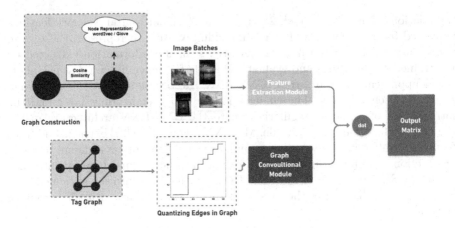

Fig. 1. Block diagram of our proposed learning model in training phase.

Graph Nodes. Each tag forms a node in the graph. These tags, denoted by t and t', are the words themselves and the word embedding vectors obtained from word2vec [18] and GloVe [19], respectively. The index of the tag in the vocabulary is represented by i.

$$t'_i = GloVe(t_i) \tag{1}$$
$$t'_i = word2vec(t_i) \tag{2}$$

Graph Edges. In the graph, the weight of each edge indicates the similarity between the nodes. According to Eq. 3, the similarity between each pair of nodes in the graph is calculated by the cosine similarity of their word embedding vectors t'_i and t'_j. The vectors are derived from the representation discussed previously. The initial weight between each pair of nodes is represented with W_{ij}, which are stored in a matrix.

$$W_{ij} = \frac{t'_i \cdot t'_j}{\|t'_i\| \times \|t'_j\|} \tag{3}$$

Reweighting Edges. First, we binarize the initial edge weights. We assign the value of zero to weights with a value less than β and one to weights with a value greater than β. We reweight them and represent them with W'_{ij} to prevent the weights from over-smoothing [14]. The diameter of the matrix is set to α, and the other values are determined by the neighborhood distribution. This re-weighting step is explained in detail in [3,14].

Quantizating Edges. A quantization function maps values from a large set (usually a continuous set) to output values in a smaller set. Many quantization functions exist, such as rounding and truncation. In this step, we utilize a scalar quantization function as illustrated in Fig. 2 to calculate the weights of graph edges.

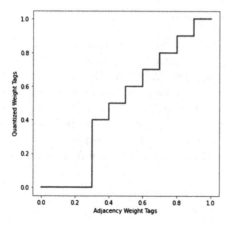

Fig. 2. Quantization function used for calculating the weights (edges) between graph nodes (vertices).

According to our previous discussion, the similarity between two nodes is calculated by cosine similarity. When the similarity of two nodes is less than a threshold, there should not be a connection between them (i.e. no edge). In the tag graph, we will ignore nodes that have a similarity score less than θ. If their weights exceed this threshold, we quantize them and round them up. The quantization function is as follows:

$$W'_{ij} = \begin{cases} 0 & W_{ij} \le \theta \\ \lceil 10W_{ij} \rceil / 10 & else \end{cases} \tag{4}$$

3.2   Architecture Design

We have already mentioned that the underlying structure of this approach is based on our previous research [3]. To extract image features, we used the last layer of a pre-trained CNN model. To integrate all features extracted from previous layers, the features are passed through one fully connected layer. One layer of graph convolution [13] is applied on the tag graph to model the relation of tags at a higher level which was explained in detail in Sect. 3.1. Afterward, the output of the graph convolution layer (i.e. tag descriptors) is multiplied by the output of the fully connected layer (image features). Multiplication forces the more relevant classes into the output matrix at the same time (e.g. if a label appears in the image, the related tag may also appear in the output). The output probability matrix is the result of this multiplication. The full structure of our architecture in training phase is represented in Fig. 1. In the annotation phase, two post-rectifying methods are applied to the final output [6,7]. The first rectifying algorithm finds unassigned tags and assigns them to the most appropriate image based on the correlation between them and the assigned ones. The second algorithm updates the output matrix based on the most similar image to the corresponding image using the cosine similarity measure.

3.3   Implementation Details

The pre-trained CNN model used in our architecture is ResNet101 [20]. We represent the tags vector in the dataset using two famous pre-trained word embedding algorithms, word2vec [18], and GloVe [19]. Our quantizing function is set to 0.3 for all datasets. We found the representation of all the tags with the GloVe embedding method. Unfortunately, some tags did not have a pre-trained vector in word2vec. Therefore, we represent them by embedding the nearest word to them. Specifically, we train our models separately for each dataset over 100 epochs. To update model parameters, we use K-Fold cross-validation.

4   Experiments

In this section, we present two datasets for automatic image annotation, followed by our evaluation metrics. To evaluate our approach, we conducted extensive experiments on the mentioned datasets. In the end, we present the quantitative results of our model.

4.1   Dataset

Similar to previous automatic annotation tasks, our model is also evaluated on two widely used datasets. There are 5,000 images in the Corel5k dataset, organized by 260 tags. The ESP Game dataset contains 20,770 images with 268 tags. Table 1 shows the general statistics of the two datasets. Each dataset is split into training, validation, and test splits, each with 80%, 10%, and 10% of the samples in the dataset. *Data Size* depicts the size of the dataset, *Train Size* the number of training samples, *Test Size* the number of test samples, VS the vocabulary size, WI the words per image, and finally IW the images per word.

Table 1. The statistics of datasets.

Dataset	Data size	Train size	Test size	VS	WI	IW
Corel5K [2]	5000	4500	500	260	3.4	58.6
ESP game [1]	20770	18689	2081	268	4.7	362.7

Our datasets are very imbalanced because of the unequal distribution of classes within each dataset, and it is one of the critical challenges every researcher is facing. Figure 3 shows the histograms of these datasets in terms of image frequency for each tag. For example, the number of images per class varies from less than ten to approximately one thousand in Corel5k. Unfortunately, we cannot learn tags with low image frequency while training the model in the benchmark datasets. In these cases, using statistical methods (post-processing) and metadata information such as tag graphs are essential.

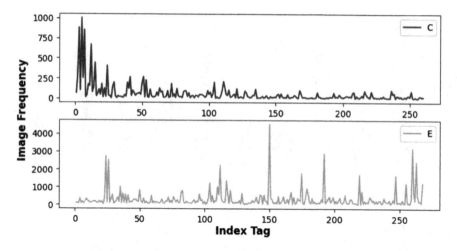

Fig. 3. Image frequencies per tag index for Corel5K (C) and ESP Game (E).

4.2   Evaluation Metrics

To evaluate our method, we use four quantitative metrics: $precision(PC)$, $recall(RC)$, F_1, and N^+. TP stands for true positive, FP for false positive, and FN for false negative. $sgn(RC_t)$ is the sign function, and RC_t is the tag recall for tag t.

$$PC = \frac{TP}{TP + FP} \tag{5}$$

$$RC = \frac{TP}{TP + FN} \tag{6}$$

$$F_1 = \frac{2 * PC * RC}{PC + RC} \tag{7}$$

$$N^+ = \sum sgn(RC_t) \quad t \in \text{Tag List.} \tag{8}$$

4.3   Results

This section compares our approach quantitatively with other state-of-the-art methods to verify its effectiveness. The state-of-the-art methods are selected based on their publication date and accuracy of results. For Corel5k, we achieved first, second, and third place with N^+ measure (with scores 229, 228, and 215, respectively) with our two reweighting methods and second and third place with F_1 measure with the quantization reweighting method. These results are presented in Table 2, and Fig. 4 in detail. By allowing the model to tag the image with as many correct tags as possible, we achieve the highest score compared to previous methods with the *recall* measure. Some concepts and tags are not fully

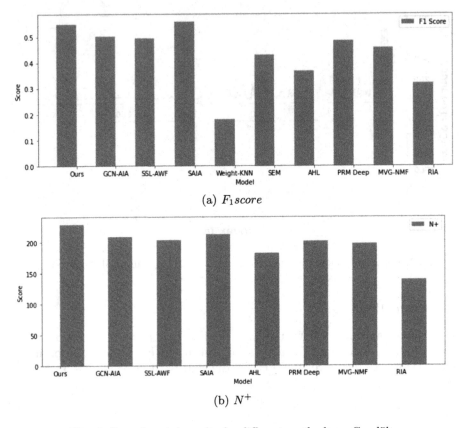

(a) $F_1 score$

(b) N^+

Fig. 4. Experimental results for different methods on Corel5k.

covered by the ground truth in the two datasets. Thus, we try to identify all the true concepts, even though it damages our *precision* score, but results show the highest score based on the N^+ measure. On ESP Game, we achieve the best performance with F_1, second and third place with N^+ and *recall*, and second place with *precision*. Results of the ESP Game are also presented in Table 3, and Fig. 5. We represent three images as the annotation samples of our proposed architecture in Fig. 6. The rightmost column represents the annotation results of the proposed architecture. Tags in black are the ground truth. However, our proposed model also generates some relevant tags shown in green that do not exist in the ground truth. Moreover, the tags in red are mispredicted. However, we had to assume the tags in green as incorrect, to provide a fair comparison to the previous state-of-the-art studies. In Table 2 and 3, green cells represent deep-based methods, and the other ones are non-deep.

Table 2. The quantitative experiments of Corel5K. RW and Q represent re-weighting and quantization, respectively. We use bold red to highlight the best and bold blue to highlight the second-best performance.

Models	PC	RC	F_1	N^+
2PKNN Deep (2016) [24]	-	-	-	-
RIA (2016) [12]	0.32	0.35	0.32	139
MVG-NMF (2017) [5]	0.44	0.475	0.456	197
PRM Deep (2018) [6]	0.453	0.5173	0.483	201
AHL (2019) [10]	0.3296	0.4007	0.3671	182
SEM (2019) [28]	0.37	0.52	0.43	-
Weight-KNN (2020) [9]	0.22	0.15	0.18	-
SAIA (2020) [7]	0.5546	0.5655	0.56	212
SSL-AWF (2021) [8]	**0.51**	0.48	0.495	203
RW-AIA (word2vec) (2021) [3]	0.3926	0.6935	0.5014	208
RW-AIA (Glove) - Proposed Method	0.3407	**0.7920**	0.4764	**228**
QAIA (word2vec) - Proposed Method	0.4181	**0.8039**	**0.5501**	229
QAIA (Glove) - Proposed Method	0.4059	0.7257	0.5206	215

Table 3. The quantitative experiments of ESP Game. RW and Q represent re-weighting and quantization, respectively. We use bold red to highlight the best and bold blue to highlight the second-best performance.

Models	PC	RC	F_1	N^+
2PKNN Deep (2016) [24]	0.52	0.27	0.36	250
RIA (2016) [12]	0.32	0.32	0.31	249
MVG-NMF (2017) [5]	0.437	0.314	0.367	254
PRM Deep (2018) [6]	-	-	-	-
AHL (2019) [10]	0.4579	0.2286	0.3108	221
SEM (2019) [28]	0.38	0.42	0.4	-
Weight-KNN (2020) [9]	0.46	0.22	0.3	-
SAIA (2020) [7]	0.6369	0.3874	**0.4817**	268
SSL-AWF (2021) [8]	0.49	0.39	0.434	261
RW-AIA (word2vec) (2021) [3]	0.3721	0.6281	0.4673	261
RW-AIA (Glove) - Proposed Method	0.3567	**0.5896**	0.4445	253
QAIA (word2vec) - Proposed Method	0.4372	0.4394	0.4383	264
QAIA (Glove) - Proposed Method	**0.5593**	0.4942	0.5248	**265**

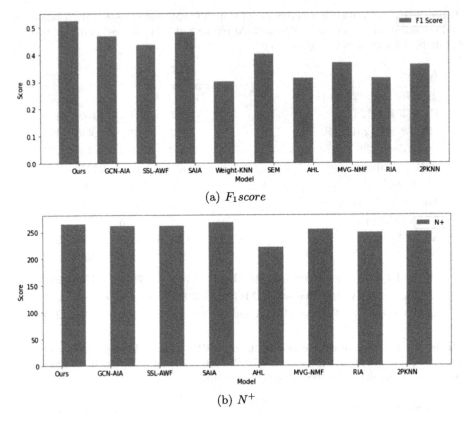

(a) $F_1 score$

(b) N^+

Fig. 5. Experimental results for different methods on ESP Game: (a) F_1 and (b) N^+.

4.4   Discussion

In Corel5k, our method was able to achieve the best results with the *recall* measure. A high *recall* in our method has resulted in lower *precision*. Moreover, it provides a very high N^+ score. Our F_1 measure is highly competitive with the state-of-the-art result. In fact since the datasets used in our experiments are highly imbalanced, obtaining the N^+ measure as the state-of-the-art and also a highly competitive F_1 proves that we have succeeded in assigning the low-frequency tags to their corresponding image. In ESP Game, we achieved the state-of-the-art score with F_1 measure and the second place with N^+ which is competitive and shows the efficiency of our model.

Dataset	Test Image	Ground Truth	Our Annotations
Corel5k		Sun Clouds Tree Sea	Sky Sun Water Clouds Tree Lake Sea Grass Birds
		Sky Jet Plane	Sky Birds Plane Runway Flight Tails F-16
		Coral Fish Ocean Reefs	Lake Sea Ships Anemone Fish Ocean Reefs Vegetation

Fig. 6. Three test samples from Corel5k with their ground truth and our model annotations. Tags printed in black are the ground truth, tags in green are our model's annotations that do not exist in the ground truth but are correct. Finally, tags in red are our model's annotations that do not exist in the ground truth and are incorrect. (Color figure online)

5   Conclusion

Our paper studies a method of automatically annotating images that is effective in tackling the problem. We use a quantization method to calculate the weights between each pair of nodes in the tag graph. To represent each node, we use two word embedding methods. Deep supervision improves the stability and efficiency of training the entire model. Modeling the relationship between tags appears to help with the image annotation problem, but there is still room for improvement. Evaluations on Corel5k and ESP Game demonstrate the efficiency of our model for automatic image annotation. In our future work, we plan to explore other connections between tags and define the graph model accordingly. Furthermore, we aim to find a new approach to construct our tag graph differently and find the optimal relationship between tags to improve its structure in our model.

References

1. Von Ahn, L., Dabbish, L.: Labeling images with a computer game. In: Proceedings of the SIGCHI Conference on Human Factors in Computing Systems, pp. 319–326, April 2004
2. Duygulu, P., Barnard, K., de Freitas, J.F.G., Forsyth, D.A.: object recognition as machine translation: learning a lexicon for a fixed image vocabulary. In: Heyden, A., Sparr, G., Nielsen, M., Johansen, P. (eds.) ECCV 2002. LNCS, vol. 2353, pp. 97–112. Springer, Heidelberg (2002). https://doi.org/10.1007/3-540-47979-1_7
3. Lotfi, F., Jamzad, M., Beigy, H.: Automatic image annotation using tag relations and graph convolutional networks. In: 2021 5th International Conference on Pattern Recognition and Image Analysis (IPRIA), pp. 1–6. IEEE, April 2021
4. Wu, B., Chen, W., Sun, P., Liu, W., Ghanem, B., Lyu, S.: Tagging like humans: diverse and distinct image annotation. In: Proceedings of the IEEE Conference on Computer Vision and Pattern Recognition, pp. 7967–7975 (2018)
5. Rad, R., Jamzad, M.: Image annotation using multi-view non-negative matrix factorization with different number of basis vectors. J. Vis. Commun. Image Represent. **46**, 1–12 (2017)
6. Khatchatoorian, A.G., Jamzad, M.: An image annotation rectifying method based on deep features. In: Proceedings of the 2nd International Conference on Digital Signal Processing, pp. 88–92, February 2018
7. Ghostan Khatchatoorian, A., Jamzad, M.: Architecture to improve the accuracy of automatic image annotation systems. IET Comput. Vision **14**(5), 214–223 (2020)
8. Li, Z., Lin, L., Zhang, C., Ma, H., Zhao, W., Shi, Z.: A semi-supervised learning approach based on adaptive weighted fusion for automatic image annotation. ACM Trans. Multimedia Comput. Commun. Appl. (TOMM) **17**(1), 1–23 (2021)
9. Ma, Y., Xie, Q., Liu, Y., Xiong, S.: A weighted KNN-based automatic image annotation method. Neural Comput. Appl. **32**(11), 6559–6570 (2019). https://doi.org/10.1007/s00521-019-04114-y
10. Tang, C., Liu, X., Wang, P., Zhang, C., Li, M., Wang, L.: Adaptive hypergraph embedded semi-supervised multi-label image annotation. IEEE Trans. Multimedia **21**(11), 2837–2849 (2019)
11. Ke, X., Zhou, M., Niu, Y., Guo, W.: Data equilibrium based automatic image annotation by fusing deep model and semantic propagation. Pattern Recogn. **71**, 60–77 (2017)
12. Jin, J., Nakayama, H.: Annotation order matters: Recurrent image annotator for arbitrary length image tagging. In: 2016 23rd International Conference on Pattern Recognition (ICPR), pp. 2452–2457. IEEE, December 2016
13. Kipf, T.N., Welling, M.: Semi-supervised classification with graph convolutional networks. arXiv preprint arXiv:1609.02907 (2016)
14. Chen, Z.M., Wei, X.S., Wang, P., Guo, Y.: Multi-label image recognition with graph convolutional networks. In: Proceedings of the IEEE/CVF Conference on Computer Vision and Pattern Recognition, pp. 5177–5186 (2019)

15. Yazici, V.O., Gonzalez-Garcia, A., Ramisa, A., Twardowski, B., Weijer, J.V.D.: Orderless recurrent models for multi-label classification. In: Proceedings of the IEEE/CVF Conference on Computer Vision and Pattern Recognition, pp. 13440–13449 (2020)
16. Feng, L., Bhanu, B.: Semantic concept co-occurrence patterns for image annotation and retrieval. IEEE Trans. Pattern Anal. Mach. Intell. **38**(4), 785–799 (2015)
17. Verma, Y., Jawahar, C.V.: Image annotation using metric learning in semantic neighbourhoods. In: Fitzgibbon, A., Lazebnik, S., Perona, P., Sato, Y., Schmid, C. (eds.) ECCV 2012. LNCS, vol. 7574, pp. 836–849. Springer, Heidelberg (2012). https://doi.org/10.1007/978-3-642-33712-3_60
18. Mikolov, T., Sutskever, I., Chen, K., Corrado, G.S., Dean, J.: Distributed representations of words and phrases and their compositionality. In: Advances in Neural Information Processing Systems, pp. 3111–3119, (2013)
19. Pennington, J., Socher, R., Manning, C.D.: Glove: global vectors for word representation. In: Proceedings of the 2014 Conference on Empirical Methods in Natural Language Processing (EMNLP), pp. 1532–1543, October 2014
20. He, K., Zhang, X., Ren, S., Sun, J.: Deep residual learning for image recognition. In: Proceedings of the IEEE Conference on Computer Vision and Pattern Recognition, pp. 770–778 (2016)
21. Hamilton, W.L., Ying, R., Leskovec, J.: Inductive representation learning on large graphs. In: Proceedings of the 31st International Conference on Neural Information Processing Systems, pp. 1025–1035, December 2017
22. Wu, B., Jia, F., Liu, W., Ghanem, B.: Diverse image annotation. In: Proceedings of the IEEE Conference on Computer Vision and Pattern Recognition, pp. 2559–2567 (2017)
23. Li, L., Wang, S., Jiang, S., Huang, Q.: Attentive recurrent neural network for weak-supervised multi-label image classification. In: Proceedings of the 26th ACM International Conference on Multimedia, pp. 1092–1100, October 2018
24. Mayhew, M.B., Chen, B., Ni, K.S.: Assessing semantic information in convolutional neural network representations of images via image annotation. In: 2016 IEEE International Conference on Image Processing (ICIP), pp. 2266–2270. IEEE, September 2016
25. Murthy, V.N., Maji, S., Manmatha, R.: Automatic image annotation using deep learning representations. In: Proceedings of the 5th ACM on International Conference on Multimedia Retrieval, pp. 603–606, June 2015
26. Liu, F., Xiang, T., Hospedales, T.M., Yang, W., Sun, C.: Semantic regularisation for recurrent image annotation. In: Proceedings of the IEEE Conference on Computer Vision and Pattern Recognition, pp. 2872–2880 (2017)
27. Niu, Y., Lu, Z., Wen, J.R., Xiang, T., Chang, S.F.: Multi-modal multi-scale deep learning for large-scale image annotation. IEEE Trans. Image Process. **28**(4), 1720–1731 (2018)
28. Ma, Y., Liu, Y., Xie, Q., Li, L.: CNN-feature based automatic image annotation method. Multimedia Tools Appl. **78**(3), 3767–3780 (2018). https://doi.org/10.1007/s11042-018-6038-x
29. Chen, J., Ma, T., Xiao, C.: FastGCN: fast learning with graph convolutional networks via importance sampling. arXiv preprint arXiv:1801.10247 (2018)
30. Zhang, J., Wu, Q., Zhang, J., Shen, C., Lu, J.: Mind your neighbours: image annotation with metadata neighbourhood graph co-attention networks. In: Proceedings of the IEEE/CVF Conference on Computer Vision and Pattern Recognition, pp. 2956–2964 (2019)

31. Datta, R., Joshi, D., Li, J., Wang, J.Z.: Image retrieval: ideas, influences, and trends of the new age. ACM Comput. Surv. (CSUR) **40**(2), 1–60 (2008)
32. Smeulders, A.W., Worring, M., Santini, S., Gupta, A., Jain, R.: Content-based image retrieval at the end of the early years. IEEE Trans. Pattern Anal. Mach. Intell. **22**(12), 1349–1380 (2000)

Interactive Process Improvement Using Simulation of Enriched Process Trees

Mahsa Pourbafrani$^{(\boxtimes)}$ and Wil M. P. van der Aalst

Chair of Process and Data Science, RWTH Aachen University, Aachen, Germany
{mahsa.bafrani,wvdaalst}@pads.rwth-aachen.de

Abstract. Event data provide the main source of information for analyzing and improving processes in organizations. Process mining techniques capture the state of running processes w.r.t. various aspects, such as activity-flow and performance metrics. The next step for process owners is to take the provided insights and turn them into actions in order to improve their processes. These actions may be taken in different aspects of a process. However, simply being aware of the process aspects that need to be improved as well as potential actions is insufficient. The key step in between is to assess the outcomes of the decisions and improvements. In this paper, we propose a framework to systematically compare event data and the simulated event data of organizations, as well as comparing the results of modified processes in different settings. The proposed framework could be provided as an analytic service to enable organizations in easily accessing event data analytics. The framework is supported with a simulation tool that enables applying changes to the processes and re-running the process in various scenarios. The simulation step includes different perspectives of a process that can be captured automatically and modified by the user. Then, we apply a state-of-the-art comparison approach for processes using their event data which visually reflects the effects of these changes in the process, i.e., evaluating the process improvement. Our framework also includes the implementation of the change measurement module as a tool.

Keywords: Process mining · Business process improvement · Process simulation · Earth mover's distance · Performance spectrum

1 Introduction

Process owners use data-driven process mining techniques to improve their processes. The discovered process models, their performance states, and hidden problems, such as deviations and bottlenecks, are critical to process improvement. The process mining techniques in the process discovery and conformance checking areas are widely used to illustrate the current states of processes and their potential problems [1]. However, before taking any action based on process mining diagnostics, one wants to have an estimation of the impact. To do so, it is required to play out the processes with the process owners' adjustments

© Springer Nature Switzerland AG 2022
H. Hacid et al. (Eds.): ICSOC 2021 Workshops, LNCS 13236, pp. 61–76, 2022.
https://doi.org/10.1007/978-3-031-14135-5_5

and then assess the effects of the actions. To improve processes in an evidence-based manner, *forward-looking* process mining techniques such as prediction and simulation are needed. They enable what-if and scenario-based analyses of business processes. However, the validity of the generated results, as well as their clear interpretation, are two determining factors when employing these techniques. The model's reliability can be improved by incorporating process mining insights, e.g., the designed simulation model is derived directly from the process's historical event data [2].

Techniques such as generating CPN models [12,14,22] and BPMN Models [3] have been proposed for generating simulation models of processes based on event logs. Simulation approaches in process mining are also useful for other applications. In [23], for example, process model simulations are used to estimate the alignment value. The gap that we aim to fill is not only providing a platform for users to easily re-run their processes using the automatically generated simulation models but also a more accurate technique for measuring improvement/changes w.r.t. the process owners' interactions with the process. The conventional comparison of two processes includes conformance checking between the event logs and the corresponding process models. In addition, for the purpose of performance comparison, general performance metrics are usually considered. Most of the current approaches are not detailed enough in both aspects, i.e., conformance checking and performance analysis. These techniques do not measure and reflect the effect of changes at the detailed level. For instance, the existing conformance checking techniques only return a value such as the fitness of two event logs, or one event log and the corresponding process model [4]. These techniques also neglect the importance of the frequency of process instances. The detailed distance between the original event log and the regenerated event log is critical for determining their similarity [19].

In this paper, we propose an approach to systematically compare the event data of a process with its simulated event data to assess the reliability of the simulation model, i.e., the accuracy of the simulation. As a result, the simulated processes in different settings can be compared. The simulation module is implemented as a new software capturing different process perspectives, in which the event logs are used to enrich the process models (trees) with existing aspects. The enriched process trees generate process behaviors in the form of event logs with/without applied changes to the process. The state-of-the-art comparison framework is then applied to the results of the simulation. It measures the effects of changes using detailed conformance and performance techniques. To demonstrate a proof of concept of the framework, we use a sample process as an example to illustrate the approach steps. Then, we employ a real-life event log to evaluate the approach.

The remainder of this paper is structured as follows. We present the related work in Sect. 2. In Sect. 3, we introduce background concepts and notations. In Sect. 4, we present our main approach. We evaluate the approach in Sect. 5 by designing simulation models, and Sect. 6 concludes this work.

2   Related Work

Process mining enables designing data-driven simulation models of processes [2]. Authors in [22] use different aspects of a process using its event data, e.g., process models, resource pooling, and performance metrics, and automatically generate simulation models. This work as a pioneer in the data-driven simulation in process mining translates insights from event data into the process simulation parameters. Other simulation approaches in process mining follow the same direction. For instance, [21] uses *stochastic Petri nets* to simulate processes and determine the duration of instances in business processes. In [18] a business simulation model is generated which is based on the user domain knowledge. Tools based on *Protos* try to reduce modeling efforts by introducing the reference process models [24]. [9] discusses how process mining insights can be exploited in the business process simulation context. As an example, the proposed tool in [3] presents the idea of combining BPMN and process mining for simulation purposes, where indicators for measuring the accuracy of the simulation results are also introduced.

In [11,15], different levels of simulating processes are proposed where all the aspects of a process are extracted at different levels, i.e., not only instance level but also higher-level, e.g., describing processes per day quantitatively. The examples of high-level simulations are presented in [16,17] with the use of the designed tool for the modeling and data extraction steps in [10]. In our approach, the enriched process models, e.g., process trees, accuracy of the performance-related aspects, effortless interaction with users, and social network analysis (resource aspects) are the main criteria for designing simulation models.

On the other side, visualization techniques are powerful tools in process mining analysis in both descriptive and predictive analyses. There are a couple of visualization techniques that are able to represent the process w.r.t. different process aspects for providing visual inspection or process comparison. For instance, the performance spectrum [5] represents the process performance behaviors in detail between every two sets of activities in the process. i.e., process segments. The stochastic conformance checking method used in [20] considers the frequency of the traces in two event logs while comparing their differences. The idea of using *Earth Mover's Distance* for conformance checking and comparing two event logs, or event logs and process models enables assessing the difference of two processes w.r.t. their behaviors in detail.

We provide a platform for regenerating a process in different settings and measure the effects of changes/results using our designed modules based on the presented ideas. The presented tool in [13] is the simulation approach taken in the current work as the intermediate tool for regenerating the process behaviors. The process trees are automatically generated and enriched with the probability and performance information and allow us to change the processes w.r.t. the activity-flow and performance aspects.

Table 1. A part of a sample event log. Each row represents an event.

	Case ID	Activity	Resource	Timestamp
e_1	1	Register request	Pete	12/30/2010 11:02
e_2	2	Register request	Mike	12/30/2010 11:32
	3	Register request	Pete	12/30/2010 14:32
...	1	Examine thoroughly	Sue	12/31/2010 10:06
	2	Decide	Sara	1/5/2011 11:22
	1	Decide	Sara	1/6/2011 11:18
	1	Reject request	Pete	1/7/2011 14:24
e_n	...	...	...	...

3   Preliminaries

In this section, we establish the basic notations for events, event logs, and process trees which are used in the framework.

Definition 1 (Event). *Let $\mathcal{A}$ be the universe of activities, $\mathcal{T}$ be the universe of timestamps, $\mathcal{R}$ be the universe of resources, and $\mathcal{C}$ be the universe of case identifier. An event e is a tuple $e = (c, a, r, t)$ where activity a at time t for case c is performed by resource r. $\mathcal{E} = \mathcal{C} \times \mathcal{A} \times \mathcal{R} \times \mathcal{T}$ is the universe of events. For each $e \in \mathcal{E}$, $\pi_D(e)$ projects e on the attribute from domain D, e.g., $\pi_\mathcal{A}(e) = a$.*

Definition 2 (Trace). *Let $\mathcal{E}$ be the universe of events, a trace $\sigma \in \mathcal{E}^*$ is a finite sequence of events. For each $\sigma = \langle e_1, ..., e_n \rangle$, $e_i \in \sigma$ happens at most once and for each $e_i, e_j \in \sigma, \pi_\mathcal{C}(e_i) = \pi_\mathcal{C}(e_j) \wedge \pi_\mathcal{T}(e_i) \le \pi_\mathcal{T}(e_j), if\ 1 \le i < j \le n$. For $\sigma = \langle e_1, ..., e_n \rangle \in \mathcal{E}^*$, $\Pi_D(\sigma) = \langle \pi_{\mathcal{D}(e_1)}, \pi_{\mathcal{D}(e_2)}, ..., \pi_{\mathcal{D}(e_n)} \rangle$ is the projection of trace σ on the attribute from domain D, e.g., $\Pi_\mathcal{A}(\sigma) = \langle \pi_{\mathcal{A}(e_1)}, \pi_{\mathcal{A}(e_2)}, ..., \pi_{\mathcal{A}(e_n)} \rangle$.*

Definition 3 (Event Log). *Let $\mathcal{E}$ be the universe of events and $\mathcal{E}^*$ be the set of possible traces, we define an event log L as a set of traces, i.e., $L \subseteq \mathcal{E}^*$.*

We denote $L_\mathcal{A} = [\Pi_\mathcal{A}(\sigma) | \sigma \in L]$ as the multiset of traces projected on the activity attribute. Furthermore, $\widetilde{L_\mathcal{A}} = \{\sigma \in L_\mathcal{A}\}$ is the set of unique traces (variants) projected on the activity attribute in the event log L. We refer to $\widetilde{L_\mathcal{A}}$ as the set of process behaviors presented in L.

Table 1 represents a part of a sample event log, where each row indicates an event, e.g., considering the first row as e_1, $\pi_\mathcal{C}(e_1) = 1$ and $\pi_\mathcal{A}(e_1) = $ *register request*. Process mining utilizes such event logs to discover running processes inside organizations. The process models are the representative ways of the discovered running processes. The

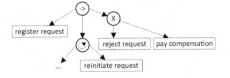

Fig. 1. A part of the discovered process tree for the sample event log.

process tree notation is one of the common approaches to present a process, where the nodes of trees are operators and leaves are activities in the process.

A part of the process tree representing the example process is shown in Fig. 1. For example, there is a choice, i.e., XOR ($\times$) as a node between activity *reject request* and *pay compensation* indicating that in the process either a request is rejected or the compensation is paid. The root node ($\rightarrow$) indicates that activity *register request* is always followed by a loop ($\circlearrowright$). A loop represents a redo of works between its children, i.e., activities in the leaves of a loop node may happen multiple times in a trace. Furthermore, the notation of τ is for silent activities which are not visible in the process but used for the representation of process trees.

Definition 4 (Process Tree). *Let L be an event log, $A_L = \{a \in \sigma | \sigma \in \widetilde{L_A}\}$ be the set of activities in L and $Op = \{\rightarrow, \times, \circlearrowright, +\}$ be the set of process operators. If $a \in A_L \cup \{\tau\}$, then $Q = a$ is a process tree. If $n \geq 1$, $Q_1, Q_2, Q_3, ..., Q_n$ are process trees, and $op \in \{\rightarrow, \times, +\}$, then $Q = op(Q_1, Q_2, ..., Q_n)$ is a process tree. If $n \geq 2$ and $Q_1, Q_2, Q_3, ..., Q_n$ are process trees, then $Q = \circlearrowright(Q_1, Q_2, ..., Q_n)$ is a process tree. For a process tree Q, we denote Q_a and Q_{op} as the set of activities and the set of operators in Q.*

For a given process tree Q, $Q_w = Q_{op} \times Q_a$ is the set of edges connecting operators to activities. For instance, ($\rightarrow$, *register request*) is an edge in the example process tree in Fig. 1 where *register request* is child of the tree under parent $\rightarrow$. Note that a process tree may also contain edges from an operator to an operator, which is not relevant in the implementation of our framework.

4   Approach

Our framework enables interactive process improvement inside organizations for designing/improving process models. The current behaviors of processes captured in the form of an event log serve as the starting point for any improvement. To enrich the discovered process models, process discovery, performance analysis, and social network analysis (resource perspective) techniques are used. We use the *Discrete Event Simulation* (DES) technique as a tool to play out the process with the current states, which results in an event log as shown in Fig. 2. The original behavior of the event log w.r.t. activity-flow (process behavior) and performance metrics are compared to ensure that the automatically designed simulation model is reliable and behaves close to reality, *Improvement Measurement* module in Fig. 2. This step allows the user to change the process parameters and re-run the process to generate the new behavior and measure the process improvement, depicted by the dotted lines in Fig. 2. These measurements are presented in a numerical format as well as in a detailed graphical format. The detailed comparative visualization increases the interaction between the framework and the user. First, we explain the automatic generation of the simulation results, including process mining techniques and enriching the process model, and continue with the *Improvement Measurement* module.

4.1   Simulating Process Trees

Enriching Process Trees. The inductive miner algorithm [7] is used to discover the process model since it is capable of capturing all the behaviors in

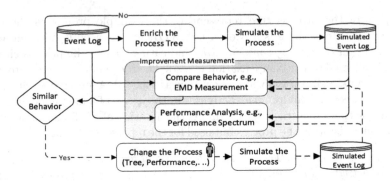

Fig. 2. The overview of the framework to improve the processes interactively. The straight lines show the path to assess the quality of the regenerated behavior by the simulation model w.r.t. activity-flow and performance metrics. The dotted lines illustrate the path that the user is able to change the process and measure and observer the improvement, i.e., the effect of changes, in the process.

a process in the form of a process model. The generated process tree by the inductive miner algorithm is able to represent the traces in the event log. The process tree's limited number of operators as defined in Definition 4 allows for easy understanding and modification of the process. To play out the process accurately, i.e., applying the new changes in the process, more information than the flow of activities provided by the process tree is required.

The tree should be enriched with the probability of activity-flows, performance information of the activities, and the corresponding resource information, e.g., organizations of the resources, the number of resources in each organization, and hand-over of activities between resources, for each activity from the real process. Therefore, the probability of the choices and the possible number of loops should be taken into account for regenerating a similar event log. Furthermore, for a process tree Q and the edge $w = (op, a) \in Q_w$, w_a represents the probability of occurrence of activity a in a generated trace from the process tree. For the edge $w = (op, a) \in Q_w$, if $op \in \{\rightarrow, +, \circlearrowright\}$, then $w_a = 1$. Note that to avoid the generation of infinite traces due to the loops in the process tree, we limit the execution of loops in the simulation with the probability of the number of occurrences of a loop on average in a trace and the maximum times that a loop happens in a trace. For all activities $a \in Q_a$, there is a binding performance metric, i.e., the average duration of each activity. Moreover, the activities are assigned to the existing automatically discovered organizations and the capacity of the resources.

For the example process shown in Fig. 1, a part of the automatically enriched tree with the activity-flow, performance, and resource information is presented in Fig. 3. For instance, for the process edge $w = (\times, reject\ request)$, $w_{rej} = 0.5$, and the shown loop in the process can be executed at most 2 times in a trace and the probability of its occurrence is 30% which is derived from the event log. Activity *register request* takes on average 43000 s to be performed,

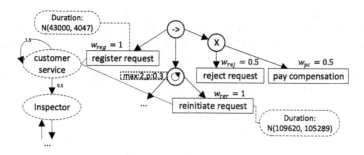

Fig. 3. Enriched tree with the probability information, resource allocation, and duration of each activity. The enriched process tree can be simulated. The hand-over of resources is shown (left) to provide more accurate simulation results (event logs) w.r.t. the resource allocation in the organizations.

Table 2. The general list of automatically discovered insights using process mining techniques to form process simulations. The top row shows what is discovered from event data. The bottom row shows what can be set or change by the user.

	Process mining											Simulation execution parameters	
	Process model (tree)	Arrival rate	Activity duration, deviation	Activities capacity	Activities unique resources (shared resources)	Waiting time	Business hours	Activity-flow probability	Process capacity (cases)	Interruption (process, cases, activities)		Start time of simulation	Number of cases
Automatically discovered	+	+	+	+	+	+	+	+	+	+		–	–
Changeable by user	+	+	+	+	+	+	+	+	+	+		+	+

and the average is used for simulating its duration using a normal distribution. Also, *register request* and *reinitiate request* belong to the *customer service* organization where the resources in this organization hand over tasks to the *inspector* organization.

The information extracted from event logs is shown in Table 2. This information, along with the discussed information for enriching process trees are the required simulation parameters. Moreover, the changeable aspects for process improvement by the user in the simulation step are specified in detail. The discovery and design of the simulation models including generating event logs as a result of the simulation models are represented in detail in [13].

4.2 Measuring the Process Improvement

To measure the changes in the newly generated process represented with an event log, we have to compare two event logs. For comparing two processes, i.e., event logs, two major aspects of the processes should be considered, activity-flow which generates the behaviors, and the performance aspects. Note that the intermediate regenerator tool can be different from the one that we use in our framework, and yet the *Measuring the Process Improvement* module can be used for measuring the effect of changes in two event logs.

Table 3. A sample example of EMD measurement for two event logs [19]. The real-location function allocates the 49 traces in L to 49 traces with activity-flow $\langle a, e, c, d \rangle$ and 1 remaining trace to $\langle a, b, c, d \rangle$ in L'. The sum of the value of the table indicates the general EMD value, i.e., the difference between the two event logs. Each cell represents the minimum cost to map its corresponding trace in the original event log (row) into the traces in the simulated event log (column).

$L_{\mathcal{A}}$	$L'_{\mathcal{A}}$			
	$\langle a,b,c,d \rangle$	$\langle a,c,b,d \rangle$	$\langle a,e,c,d \rangle^{49}$	$\langle a,e,b,d \rangle^{49}$
$\langle a,b,c,d \rangle^{50}$	$\frac{1}{100} \times 0$	0×0.5	$\frac{49}{100} \times 0.25$	0×0.5
$\langle a,c,b,d \rangle^{50}$	0×0.5	$\frac{1}{100} \times 0$	0×0.5	$\frac{49}{100} \times 0.25$

Activity-Flow Behaviors. The fact that process trees include silent transitions, loops, and XOR operators makes generating more behavior (new traces) than the existed ones in the original log possible. Therefore, the similarity of behaviors is one of the main indicators in the comparison step.

Given two event logs, the original event log L and the simulated event log L', we show the presented behaviors in each event log using their set of unique traces, i.e., $\widetilde{L}_{\mathcal{A}}, \widetilde{L}'_{\mathcal{A}}$. The new generated behaviors in the simulated event log, i.e., not existing in the original event log, and the removed behaviors from the original process are calculated as $\widetilde{L}'_{\mathcal{A}} \backslash \widetilde{L}_{\mathcal{A}}$, and $\widetilde{L}_{\mathcal{A}} \backslash \widetilde{L}'_{\mathcal{A}}$, respectively. Therefore, $\frac{|\widetilde{L}'_{\mathcal{A}} \backslash \widetilde{L}_{\mathcal{A}}|}{|\widetilde{L}_{\mathcal{A}} \cup \widetilde{L}'_{\mathcal{A}}|}$ and $\frac{|\widetilde{L}_{\mathcal{A}} \backslash \widetilde{L}'_{\mathcal{A}}|}{|\widetilde{L}_{\mathcal{A}} \cup \widetilde{L}'_{\mathcal{A}}|}$ are the fraction of the new and removed behaviors, respectively.

These metrics represent the pairwise difference between two event logs. They evaluate whether the simulation of the original log is close to reality, as well as capturing any different behavior added/removed due to the changes in a process tree (flow of activities). In the example process presented in Fig. 1, after regenerating the process without any change multiple times, on average 22% of the generated variants (unique traces) in the simulated logs are newly generated. The sample event log for further experiments with the tools is publicly available[1]. Furthermore, the precise comparison of two event logs should be based on their behavior, taking into account the frequency of the behavior. To determine the difference between the original and the simulated event logs, we employ a stochastic conformance checking approach.

Earth Mover's Distance Conformance Checking. To accurately compare two event logs' behaviors, we use the probability distance of each two traces in two event logs based on Earth Mover's Distance (EMD). To calculate the EMD measurement between two event logs, we use the conformance techniques presented in [6]. For every trace in the original log, we calculate the movement of its frequency to all the traces in the simulated event log using the reallocation function. As the next step, the cost of the movement is considered using the trace distance function.

---

[1] https://github.com/mbafrani/VisualComparison2EventLogs.

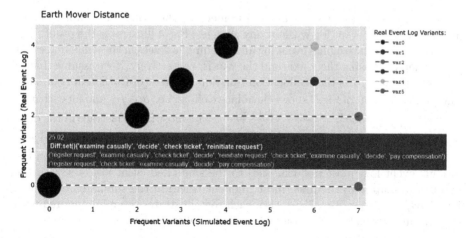

Fig. 4. The detailed comparison of two event logs for the sample process, i.e., the results of EMD measurement. It is the results of the EMD *reallocation* and *trace distance* functions in the form of a table such as Table 3. The points are the proportional cost of moving every trace in one event log to the simulated event log. Each row (color) indicates a trace in the original event log. The black points are similar traces in both event logs. The sizes of the points are the relative costs of movement for each variant (unique traces) in the original event logs.

Reallocation. Let L and L' be the original and the simulated event logs, respectively. Function $r \in \widetilde{L}_{\mathcal{A}} \times \widetilde{L}'_{\mathcal{A}} \to [0,1]$ returns the relative frequency of $\sigma \in \widetilde{L}_{\mathcal{A}}$ that should be transformed to $\sigma' \in \widetilde{L}'_{\mathcal{A}}$, i.e., $r(\sigma,\sigma')$. Note that for all $\sigma \in \widetilde{L}_{\mathcal{A}}$, $\frac{L_{\mathcal{A}}(\sigma)}{|L_{\mathcal{A}}|} = \sum_{\sigma' \in \widetilde{L}'_{\mathcal{A}}} r(\sigma,\sigma')$, i.e., the frequency of each $\sigma \in \widetilde{L}_{\mathcal{A}}$ is considered properly. The same should be considered for each $\sigma' \in \widetilde{L}'_{\mathcal{A}}$.

Trace Distance. The distance between each two traces in the original log and the simulated logs is calculated based on *normalized string edit distance* (Levenstein) [8]. Function $d \in \mathcal{A}^* \times \mathcal{A}^* \to [0,1]$ calculates the distance between two traces, where for two similar traces the value is 0 and $d(\sigma,\sigma') = d(\sigma',\sigma)$.

To represent the algorithm clearly, we reduced the sample process and presented a couple of traces in Table 3. The EMD measurement of the two event logs is $EMD(L_{\mathcal{A}}, L'_{\mathcal{A}}) = \min_{r \in R} r.d = \sum_{\sigma \in \widetilde{L}_{\mathcal{A}}} \sum_{\sigma' \in \widetilde{L}'_{\mathcal{A}}} r(\sigma,\sigma')d(\sigma,\sigma')$ where R is the universe of reallocation functions. Table 3 represents a sample EMD measurement for two sample event logs L and L'. For instance, for $\langle a,b,c,d \rangle^{50}$ in L and $\langle a,e,c,d \rangle^{49}$ in L', the trace distance value is 0.25 given the differences between two traces using *normalized string edit distance* (Levenstein). The reallocation value is 0.49, i.e., 49 of 100 traces in L are reallocated to 49 traces with the sequence $\langle a,e,c,d \rangle$ in L'. Therefore, the minimum effort of mapping the one trace to the second one is $0.49 * 0.25 = 0.122$. Besides the EMD value of two event logs that indicates how two event logs are different, we are interested in the required effort for every trace in the original event log to be mapped/transformed into the simulated event log for accurate comparison of the simulation results.

Applying the designed EMD measurement to the complete sample process and its simulated event log without any changes, Fig. 4 illustrates the result. The unique traces in the original event log and the unique traces in the simulated event log are depicted using the x-axis and the y-axis, respectively. If we assume that in our example, $r \in R$ is the reallocation function, the cost of EMD (effort of mapping) for each point of Fig. 4 shows the relative effort, i.e., $\sigma \in \widetilde{L}_A$ and for each $\sigma'_i \in \widetilde{L}'_A$, $effort_{L_A,L_A}(\sigma, \sigma'_i) = \frac{d(\sigma,\sigma'_i).r(\sigma,\sigma'_i)}{\sum_{\sigma' \in \widetilde{L}'} d(\sigma,\sigma').r(\sigma,\sigma')}$. The most frequent trace in the original event log (first row) will be converted to the $(74.98\%, 0, 0, 0, 0, 0, 0, 25.02\%)$, i.e., points in the first row. The values indicate that to map the first trace in the original event log (most frequent one) to the simulated event log 74.98% of the effort is to map it to the first (most frequent trace) in the simulated event log, i.e., $effort_{L_A,L_A}(\sigma_1, \sigma'_1) = 75\%$. Also, each row illustrates the minimum required effort to map/transform the traces into the simulated event log.

Performance Behaviors. Performance is the second factor to consider when assessing improvement/changes. However, because the times are abstracted from the real data in prediction and simulation techniques, exact measurements are impossible. It is worth noting that in many cases, time-related parameters such as the duration of simulation events are generated using a random function, e.g., normal distribution in our case. General performance KPIs at a high level of aggregation, e.g., the average waiting time of traces, or average service time are too abstract to represent the effects of the changes in the process. Therefore, besides the usual metrics, we use the performance spectrum, which relies on the structure of the process and directly reflects the effects of changes in specific parts of the process on others. For instance, changing the current service time of the activity *examine thoroughly* in the example process has an impact not only on the overall metrics but also on the duration of the later activities in the traces, e.g., *decide* or *reinitiate request*.

Aggregated Performance Spectrum. Performance Spectrum is a concept introduced to visualize the performance of process steps at the detailed level. A process segment in event log L is a step from activity a to activity b, i.e., $(a, b) \in A_L \times A_L$ is a process segment in L where $A_L = \{a \in \sigma | \sigma \in \widetilde{L}_A\}$. Each occurrence of a segment in a trace allows measuring the time between occurrences of a and b [5]. We define the set of all tuples of events that are directly followed in the traces in L as $SEG^L = \{(e_i, e_{i+1}) | \exists_{\sigma = \langle e_1, e_2, ..., e_n \rangle \in L} e_i, e_{i+1} \in \sigma\}$. The projection of the events in SEG^L on their activity attribute provides the multiset of process segments, i.e., $SEG^L_A = [(\pi_A(e_1), \pi_A(e_2)) | (e_1, e_2) \in SEG^L]$. For instance, $[(examine\ thoroughly, decide)^{17}, (examine\ thoroughly, reinitiate\ request)^{20}]$ is the part of the multiset of segments in our example.

We consider two aspects for representing a process segment in an event log: *average time* of the segment and *frequency* of the segment. For $seg = (a, b) \in SEG^L_A$, function $PS(seg, L) = (AvgTime(seg, L), Freq(seg, L))$ represents the frequency of the process segment seg and the corresponding average time difference for the segment. For $seg = (a, b) \in SEG^L_A$, we define

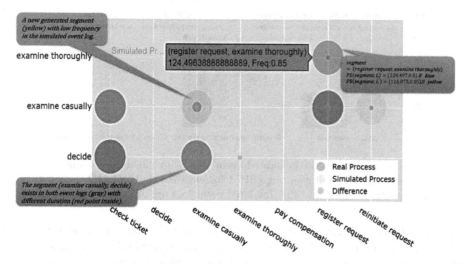

Fig. 5. Part of the performance measurement for the example process based on the aggregated performance spectrum. Each event log is represented by a different color, i.e., blue for the original and yellow for the simulated one. Overlapping segments are represented by the gray color (same duration between segments). Each point's transparency and size indicate the frequency and duration of the segment in the event logs. (Color figure online)

$AvgTime(seg, L) = Avg(\{\pi_T(e_2) - \pi_T(e_1)|(e_1, e_2) \in SEG^L \wedge \pi_A(e_1) = a \wedge \pi_A(e_2) = b\})$ and $Freq(seg, L) = SEG_A^L((a, b))$.

Figure 5 is the result of the introduced performance measurement (PS) for the example process and the regenerated event log. In order to represent different aspects of the results, e.g., new/eliminated segments and different duration, we performed the simulation based on the changed process. For instance, given L and L' as the original and simulated event logs, each segment's colors refer to an event log, the size refers to the average time difference between the segments, and the transparency indicates the frequency (darker means more frequent). The gray color represents the overlapped segment in two event logs with similar performance metrics, and the yellow points represent the new segment generated in the simulated event log as a result of process tree choices. The implementation also includes the option to display only the difference (red points).

5 Evaluation

A real event log representing the process of taking loans by customers inside a financial company, known as the *BPI challenge 2012*, is used in this section. First, we simulate a similar process with different configurations and assess how close they are to the original event log. Following that, we alter the activity-flow of the process model in order to improve the process and evaluate the effect of the applied changes. Having both simulated and original behaviors of the

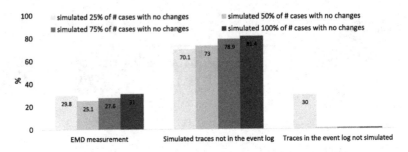

Fig. 6. The comparison of the generated event logs using simulating a specific number of traces in the original event log (BPI Challenge 2012). The EMD measurement indicates, how the original and the simulated event logs are different.

process (with or without modifications) the possibility of comparing between two processes is easily provided. To do so, we used our tool $SIMPT$[2] for simulating the process, and our developed modules for comparing two event logs w.r.t. the detailed performance and control flow aspects[3]. The provided tools make it possible to evaluate the framework for the interactive improvement of different processes for different event logs.

We start with automatically discovering and enriching the underlying process tree before regenerating the process, where the similarity of the two event logs indicates the possibility of using the simulation models for further investigation. Therefore, we simulated the event log multiple times without applying any changes. As shown in Fig. 6, we took a specific percentage of the total number of traces in the process for each round of simulation of the original process. As expected, when the number of simulated traces is small, there is a chance of missing specific process behaviors, e.g., using 25% of the number of traces, we lost 30% of the behaviors (unique traces). On the other hand, increasing the number of simulated traces increases the number of new behaviors. Since the generation of the traces (activity-flow) is based on probability and the process tree includes both XOR choices and silent transitions, the new behaviors are expected to be generated.

Afterward, in the process tree of the original process, we changed the optional activity *preaccepted* to be a mandatory activity for all the traces that are going to be *accepted* in Fig. 7. The structure of the process tree (activity-flow) is changed from $\rightarrow$ $(submitted, partlysubmitted, \times(\tau, preaccepted), \times(\tau, accepted), \times(\tau, finalized), \times(declined, cancelled))$ to $\rightarrow$ $(submitted, partlysubmitted, \times(\tau, \rightarrow (preaccepted, accepted)), \times(\tau, finalized), \times(declined, cancelled))$. Note that these changes are possible in different aspects of the process such as the process model, performance metrics, e.g., activity duration, arrival rate of the traces, or capacity of the resources.

Based on the shown results of simulating the original event log without any changes in Fig. 8, we simulated the changed process model with 50% of original

---

[2] https://github.com/mbafrani/SIMPT-SimulatingProcessTrees.

[3] https://github.com/mbafrani/VisualComparison2EventLogs.

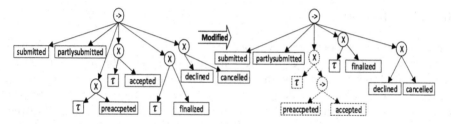

Fig. 7. The process tree for handling application in the BPI challenge 2012 event logs (left). To evaluate the approach, the optional activity *preaccepted* is changed to be mandatory in the flow of activities for all the traces (right) in the process. Dotted lines indicate the parts of the tree that have changed.

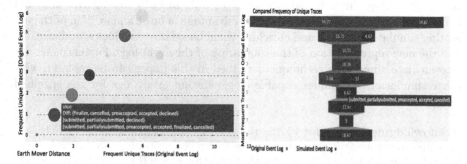

Fig. 8. The detailed comparison of the changed process and the original process model. The detailed EMD diagram (left) shows the differences of the two event logs w.r.t. the activity-flow and the comparing frequent chart (right) represents the preserved and removed behavior in the simulated process as the effect of the applied changes.

traces. In the proposed scenario (changed process tree), 63.6% of generated behaviors (unique traces) are new. However, it is less than the behaviors in the simulated event log without any modifications, since we removed one of the XOR choices limiting the possibilities of producing new behaviors. On the other hand, 23% of the behaviors due to the change in the process tree are eliminated, i.e., the traces that skipped the activity *preaccepted* in the original process. Also, in Fig. 6, the pairwise comparison of the traces (right), as well as the detailed EMD companions for the cost of the mapping of two event logs (left) after the changes, are shown. The applied changes in the process model not only affected the process behavior but also these changes affected the performance of the later segments in the process, e.g., the duration for the process segment *accepted* and *finalized* increased while activity *finalized* was not changed. The provided detailed comparison along with the intermediate simulation tool enables the possibility of capturing these types of unexpected insights. Note that the reliability of the simulation techniques such as the presented ones in Sect. 2 can be assessed using the measurement modules.

6 Conclusion

Process mining supports organizations in finding running processes, as well as identifying challenges or possible areas for improvement. The process improvement should be supported with process knowledge. We use process mining insights and simulation models as an intermediate method to regenerate processes in various scenarios. The framework begins with an event log, discovers a process tree, and enriches it with all the knowledge needed to regenerate the process. The similarity of the simulated results and the original process behavior in the form of an event log is then measured in the next step. The degree of similarity reflects the accuracy of our model. As a result, the improvement of the change in the process can be played out, and the impact of changes can be tracked using the same measurement module in both the activity-flow and performance aspects of the process. The advantage of our framework in both generating simulation models and enriching them based on event logs automatically, and the new representation of the comparing of the event logs. Furthermore, the intermediate simulation technique described in this paper can be replaced with other simulation techniques capable of generating event logs for the specified changes.

Acknowledgments. Funded by the Deutsche Forschungsgemeinschaft (DFG, German Research Foundation) under Germany's Excellence Strategy-EXC-2023 Internet of Production - 390621612. We also thank the Alexander von Humboldt (AvH) Stiftung for supporting our research.

References

1. van der Aalst, W.M.P.: Process Mining - Data Science in Action, 2nd edn. Springer, Heidelberg (2016). https://doi.org/10.1007/978-3-662-49851-4
2. van der Aalst, W.M.P.: Process mining and simulation: a match made in heaven! In: Computer Simulation Conference, pp. 1–12. ACM Press (2018)
3. Camargo, M., Dumas, M., Rojas, O.G.: Simod: a tool for automated discovery of business process simulation models, pp. 139–143 (2019)
4. Carmona, J., van Dongen, B.F., Solti, A., Weidlich, M.: Conformance Checking - Relating Processes and Models. Springer, Cham (2018). https://doi.org/10.1007/978-3-319-99414-7
5. Denisov, V., Fahland, D., van der Aalst, W.M.P.: Unbiased, fine-grained description of processes performance from event data. In: Weske, M., Montali, M., Weber, I., vom Brocke, J. (eds.) BPM 2018. LNCS, vol. 11080, pp. 139–157. Springer, Cham (2018). https://doi.org/10.1007/978-3-319-98648-7_9
6. Leemans, S.J.J., Syring, A.F., van der Aalst, W.M.P.: Earth movers' stochastic conformance checking. In: Hildebrandt, T., van Dongen, B.F., Röglinger, M., Mendling, J. (eds.) BPM 2019. LNBIP, vol. 360, pp. 127–143. Springer, Cham (2019). https://doi.org/10.1007/978-3-030-26643-1_8
7. Leemans, S.J.J., Fahland, D., van der Aalst, W.M.P.: Discovering block-structured process models from incomplete event logs. In: Ciardo, G., Kindler, E. (eds.) PETRI NETS 2014. LNCS, vol. 8489, pp. 91–110. Springer, Cham (2014). https://doi.org/10.1007/978-3-319-07734-5_6

8. Levenshtein, V.I.: Binary codes capable of correcting deletions, insertions, and reversals. In: Soviet Physics Doklady, vol. 10, pp. 707–710. Soviet Union (1966)

9. Martin, N., Depaire, B., Caris, A.: The use of process mining in business process simulation model construction. Bus. Inf. Syst. Eng. **58**(1), 73–87 (2016). https://doi.org/10.1007/s12599-015-0410-4

10. Pourbafrani, M., van der Aalst, W.M.P.: PMSD: data-driven simulation in process mining. In: Proceedings of the Demonstration Track at BPM 2020 co-located with 18th International Conference on Business Process Management, BPM, pp. 77–81 (2020). http://ceur-ws.org/Vol-2673/paperDR03.pdf

11. Pourbafrani, M., van der Aalst, W.M.P.: Extracting process features from event logs to learn coarse-grained simulation models. In: La Rosa, M., Sadiq, S., Teniente, E. (eds.) CAiSE 2021. LNCS, vol. 12751, pp. 125–140. Springer, Cham (2021). https://doi.org/10.1007/978-3-030-79382-1_8

12. Pourbafrani, M., Balyan, S., Ahmed, M., Chugh, S., van der Aalst, W.M.P.: GenCPN: automatic generation of CPN models for processes (2021)

13. Pourbafrani, M., Jiao, S., van der Aalst, W.M.P.: SIMPT: process improvement using interactive simulation of time-aware process trees. In: Cherfi, S., Perini, A., Nurcan, S. (eds.) RCIS 2021. LNBIP, vol. 415, pp. 588–594. Springer, Cham (2021). https://doi.org/10.1007/978-3-030-75018-3_40

14. Pourbafrani, M., Vasudevan, S., Zafar, F., Xingran, Y., Singh, R., van der Aalst, W.M.P.: A python extension to simulate petri nets in process mining. CoRR abs/2102.08774 (2021)

15. Pourbafrani, M., van Zelst, S.J., van der Aalst, W.M.P.: Semi-automated time-granularity detection for data-driven simulation using process mining and system dynamics. In: Dobbie, G., Frank, U., Kappel, G., Liddle, S.W., Mayr, H.C. (eds.) ER 2020. LNCS, vol. 12400, pp. 77–91. Springer, Cham (2020). https://doi.org/10.1007/978-3-030-62522-1_6

16. Pourbafrani, M., van Zelst, S.J., van der Aalst, W.M.P.: Supporting automatic system dynamics model generation for simulation in the context of process mining. In: Abramowicz, W., Klein, G. (eds.) BIS 2020. LNBIP, vol. 389, pp. 249–263. Springer, Cham (2020). https://doi.org/10.1007/978-3-030-53337-3_19

17. Pourbafrani, M., van Zelst, S.J., van der Aalst, W.M.P.: Supporting decisions in production line processes by combining process mining and system dynamics. In: Ahram, T., Karwowski, W., Vergnano, A., Leali, F., Taiar, R. (eds.) IHSI 2020. AISC, vol. 1131, pp. 461–467. Springer, Cham (2020). https://doi.org/10.1007/978-3-030-39512-4_72

18. Pufahl, L., Weske, M.: Extensible BPMN process simulator. In: Proceedings of the BPM Demo Track and BPM Dissertation Award co-located with 15th International Conference on Business Process Modeling (BPM) (2017)

19. Rafiei, M., van der Aalst, W.M.P.: Towards quantifying privacy in process mining. In: Leemans, S., Leopold, H. (eds.) ICPM 2020. LNBIP, vol. 406, pp. 385–397. Springer, Cham (2021). https://doi.org/10.1007/978-3-030-72693-5_29

20. Rafiei, M., Schnitzler, A., van der Aalst, W.M.P.: PC4PM: a tool for privacy/confidentiality preservation in process mining. In: Proceedings of the Demonstration Track at BPM co-located with 19th International Conference on Business Process Management (BPM), vol. 2973, pp. 106–110. CEUR-WS.org (2021)

21. Rogge-Solti, A., Weske, M.: Prediction of business process durations using non-Markovian stochastic petri nets. Inf. Syst. **54**, 1–14 (2015)

22. Rozinat, A., Mans, R.S., Song, M., van der Aalst, W.M.P.: Discovering simulation models. Inf. Syst. **34**(3), 305–327 (2009)

23. Sani, M.F., Gonzalez, J.J.G., van Zelst, S.J., van der Aalst, W.M.: Conformance checking approximation using simulation. In: 2020 2nd International Conference on Process Mining (ICPM), pp. 105–112 (2020)
24. Verbeek, E., van Hattem, M., Reijers, H., de Munk, W.: Protos 7.0: simulation made accessible. In: Ciardo, G., Darondeau, P. (eds.) ICATPN 2005. LNCS, vol. 3536, pp. 465–474. Springer, Heidelberg (2005). https://doi.org/10.1007/11494744_27

A Reproducible Approach for Mining Business Activities from Emails for Process Analytics

Raphael Azorin, Daniela Grigori$^{(\boxtimes)}$, and Khalid Belhajjame

PSL, Université Paris-Dauphine, LAMSADE, Paris, France
{raphael.azorin,daniela.grigori,khalid.belhajjame}@dauphine.fr

Abstract. Emails are more than just a means of communication, as they are a valuable source of information about undocumented business activities and processes. In this paper, we examine a solution that leverages machine learning to i) extract business activities from emails, and ii) construct business process instances, which group together these activities involved in achieving a common goal. In addition, we examine how relational learning can exploit the relationship between sub-problems (i) and (ii) to further improve their results. The research results presented in this paper are reproducible, and the recipe and data sets used are freely available to interested readers.

1 Introduction

To facilitate personal email management, several research proposals have been made, see e.g., [5, 8, 23]. For example, Corston et al. [8] exploit emails to identify actions (tasks) that can be added to the user's "to-do" list. While useful, current emails management tools lack the ability to recast emails into *business activity-centric resources*. In particular, our work targets people that would like to apply analytics on emails upon the extraction of their business activities and their organization into instances. Thus, the elicitation of business activities from a set of emails opens up the door to activity analytics by leveraging emails to answer queries such as: *What are the business activities executed by a specific employee?* In addition to activities, their relationships and organization into processes are also useful to answer analytical queries like: *What is the average duration of a business process?* This can be computed by averaging the time taken by all process instances of the same process.

An email folder will usually contain several executions of the same process, which are called cases or instances. While threads can be an indication of such related activities inside a process instance, this information is not sufficient, as the topic inside a thread can drift and people can deal with several processes inside a single email or thread. The first contribution we make in this paper is a solution to extract activities and process instances from an email log. The underlying issue is the identification of the relevant features that are crucial for the

© Springer Nature Switzerland AG 2022
H. Hacid et al. (Eds.): ICSOC 2021 Workshops, LNCS 13236, pp. 77–91, 2022.
https://doi.org/10.1007/978-3-031-14135-5_6

quality of the extraction operation. The solution we adopted for this is to sweep and empirically identify the email characteristics (i.e., subject, sender, recipient, email body, named entities as well as temporal information) that can be used for activity extraction on one hand, and for instance identification on the other. Our solution transforms an email corpus into a process event log allowing process discovery and analysis. The operation of mining activities and the operation of mining instances are not independent of each other. The second contribution we make is to study the interplay between these two problems. Specifically, we examined whether activity discovery can be leveraged to effectively improve process instance discovery, and vice versa, by using relational learning techniques [18]. In particular, we examine the extent to which relational learning can be used to improve the baseline results obtained using traditional machine learning techniques.

This paper is organised as follows. We present the framework we elaborated for email and process instance discovery in Sect. 2. We then study in Sect. 3 the relationship between activity and instance discovery. Experiments and results of our approaches are provided in Sect. 4 using a real-world email log collected in the context of an open source project development [2]. It is worth noting that the results presented are reproducible as we make the documented recipe (notebook) and datasets used in our research freely available online[1]. We discuss related work and present concluding remarks in Sects. 5 and 6, respectively.

2 Framework for Process Discovery and Analysis from Email

Figure 1 presents the modules of the framework we elaborated to transform an email corpus into a process-oriented event log that can be used for process discovery and analysis.

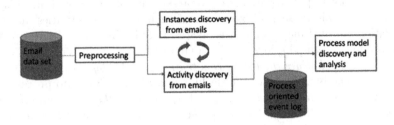

Fig. 1. Overview of the approach

There is a plethora of process mining tools and approaches (see e.g., [19] for a survey) that can be utilized to discover process information from logs. However, such tools require as input an event log that explicitly contains information

---

[1] https://github.com/Raphaaal/icpm_emails_process_mining.

about the activities that have taken place, the process instances (also known as cases) to which those activities belong, and their process model. Table 1 shows an example of the structure of such event log that is required by process mining tools. When dealing with emails, the challenge lies in creating this event log to be used as input by process mining tools. The framework illustrated in Fig. 1 is designed for this purpose. After preprocessing the emails with natural language processing techniques in the first step (preprocessing), activities and instances are identified. The remainder of this section presents the approach we adopted to discover activities and process instances.

Table 1. Event log example

Process ID	Process instance ID	Activity name	Timestamp
1	1	Request	2016-05-03 12:45
1	1	Send	2016-05-04 14:40
1	1	Confirm	2016-05-04 16:56
1	2	Send	2016-05-11 12:46
1	2	Confirm	2016-05-12 23:22
1	2	Request	2016-05-10 10:03
1	3	Request	2017-03-03 12:09
1	3	Send	2017-03-03 13:10
1	3	Notify	2017-03-04 15:38

2.1 Email Activities Extraction

Activity extraction is formulated as a classification problem where a training instance is composed of the embedding of the email and its sender's domain, the label being the activity present in the email. The label is one of the categorical labels representing each possible activity. We use words embeddings (Sense2Vec 2019 [20]) and average them to compute an embedding for the subject and for the body. Both of these embeddings are then averaged to obtain the email final embedding. We also extract the sender email domain that we encode using one-hot encoding. In this work, we assume that each email contains one activity and belongs to one instance.

2.2 Instances Discovery

To identify process instances, we start by determining if a pair of emails belong or not to the same instance. To build the binary classifier that allows us to do so, we were guided by the following observation. Emails belonging to the same instance are likely to have a sender or receiver in common, are close in time and talk about the same process (i.e., they have similar text and contain the same named entities). Therefore, we used the following features to describe a pair of emails: i) the (cosine) similarity of the embeddings of the email subjects and the (cosine) similarity of the embeddings of the email bodies, ii) the timestamp difference between the two emails, and iii) the number of common recipients (senders and receivers) between emails. Moreover, we used the cosine similarity of the embeddings of the named entities found in the two emails. These features compose a training instance and the label is 1 if the two emails belong to the same process instance, 0 otherwise.

Once the email pairs are identified as belonging or not to the same process instance, we then construct the full process instances. We have developed the Algorithm 1 for this purpose. Due to space limitations, we cannot explain the algorithm in detail. However, it is sufficient to present the essence of the algorithm, which proceeds as follows. It considers the results of the binary classifier presented above as votes: two emails classified as belonging to the same process instance are considered as a vote for an instance containing both emails, and two emails found not to belong to the same process instance are considered as a vote for separate instances. These votes are then aggregated to construct process instances, with each email being associated with the process instance to which it is most likely to belong based on the votes counted for that email.

Alternatively, we also developed an unsupervised approach based on hierarchical clustering to directly group emails into process instances. For each email, we used the following features: i) subject line embedding, ii) body embedding, and iii) date embedding. These features were normalized and their cosine similarity was used to guide the generation of the clusters. We used the average distance between emails as a stopping criterion, i.e., the maximum distance between emails in a cluster has to be inferior to the average similarity.

2.3   Workflow Discovery and Analysis

After applying the previous approaches, the email corpus has been transformed into an event log in which each email is associated with an activity name and an instance ID. The email timestamp is assumed to be the activity timestamp.

Therefore, existing process discovery technique can now be applied on this event log. For structured processes, techniques discovering the whole model can be applied. For unstructured processes, behavioral patterns (i.e., small group of activities and their relationships) can be extracted using specific approaches for behavioral patterns discovery (see e.g., [1]).

3   Relational Approach for Activity Discovery

Can we exploit information about emails process activities in order to group these emails into process instances? Conversely, can we leverage information about which email belongs to which process instance to discover process activities? In this section, we investigate these hypothesis by leveraging relational learning.

As an example, suppose that a candidate email is about purchasing an item P and another candidate email is the confirmation of the purchase of item P.

ConstuctInstances
Record votes ;
Data: counter =0
Data: EmailsByInstances = {}
for each pair $(email_1, email_2)$ **do**
    if pair predicted as same instance **then**
        if one email of the pair is already accepted in EmailsByInstances at the key InstanceId **then**
            append both emails to EmailsByInstances[InstanceId][accepted];
        else
            create entry EmailsByInstances[counter][accepted] = $[email_1, email_2]$;
            counter = counter + 1;
            end if
        end
    end
    else
        if one email (A) of the pair is already accepted in EmailsByInstances at the key InstanceId **then**
            append the other email (B) to EmailsByInstances[InstanceId][rejected];
        else
            create entry EmailsByInstances[counter][accepted] = [A];
            create entry EmailsByInstances[counter][rejected] = [B];
            counter = counter + 1;
        end
        end
    end
    end
end
Count votes;
InstanceIdentifiers = {};
for each email in the emails set **do**
    EmailScore = {};
    for each InstanceId in EmailsByInstances **do**
        EmailScore[InstanceId] = 0;
        if email is accepted in EmailsByInstances at the key InstanceId **then**
            EmailScore[InstanceId] = EmailScore[InstanceId] + 1;
        end
        if email is rejected in EmailsByInstances at the key InstanceId **then**
            EmailScore[InstanceId] = EmailScore[InstanceId] - 1;
        end
    end
    Take the instance with the maximum score as InstanceIdentifiers[email] ;
    This is the instance ID for the considered email
end

Algorithm 1: Reconstructing instances from emails pairs

Both emails will contain data about P with names of the selling company or the name of the buyer or attachments about the item characteristics or the bill. Using such information about the activities "Purchase Item" and "Confirm Purchase" can help in associating both emails with the same process instance.

On the other hand, information about process instances can be used to help extract email process activities. Consider again the above example, and consider that we know that emails e_1 and e_2 belong to the same process instance I where e_2 is a response to e_1. Taking into consideration that e_1 contains the activity "Purchase Item" and that e_2 is the email following e_1 in the process instance I, we could build a classification model that could successfully predict that e_2 is likely to contain the activity "Confirm Purchase" (based on old occurrences of emails similar to e_1 and e_2).

To investigate the above hypothesises, we make use of relational learning techniques [17]. Accordingly, we reformulate the problem into a relational problem, that consists of the following four phases:

- **Phase 1**: Extracting process activities from emails using their content (i.e. without using information about emails process instances).
- **Phase 2**: Identifying process instances from emails using a similarity computation of their content (i.e. without using any information about email process activities).
- **Phase 3**: Using the extracted email activities, can we improve the identification of process instances in emails?
- **Phase 4**: Using the links between emails belonging to the same process instances, can we improve the extraction of emails process activities?

We have already presented solutions for Phases 1 and 2 in the previous section. We focus on Phases 3 and 4 in the rest of this section.

Phase 3: Instance Discovery Using Extracted Activities Information. In this phase, the problem is to predict whether two emails belong or not to the same instance using information coming from the activity discovery step (in addition to email intrinsic features). For each pair of emails, in addition to the intrinsic features used in the baseline approach (see the previous section about **Phase 2**), a training instance is now composed of the one hot encoding of each email's activity. The label is 1 if the two emails belong to the same instance, 0 otherwise.

Phase 4: Activity Discovery Using Instances Information. In this phase, the problem is to discover email activities using information about process instances in addition to email intrinsic features. For each email, to guide the classification of its activity, we use supplementary information about the previous and next activity labels present in its process instance. A training instance is now composed of the following features:

- its subject line and body embedding (intrinsic features, see the previous section about **Phase 1**)
- for a predefined number of generations i:

- the i-th previous activity in its process instance (one hot encoding), its subject line embedding, its email domain embedding (one-hot encoding) and its timestamp difference with the current email
- the i-th next activity in its process instance (one hot encoding), its subject line embedding, its email domain embedding (one-hot encoding) and its timestamp difference with the current email

The number of generations i is defined during training as the maximum number of activities in one instance. The label is one of the categorical labels representing each possible activity.

3.1   Relational Algorithm

Algorithm 2 lists the main steps of our iterative relational classification approach. The results of **Phase 1** and **Phase 2** are used to solve **Phase 3** and **Phase 4**.

First, we have to check if there is some potential for improvement with a relational approach. For this, we apply phases 3 and 4 on the training set while using the ground truth information (i.e., the real instances and real activity labels as extrinsic features). If improvement is observed, we put all the phases together into an iterative and relational approach. If only one relational model shows potential for improvement, it will be the only one to be included in the loop.

In the relational approach, predictions with high confidence deduced from **Phase 3** are fed back to **Phase 4** to improve the accuracy of its results and vice versa. This is an iterative process, so inferences are dynamically changing between iterations until convergence.

---

1. Baseline approach for email activities extraction (**Phase 1**).
2. Baseline approach for process instances extraction (**Phase 2**).
for Iteration $i = 1...K$ **do**
  /* If RelationalInstances showed potential */
  3. Use extracted email activities to identify process instances (**Phase 3**).
  Accept predictions for which Confidence > Threshold and update instances
  /* If RelationalActivities showed potential */
  4. Use extracted process instances to identify email activities (**Phase 4**).
  Accept predictions for which Confidence > Threshold and update activities
  Threshold = Threshold/2
end

Algorithm 2: Iterative Relational Classification Approach.

---

4   Experiments

In this section, we present the results of the empirical evaluations we conducted to assess the effectiveness of the solutions presented in Sects. 2 and 3. We used an open labelled data corpus [2] that was constructed from a publicly available email dataset to facilitate automated workflow analysis by providing ground truth. This dataset associates each email with an activity and a process instance. There are 19 possible activity labels in total. In addition, it contains 6 process models that act as specifications of the process instances and specify the dependencies between the activities. Our objective is to empirically examine how close the solutions presented in this article for the automatic discovery of activities and process instances are to the activities and process instances that have been manually labeled in [2].

4.1 Implementation

We implemented the solutions presented in this paper as a Python Notebook that is available online[2]. For word embeddings we used Sense2Vec 2019 [20], an advanced version of Word2Vec. For the classification tasks, we used XGBOOST, a decision-tree-based ensemble algorithm that uses a gradient boosting framework, which is able to handle a mixture of weak and strong features automatically and works well on tabular data. We used the output probability prediction as a measure of confidence (e.g., if the predicted class has a probability > 0.9).

For preprocessing purposes, we applied classical cleaning steps when dealing with text: removing special characters, spaces, etc. Then the embedding of each subject line and body has been obtained using Sense2Vec 2019. Named entities have been extracted in the same way. We have also filtered out process instances containing a single activity, since these patterns presents little interest, especially for relational learning. At the end, we obtained a cleaned dataset containing and characterizing 157 emails.

4.2 Building Classification Models

When splitting the dataset into training and testing sets, we made sure that emails from to the same process instances belong (collectively) to either the training or the testing dataset. Moreover, the Algorithm 3 ensures that the two sets contains the same activities and a balanced number of instances per activity. Finally, this split algorithm is randomized and reproducible thanks to the seed used.

The hyperparameters of XGBOOST were set via grid search. More information on the hyperparameters used in our experiments can be found in the Python Notebook.

In training mode, as we have access to the ground truth activity label and instance identifier for each email, we know each activity composing a process instance. However, when testing the relational approach, we will use the predicted activities and instances instead. On the test dataset, Algorithm 2 iteratively uses high confidence predictions from the relational instances classifier to update the current assignments and then feed them to the relational activity classifier (and vice versa).

---

[2] https://github.com/Raphaaal/icpm_emails_process_mining.

Train/Test dataset split
Data: exclusion list = [];
for <u>each activity</u> **do**
| count the number of instances containing it.
end
for <u>for each activity, ordered by increasing number of instances</u> **do**
| instances = activity instances that are not in exclusion list;
| shuffle instances;
| separate instances btw. the two datasets (50% - 50%) ;
| append instances to exclusion list;
end
Algorithm 3: Algorithm for splitting the data set into train/test set

4.3 Results

All the presented results are computed as an average on 10 experiments. An experiment is a full pipeline evaluation starting with the generation of pseudo random train/test sets.

Instances Discovery. To evaluate the instances discovery (baseline and relational approaches), we considered all the pairs of emails in the test set and computed precision, recall and F-score in a binary fashion. That is to say, we checked whether each pair of emails belong or not to the same instance (ground truth vs prediction). For the relational approach, this evaluation is made at the end of each collaborative learning iteration. The Algorithm 2 stabilized after 5 iterations (afterwards the variations of the F-score was less than 1%). Figure 2a summarizes the results obtained across the 10 experiments for instance discovery. The first bar shows the F-score for the baseline solution. The second bar shows the F-score obtained using relational learning that is based on the ground truth. Whereas the remaining bars show the results obtained using relational learning at different iterations of Algorithm 2. The figure shows that the different flavors of our solution achieve an F-score greater than 60%. The use of relational learning allows us to achieve some improvement albeit minor (up to 2%). This can be explained, at least partly, by the fact that the activity discovery was not instrumental in distinguishing between instances due to the unstructured nature of the underlying business processes.

Activities Discovery. To evaluate the activity discovery (baseline and relational approaches), we considered all emails in the test set (ground truth vs prediction) and computed precision, recall and F-score in a micro-average fashion (i.e., calculating metrics globally by counting the total true positives, false negatives and false positives). Figure 3 shows the evolution of the F-score for various activities during the iterations of the relational approach. It also shows the baseline approach score and the relational approach score when used with ground truth labels (to identify the potential for improvement). Note that the results reported as "relational" in the figure correspond to a variant of Algorithm 2 where the predictions of the instances are not changed (i.e., remain the same) across the

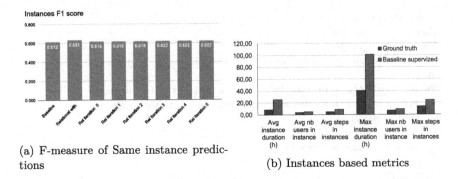

(a) F-measure of Same instance predictions

(b) Instances based metrics

Fig. 2. Instance discovery results

iterations. This version of the algorithm performed best because the relational instances classification model did not show significant potential for improvement. The support (i.e., number of emails containing the respective activity in the dataset) is also indicated. We can see that for activities with a significant support like "issue comment" and "issue update", the relational approach leads to an improvement of 0.119 and 0.024, respectively. For the activity "commit changes", with a support of 62, the potential for improvement for the relational approach is quasi null: in this case, the relational approach with ground truth is 0.001 greater than the baseline. Some activities have a very small support and are depicted with lighter colors in the figure. Also, note that, following [2], some pairs of activities (like "issue comment" and "issue update") are less distinguishable because they had a large overlap of their label-email keyword frequency distributions and had a weaker consensus of annotators. Overall, the relational approach leads to an increase of 0.022 in F-score.

These experiments suggest that the relational approach could be beneficial, but more experiments should be done to asses its advantage. Our dataset was limited in terms of number of activity occurrences and number of instances for different processes. Also, the business processes are unstructured and this lack of regularity may be detrimental to relational learning.

Business Metrics and Model Mining. Figure 2b shows some business metrics based on the discovered instances. It shows that average metrics (like average duration, average number of users involved in one instance, etc.) can be derived with good accuracy from the discovered instances. Obviously, metrics like maximum are sensible to prediction errors. The same conclusion can be drawn form Fig. 4 which presents metrics based on the discovered activities: average activity duration, average number of users executing a specific activity, etc.

Figure 5 shows the models mined using FHM (Flexible Heuristic Miner, [22]) from the generated event log vs the event log based on the ground truth labels. We can see that 7 relationships have been correctly identified among the 9 relationships existing between recognized activities.

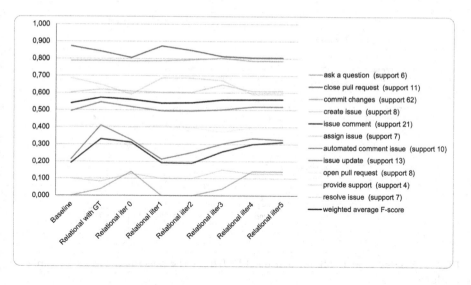

Fig. 3. F-measure of activities discovery

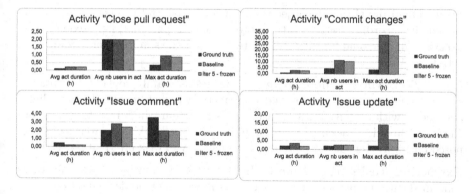

Fig. 4. Activity based metrics

These experiments show the feasibility of a full analytics pipeline that starts with a raw email corpus and outputs analytics about the executed process. However, in order to fully assess the advantage of a relational approach versus baseline approaches, a richer email dataset should be used (i.e. containing more process instances). This would allow the clear identification of cases where the relational approach should be used, for example when processes have some structure and when some individual activities are difficult to distinguish.

5   Related Work

We categorize the related works into the following categories:

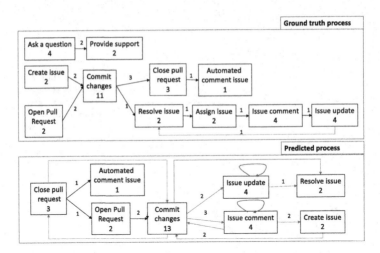

Fig. 5. Process models obtained using FHM from the generated log vs the ground truth log

Email Organization. The objective of the proposals in this category is to classify emails according to their nature (professional or personal) [3], predefined set of classes [4], personal prioritization criteria [24] or the similarity of encompassed activities [16].

Email Task Management. Several applications mainly consider the process of associating manually emails and their metadata such as attachments, links, and actors with activities. TaskMaster [6] is a system which recasts emails into Thrasks (thread + task). In Gwizdka et al. [14], the TaskView interface is proposed for improving the effectiveness and efficiency of task information retrieval.

Extracting Tasks from Emails. In the work of Faulring et al. [13], tasks contained within sentences of emails are classified using 8 predefined set of classes of tasks. In the same context, Cohen et al. [7] use text classification methods to detect "email speech acts" (e.g., Request, Deliver, Propose, Commit). Another work by Corston-Oliver et al. [8] identifies action items (tasks) in email messages that can be added by the user to his/her "to-do" list. Syntactic and semantic features are used to define the sentences of the emails to be classified using SVM as containing tasks or not.

Extracting Business Process Information from Emails. The EMailAnalyzer tool [21] assists users in analyzing and transforming e-mail messages to a process event log format that can be used by process mining tools. The user can search for process instances using several simple options such as linked contact, thread, sender, receiver or instance ID (if present in the message topic), thread, etc. [19] extracts business process information from emails and organizes this information as a mind map, a diagram meant to be used for knowledge management, showing process participants, exchanged artifacts, etc. The approach in [9] makes use of object matching algorithms to obtain clusters of related email conversations (process instances). For that purpose, they use the distances between names of attachments, body (as a single string) and sender/receiver of each email. To classify emails into process instances, Drezde et al. [10] propose a model combining three learners: two of them for comparing people involved in a process against the recipients of the message, and the third one that uses a form of Latent Semantic Analysis to compare emails' contents. Approaches for extracting multiple business process activities from each email, their associated data and actors are proposed in [11,12,15]. The approach in [15] uses a classifier that extracts sentences containing business process activities from each email, identifies verb-noun pairs as activity names candidates using also a thesaurus and then clusters similar activities. An unsupervised method for extracting multiple business activities from each email and their metadata is proposed in [11] based on the frequency of activity names and their synonyms. Afterwards, these activities are associated to actors and speech acts [12].

Compared with prior work, our goal was to check the feasibility of a full and automated pipeline starting from a user-labelled corpus of emails to analytics. In a practical scenario, the activity label could come from the tasks added to the user task list and instances could be initially approximated by threads. A second goal was to test the relationship between activity and instances discovery. Our relational learning approach is inspired by the work of Khoussainov et al. [17], which proposes an iterative approach to identify task and relations between individual messages in a task. Each message is a associated to one of 5 predefined speech acts. Our approach allows to use an open list of activities, as available in the user labelled dataset. Moreover, we apply the relational approach on a public open dataset and publish the code thus contributing to further developments in this area.

6   Discussion and Future Work

We presented in this paper a reproducible approach that takes as input an email log and extracts business activities and their organization into process instances. The resulting process event log can be used for process discovery and analysis. We applied our approach to an open data set and we provided an activity-centric analysis of emails.

The approach and its validation can be extended in several ways. As the dataset has only one activity per email and each email belongs to one instance, our models are restricted to these assumptions. However, in other applications, an email may contain several activities and these activities may belong to different instances. We envisage to extend our models to cover these cases. The experiments were limited to the studied dataset which contains a small numbers of emails. The quality of the results may be affected by the scarcity of the data. Testing the approach on a different dataset containing more structured processes (like orders processing on e-commerce websites) would allow us to confirm our intuition that the relational approach works better on more structured processes.

As future work we envisage to propose active learning approaches by implying users in a small number of tasks in order to improve the result. User involvement could assure the production of "islands of truth" that could improve the relational approach [18]. Asking the right questions to users would allow to improve the classification and consequently the analytics metrics.

References

1. Acheli, M., Grigori, D., Weidlich, M.: Efficient discovery of compact maximal behavioral patterns from event logs. In: Giorgini, P., Weber, B. (eds.) CAiSE 2019. LNCS, vol. 11483, pp. 579–594. Springer, Cham (2019). https://doi.org/10.1007/978-3-030-21290-2_36
2. Allard, T., Alvino, P., Shing, L., Wollaber, A., Yuen, J.: A dataset to facilitate automated workflow analysis. PLoS One **14**(2), e0211486 (2019)
3. Alsmadi, I., Alhami, I.: Clustering and classification of email contents. J. King Saud Univ. Comput. Inf. Sci. **27**(1), 46–57 (2015)
4. Bekkerman, R.: Automatic categorization of email into folders: Benchmark experiments on Enron and Sri corpora. Technical report, UMass CIIR (2004)
5. Bellotti, V., Ducheneaut, N., Howard, M., Smith, I., Grinter, R.E.: Quality versus quantity: e-mail-centric task management and its relation with overload. Human-Comput. Interact. **20**(1), 89–138 (2005)
6. Bellotti, V., Ducheneaut, N., Howard, M., Smith, I.: Taskmaster: recasting email as task management. In: PARC, CSCW, vol. 2 (2002)
7. Cohen, W.W., Carvalho, V.R., Mitchell, T.M.: Learning to classify email into speech acts. In: Proceedings of Empirical Methods in Natural Language Processing (2004)
8. Corston, S,O., Ringger, E., Gamon, M., Campbell, R.: Task-focused summarization of email. In: Association for Computational Linguistics, July 2004
9. Di Ciccio, C., Mecella, M., Scannapieco, M., Zardetto, D., Catarci, T.: MailOfMine – analyzing mail messages for mining artful collaborative processes. In: Aberer, K., Damiani, E., Dillon, T. (eds.) SIMPDA 2011. LNBIP, vol. 116, pp. 55–81. Springer, Heidelberg (2012). https://doi.org/10.1007/978-3-642-34044-4_4
10. Dredze, M., Lau, T., Kushmerick, N.: Automatically classifying emails into activities. In: 11th International Conference on Intelligent User Interfaces, pp. 70–77. ACM (2006)
11. Elleuch, M., Ismaili, O.A., Laga, N., Assy, N., Gaaloul, W.: Discovery of activities' actor perspective from emails based on speech acts detection. In: 2nd International Conference on Process Mining, ICPM 2020, pp. 73–80. IEEE (2020)
12. Elleuch, M., Ismaili, O.A., Laga, N., Gaaloul, W., Benatallah, B.: Discovering activities from emails based on pattern discovery approach. In: Fahland, D., Ghidini, C., Becker, J., Dumas, M. (eds.) BPM 2020. LNBIP, vol. 392, pp. 88–104. Springer, Cham (2020). https://doi.org/10.1007/978-3-030-58638-6_6
13. Faulring, A., et al.: Agent-assisted task management that reduces email overload. In: International Conference on Intelligent User Interfaces, pp. 61–70. ACM (2010)
14. Gwizdka, J.: Taskview: design and evaluation of a task-based email interface. In: Proceedings of the 2002 conference of the Centre for Advanced Studies on Collaborative research, p. 4. IBM Press (2002)
15. Jlailaty, D., Grigori, D., Belhajjame, K.: Email business activities extraction and annotation. In: Kotzinos, D., Laurent, D., Spyratos, N., Tanaka, Y., Taniguchi, R. (eds.) ISIP 2018. CCIS, vol. 1040, pp. 69–86. Springer, Cham (2019). https://doi.org/10.1007/978-3-030-30284-9_5

16. Jlailaty, D., Grigori, D., Belhajjame, K.: Mining business process activities from email logs. In: ICCC, pp. 112–119. IEEE (2017)
17. Khoussainov, R., Kushmerick, N.: Email task management: an iterative relational learning approach. In: CEAS 2005 - Second Conference on Email and Anti-Spam, 21–22 July 2005, Stanford University, California, USA (2005)
18. Neville, J., Jensen, D.D.: Iterative classification in relational data. In: AAAI 2000 Workshop Learning Statistical Models from Relational Data (2000)
19. Soares, D.C., Santoro, F.M., Baião, F.A.: Discovering collaborative knowledge-intensive processes through e-mail mining. J. Netw. Comput. App. **36**(6), 1451–1465 (2013)
20. Trask, A., Michalak, P., Liu, J.: sense2vec - A fast and accurate method for word sense disambiguation in neural word embeddings. CoRR, abs/1511.06388 (2015)
21. van der Aalst, W.M.P., Nikolov, A.: Emailanalyzer: an e-mail mining plug-in for the prom framework. BPM Center report BPM-07-16, BPMCenter.org (2007)
22. Weijters, A.J., Ribeiro, J.T.: Flexible heuristics miner (FHM). In: CIDM 2011, pp. 310–317. IEEE (2011)
23. Whittaker, S., Bellotti, V., Gwizdka, J.: Problems and possibilities. Commun. ACM, Email and Pim (2007)
24. Yoo, S., Yang, Y., Lin, F., Moon, I.-C.: Mining social networks for personalized email prioritization. In: Proceedings of ACM SIGKDD, pp. 967–976. ACM (2009)

Human-in-the-Loop Optimization for Artificial Intelligence Algorithms

Helia Farhood[1]([⊠]), Morteza Saberi[2]([⊠]), and Mohammad Najafi[3]([⊠])

[1] School of Computing, Macquarie University, Sydney, Australia
helia.farhood@mq.edu.au
[2] School of Information, Systems and Modelling, University of Technology Sydney, Sydney, Australia
morteza.saberi@uts.edu.au
[3] School of Electrical and Data Engineering, University of Technology Sydney, Sydney, Australia
mohammad.najafi@uts.edu.au

Abstract. Numerous organisations use artificial intelligence algorithm-based products in their different activities. These solutions help with a wide range of jobs, from operational task automation to augmentation-based strategic decision making. The users' trust in the truth and fairness of a product's outputs must be built before it can be completely integrated and embedded in an organization's daily functioning. They would be burdened with more work if Artificial Intelligence (AI) products did not have this feature. A human-in-loop decision-making process is important for building confidence and producing a successful AI-powered solution. In this research, a novel interactive system was created to explore the behaviour of AI-powered products. When designing our framework, we considered the necessity of integrating a human-in-the-loop technique in the design stage, something that had been missed in prior research of a comparable scale. The proposed software can optimise and monitor the AI-powered product process and outputs to involve people directly in the optimisation loop to identify and avoid likely and diverse failures. The Local Interpretable Model-agnostic Explanations (LIME) heatmap was utilised to illustrate decision-making features and mistake details more effectively throughout the improvement phase. The literature highlights the need of taking these issues into account throughout the design stage of an AI-powered product. This article describes how a human-in-loop AI-powered product is created by combining technologies from the AI, risk management, and human-computation domains. The designed system is based on deep learning as its decision-making engine, LIME as its approach explanation module, and the human aspect of knowledge workers. For real-world applications, we show how the created system improves product dependability and understandability by using real data and benchmark datasets.

Keywords: Human-in-the-loop optimization · Artificial intelligence improvement · LIME · Knowledge workers

© Springer Nature Switzerland AG 2022
H. Hacid et al. (Eds.): ICSOC 2021 Workshops, LNCS 13236, pp. 92–102, 2022.
https://doi.org/10.1007/978-3-031-14135-5_7

1 Introduction

Task automation vs. enhancement is a key factor when developing AI-powered solutions. When deciding on the best form of transportation, a number of variables must be carefully examined. When it comes to safety, automation is the way to go. However, augmentation is preferable in high-stakes situations like health and financial. People do not want to conduct repeated and dangerous activities; thus, automated solutions are preferable. However, if they are in charge of the task's result, they may opt to employ an AI-powered system as a decision support system. The company should ensure that the AI-powered product is efficient and compatible in its environment for both forms of task automation and augmentation. This is due to the fact that AI models might make mistakes. AI-based classifiers have matured to the point where they can achieve remarkable accuracy using advanced deep learning algorithms [7].

For binary classification, these sophisticated classifiers keep an eye out for two sorts of errors: false positives and false negatives. For the implication phase, this, however, is not enough information. Companies may not want genuine positive or actual negative outcomes. Deep learning, for example, cannot accurately identify a swan image if there is no "black swan" instance image in the training set (concept drift issue) [11].

This is a massive win for the deep learning designer. However, seen from the perspective of an organisation, the tale is quite different, and the result is deemed a failure. To put it another way, even when AI classifiers have been tuned to maximise their reward function and use just training data, there is always the possibility that something goes wrong without the user being aware. In literature, this is referred to as the "unknown unknown" or "weak spots" [12].

For many causes, like not having access to the necessary domain expertise, utilising biassed data or an imbalanced data collection, this unknown unknown/weak spots risk develops. This might result in serious consequences, and particularly in a high-stakes environment, the company may wish to monitor the AI-powered product's process and outputs in order to discover and prevent probable and diverse failures, particularly the unknown. It should be emphasised that while academics strive to develop AI models using a variety of qualitative approaches, this is insufficient for the real-world consequences. Including a person in the loop may be an alternate method of mitigating these difficulties [6].

In other words, people should be included in the loop between the AI algorithm and the context in which the system operates [18]. While AI accuracy is a critical aspect of any AI-powered product, its user mental model is as critical, while being the most overlooked. For instance, while a 95% accuracy rate may be considered a success from an AI standpoint, a 5% mistake rate may be a big concern for a beginner or strict user. If a user expects AI to offer perfectly correct solutions, a 5-% error margin may sound strange. Thus, through conducting coordination sessions, knowledge workers should have a better understanding of the system, with the goal of bringing their mental model in line with the AI model's actual capabilities. This is a critical stage because it enables the company to harness the collective wisdom of its knowledge employees.

In response to this practical challenge and knowledge gap, we developed a monitoring system that enables the organisation to improve the detectability rate of unknown problems. The purpose of this article is to describe the created solution for human-in-the-loop enhancement of AI-powered products. The suggested solution combines the strength of several strong tools across interrelated disciplines to enable organisations to control their AI-powered products. By involving knowledge workers in the loop between AI models and their output, the system establishes a scalable and realistic quality assurance method. By including an explanation feature in the AI model, the process becomes more flexible and efficient [5]. It should be emphasized that while the suggested system was created and validated for image classification, it can be modified to perform other tasks such as forecasting and text analytics by incorporating additional data types such as text and tabular datasets. Additionally, while image categorisation is often a low-stakes work, system implications might elevate it to a high-stakes activity [20].

This research's primary contribution is the development of a scalable and practical framework for embedding humans with appropriate domain knowledge within an AI-powered solution. This enables organisations to monitor, analyse, and enhance the product's performance. We developed a novel framework for realising such a system incorporates techniques from human computation, explainable AI, and risk management. Additionally, our platform is backed up by innovative software engineering techniques to make it more user-friendly.

The remainder of the article is structured as follows. Section 2 presents the motivating scenario. We examine related work and added value in Sect. 3. Section 4 discusses the proposed framework. Section 5 demonstrates experimental results. Section 6 summarises the paper's findings and suggests areas and future investigations.

2   Motivating Scenario

The development of new learning algorithms and concepts, the continual expansion of data and low-cost processing have fuelled recent machine learning advances. However, when applied to complicated real-world problems, machine learning systems frequently give disappointing results. The human-in-the-loop approach could aid in resolving these challenges. According to a widespread notion in neural network models with users in the loop, individuals can provide expertise about complex tasks that datasets cannot capture. In some machine learning applications, tasks are exceedingly complicated, and machine learning algorithms are incapable of resolving the issue successfully. Human input and involvement can be beneficial in certain instances.

Interpretability and explicability are mutually exclusive in machine learning. The issue is that these products usually function in a mysterious black box, leaving consumers in the dark about why these algorithms make the decisions they do [17]. As a result, the development of interpretable structures for communication by a human is one solution for this problem. Our study in this paper is focused on developing a monitoring system that can assist an organisation in identifying previously unidentified issues. The purpose of this article is to present a recently built human-in-the-loop AI product improvement system. This enables companies to monitor, analyse, and enhance the performance of their products. Since this new technique allows knowledge workers in a particular domain (such as government or finance) to automate and facilitate their operations, it may be delivered as a service.

3   Related Work

Work on improving deep learning models can be categorised into two key areas: using proxies to learn interpretable structures and learning from human interference.

3.1   Using Proxies to Learn Interpretable Structures

Interpretable frameworks may be learned using many different techniques, some of which focus on improving proxies that are directly derived from the system. Decision tree depth [8,9] and the amount of integer linear regression [21] a couple of instances. On any model structure, lage et al. [14] offered a method that reduces the amount of user studies needed to discover predictions that are both effective and easy to understand. According to Nada et al. [15] physicians favour larger decision trees than shorter sides, indicating that these proxy measures do not reflect what it takes for a system to be interpretable throughout all circumstances. It is also possible to see optimising a proxy as MAP prediction under such a traditionally adopted even before in some circumstances [2]. Zhao et al. [23] offered a visible embedding learning technique based on deep interpretable machine learning. Their structural matching method computes an optimum matching flow among feature maps from the two pictures, which directly coordinates the spatial embeddings. These models allow deep networks to learn measurements in a way that is more human-friendly by decomposing the correlation between different pictures and their implications to the total similarity. In these proxy-based techniques, it is assumed that a priori knowledge of the interpretability quality may be expressed as a functional feature of the model.

3.2   Human Feedback Enables Learning

Taking into account the needs of the user during the process of system, using brain electroencephalogram signals to record prefered design aspects, Pan et al. [22] devised a system that reports findings from a human in-the-loop approach. They created a system that uses an encoder to identify electroencephalogram (EEG) characteristics from signals coming collected from individuals as they viewed photos from ImageNet. Conversational agents must have the capacity to learn from their errors and adapt via interacting with people. This is an essential consideration when creating conversational agents. When it comes to relevance feedback, Li et al. [16], looked into this area in a scenario where the chatbot gets feedback from an instructor after it generates answers to questions. In a chaotic manner, they developed a simulator to evaluate different components of this training and to propose models that operate in this domain. Human in-the-Loop (HitL) network processing paradigm and dataset termed HOOPS for the job of a knowledge graph inspired verbal evaluation are proposed by Fu et al. [10] in the field of conversational recommendation. To be more specific, they create a knowledge tree that interprets various user actions and identifies relevant attribute entities for each user-item combination. A knowledge graph organization is traced by simulating speech turns. This reflects the human decision-making processes. Due to the fact that algorithms and people both produce and acquire the characteristics and relevant data for training suggestions, biases might creep into the data preparation and training stages. Beheshti et al. [4] used a combination of crowdsourcing approaches and rule-based algorithms to build feedback loops that may be used over the duration to deal with biases [1,3].

3.3 Problems of Existing Approaches and Our Added Value

Each of the approach types discussed has some limits when it comes to improving AI products. For end-users, being able to provide feedback to the AI model makes an AI-powered product more trustworthy than utilising proxies to develop interpretable structures for communication. Also, the AI model may be customised to better suit the demands of individual users. However, in some high-stakes situations, it may be too late to ask for such input because of the end-rigid user's mental model of the situation. For example, if a medical professional does not trust an AI-based health diagnostic, even a small mistake might jeopardise the efficacy of the system. Since input should be sought sooner in AI-powered product development, or consumers should be engaged with a more inviting mental model at the same time, it would be preferable.

One way to address this issue and its natural risk is to utilize the human-in-the-loop mechanism during the design stage, which has been considered in the design of our proposed framework, of which no attention was made to earlier comparable research. In addition, we used LIME's heatmap to highlight decision making features and mistake details more precisely during the improvement process.

4 Proposed Framework

Our developed system is a simple yet powerful system that uses knowledge workers as its AI auditors. The main task of these workers is to create golden human intelligence tasks (HIT). It should be noted that the way proposed framework uses workers is different to most typical crowdsourcing projects where the workers complete HIT rather than making them. Table 1 depicts the settings of the human - computation model of the system.

Table 1. Human-computation model setting of developed framework

Requester	HIT	HIT creator	Workers
Organization which uses an AI-powered product	Live image, offline image (benchmark database)	Knowledge workers	AI algorithm

The suggested framework relies on two connected modules. In the first, the accuracy of the deep learning network is tested through a golden HITs. The second module will redo this assessment using noisy golden HITs. Figure 1 depicts the process involved in the first module. The knowledge workers do the assessment either by uploading a live image using a smartphone or by retrieving it from a benchmark database. When a user uploads a live image, we assume they know the label of the image and are able to verify the model's output. For an image from a benchmark dataset, verification is conducted through a comparison with the model's output and the database's predefined label. Developed model then visually explains how the deep learning model arrived at the label by applying LIME in an online manner [19]. LIME's heatmap is visualized for the user to this end. Either the system or user can verify whether the model made a correct decision. If the decision is wrong, the user is able to report this to the data science team by sending the instance and the heatmap. Otherwise, the second module is initiated.

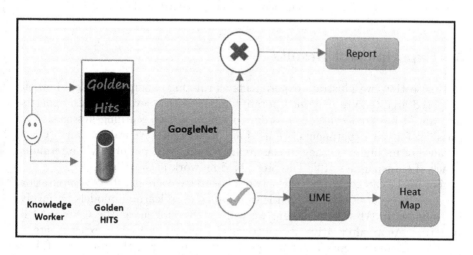

Fig. 1. First model of our proposed model

The second module's goal is to pinpoint the network's weak spot. By utilising image modification techniques, the created system allows the user to do this action automatically. When it comes to knowledge workers, they have two options: delete the essential section of the image or blur it. AI model reacts as follows: (a) model failure, (b) model accuracy decline, and (c) no change in the model performance. The system immediately recognises these three situations. In the first two scenarios, the user can inform the data science team about the failure of the system or the decrease in performance. Figure 2 shows the process involved in the second module.

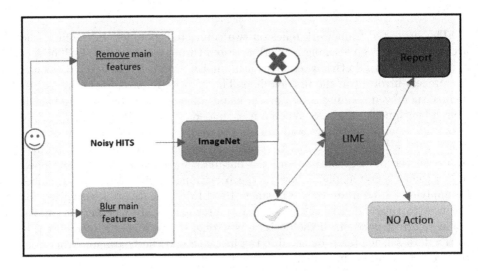

Fig. 2. Second model of our proposed model

5   Experimental Results

In this section, we illustrate experiments as running examples to show how the proposed architecture may incorporate humans with relevant domain expertise into an AI-powered solution in order to enhance deep learning systems. As a classifier for our experiments, we used a Convolutional Neural Network (CNN) because of its high accuracy in classifying complicated pictures while retaining a lack of transparency [13]. The pretrained network is GoogLeNet.

For image categorisation, the network has to have a high rate of computation. Because the goal of this research is to improve deep learning models by involving humans in AI system design, we will not be looking at the CNN model's structure. As an alternative, it will serve as a black box for our experiments.

This system accepts input images in two ways: first, the user may take a picture with their mobile phone and upload it as an input to the system (live image). The second method allows users to import images from absolutely any database (offline image). There are zero means in the pretrained network; thus, the size is 224 * 224 in the RGB colour space. The suggested method will next resize the sample picture to fit the network's input size. Once the test picture had been classified, the results were shown, starting with the highest classification score on display. This stage evaluates the deep learning network's accuracy by comparing it against a set of gold HITs.

For the sake of our model, we will assume that every user who uploads a live image already knows what it is supposed to be. Verification is done by comparing the image's label to the database's specified label and the model's output. Once the developed model is complete, it shows how LIME was used to achieve at the label. As a unique explanatory approach for dealing with the predictions of complicated machine learning models like deep neural networks, the LIME

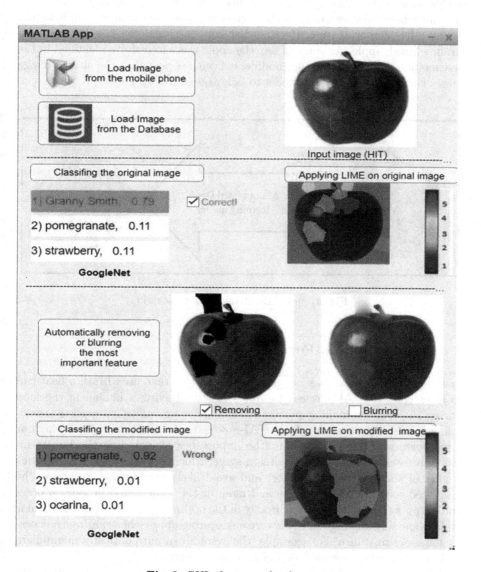

Fig. 3. GUI of proposed software

framework emerged. Using a LIME model, researchers can gain valuable insight into how visuals and words influence decision-making. To discover a weak spot in the network, the created system will automatically modify the input image in the following stage. It is possible to achieve this by offering the user two options: either remove or blur essential elements (depending on the LIME heatmap).

These three outcomes are possible while using the system: model failure, a decrease in the accuracy of the models, and the model's performance remain unchanged. The system immediately recognises these three situations. Because of this, users would inform the data science team about system failures and performance drops. Figure 3 depicts the various components of our system in action during a live demonstration. Here is what happens when you give the system a Granny Smith apple. As you can see, the apple was successfully identified in the first module. But in the second module, it becomes clear that the altered image causes the image recognition model to fail, as seen in Fig. 4.

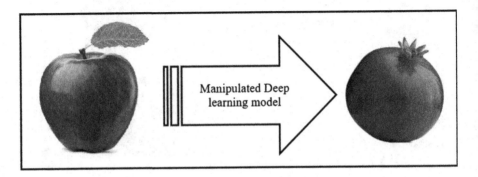

Fig. 4. Shows the manipulation scenario

6   Conclusion and Future Work

In this study, we created a new interactive system that uses LIME's heatmap to investigate how AI-powered products behave. Having a human in the loop during the decision-making process is critical for providing a supportive environment and delivering an effective AI-powered solution. When developing our framework, we took into account the importance of including the human-in-the-loop mechanism during the design stage, which was overlooked in previous studies of similar scope. To detect and avoid likely and diversified failure, the suggested software may optimise and monitor the AI-powered product process and outputs to include humans directly in the optimisation loop. Our platform is underpinned by cutting-edge software engineering approaches that make it easier for users to utilise. For example, the system can automatically manipulate an image to reduce the workload of knowledge workers.

We also used LIME's heatmap during the enhancement phase to highlight decision-making characteristics and error details more clearly. This monitoring system allows the organisation to enhance the detectability rate of unknown issues. In order to provide organisations control over their AI-powered products, a solution has been proposed that combines the strengths of many powerful tools from several fields. Quality assurance may be scaled and realistically achieved by including knowledge workers in the loop between AI models and their output. The procedure may be made more adaptable and efficient by incorporating an explanation feature in the AI model. In addition, the proposed system can be extended to do other tasks such as forecasting and text analytics by adding new data types such as text and tabular datasets, which were produced and verified for image classification. Future research can go in a lot of different areas. Eventually, we hope to extend this framework to analyse and monitor clinical data and text to make more informed recommendations on improving the medical system.

References

1. Allahbakhsh, M., Ignjatovic, A., Benatallah, B., Beheshti, S., Bertino, E., Foo, N.: Reputation management in crowdsourcing systems. In: Pu, C., Joshi, J., Nepal, S. (eds.) 8th International Conference on Collaborative Computing: Networking, Applications and Worksharing, CollaborateCom 2012, Pittsburgh, PA, USA, October 14–17, 2012, pp. 664–671. ICST/IEEE (2012)

2. Bach, F.: Structured sparsity-inducing norms through submodular functions. arXiv preprint arXiv:1008.4220 (2010)

3. Beheshti, A., Vaghani, K., Benatallah, B., Tabebordbar, A.: CrowdCorrect: a curation pipeline for social data cleansing and curation. In: Mendling, J., Mouratidis, H. (eds.) CAiSE 2018. LNBIP, vol. 317, pp. 24–38. Springer, Cham (2018). https://doi.org/10.1007/978-3-319-92901-9_3

4. Beheshti, A., Yakhchi, S., Mousaeirad, S., Ghafari, S.M., Goluguri, S.R., Edrisi, M.A.: Towards cognitive recommender systems. Algorithms 13(8), 176 (2020)

5. Beheshti, S.-M.-R., Benatallah, B., Motahari-Nezhad, H.R., Sakr, S.: A query language for analyzing business processes execution. In: Rinderle-Ma, S., Toumani, F., Wolf, K. (eds.) BPM 2011. LNCS, vol. 6896, pp. 281–297. Springer, Heidelberg (2011). https://doi.org/10.1007/978-3-642-23059-2_22

6. Benedikt, L., Joshi, C., Nolan, L., Henstra-Hill, R., Shaw, L., Hook, S.: Human-in-the-loop AI in government: a case study. In: Proceedings of the 25th International Conference on Intelligent User Interfaces, pp. 488–497 (2020)

7. Cohen, N., Sharir, O., Shashua, A.: On the expressive power of deep learning: a tensor analysis. In: Conference on Learning Theory, pp. 698–728. PMLR (2016)

8. Farhood, H., He, X., Jia, W., Blumenstein, M., Li, H.: Counting people based on linear, weighted, and local random forests. In: 2017 International Conference on Digital Image Computing: Techniques and Applications (DICTA), pp. 1–7. IEEE (2017)

9. Freitas, A.A.: Comprehensible classification models: a position paper. ACM SIGKDD Explor. Newsl. 15(1), 1–10 (2014)

10. Fu, Z., et al.: Hoops: human-in-the-loop graph reasoning for conversational recommendation. In: Proceedings of the 44th International ACM SIGIR Conference on Research and Development in Information Retrieval, pp. 2415–2421 (2021)

11. Gama, J., Žliobaitė, I., Bifet, A., Pechenizkiy, M., Bouchachia, A.: A survey on concept drift adaptation. ACM Comput. Surv. (CSUR) **46**(4), 1–37 (2014)

12. Heaven, D.: Why deep-learning AIS are so easy to fool (2019)

13. Khatami, A., Nazari, A., Beheshti, A., Nguyen, T.T., Nahavandi, S., Zieba, J.: Convolutional neural network for medical image classification using wavelet features. In: 2020 International Joint Conference on Neural Networks, IJCNN 2020, Glasgow, United Kingdom, July 19–24, 2020, pp. 1–8. IEEE (2020)

14. Lage, I., Ross, A.S., Kim, B., Gershman, S.J., Doshi-Velez, F.: Human-in-the-loop interpretability prior. Adv. Neural Inf. Process. Syst. **31**, 1–10 (2018)

15. Lavrač, N.: Selected techniques for data mining in medicine. Artif. Intell. Med. **16**(1), 3–23 (1999)

16. Li, J., Miller, A.H., Chopra, S., Ranzato, M., Weston, J.: Dialogue learning with human-in-the-loop. arXiv preprint arXiv:1611.09823 (2016)

17. Maadi, M., Akbarzadeh Khorshidi, H., Aickelin, U.: A review on human-AI interaction in machine learning and insights for medical applications. Int. J. Environ. Res. Public Health **18**(4), 2121 (2021)

18. Rezvani, N., Beheshti, A.: Attention-based context boosted cyberbullying detection in social media. J. Data Intell. **2**(4), 418–433 (2021)

19. Ribeiro, M.T., Singh, S., Guestrin, C.: "Why should i trust you?" Explaining the predictions of any classifier. In: Proceedings of the 22nd ACM SIGKDD International Conference on Knowledge Discovery and Data Mining, pp. 1135–1144 (2016)

20. Rudin, C.: Stop explaining black box machine learning models for high stakes decisions and use interpretable models instead. Nat. Mach. Intell. **1**(5), 206–215 (2019)

21. Ustun, B., Rudin, C.: Optimized risk scores. In: Proceedings of the 23rd ACM SIGKDD International Conference on Knowledge Discovery and Data Mining, pp. 1125–1134 (2017)

22. Wang, P., et al.: Human-in-the-loop design with machine learning. In: Proceedings of the Design Society: International Conference on Engineering Design, vol. 1, pp. 2577–2586. Cambridge University Press (2019)

23. Zhao, W., Rao, Y., Wang, Z., Lu, J., Zhou, J.: Towards interpretable deep metric learning with structural matching. In: Proceedings of the IEEE/CVF International Conference on Computer Vision, pp. 9887–9896 (2021)

A Recommender System and Risk Mitigation Strategy for Supply Chain Management Using the Counterfactual Explanation Algorithm

Amir Hossein Ordibazar[1]([✉]), Omar Hussain[1], and Morteza Saberi[2]

[1] School of Business, UNSW Canberra, Canberra, ACT, Australia
a.ordibazar@adfa.edu.au
[2] School of Computer Science, University of Technology Sydney,
Sydney, NSW, Australia

Abstract. Supply chain management (SCM) and its disruptions and risks have been the focus of many researchers in recent times. It is important how to identify these disruptions and risks to avoid them, therefore many risk mitigation strategies have been developed. Artificial intelligence (AI) is a powerful tool to identify and predict the occurrence of risks, it is also important that the solutions to avoiding risks must be explainable for risk managers. Recently, making transparent and explainable AI models has been the focus of a large number of research studies and many post-hoc algorithms such as counter-factual explanation (CE) algorithms have been developed. In this paper, first we propose an optimization problem to design a transportation schedule for the supply chain network (SCN), then to increase the resiliency and transparency of the designed schedule, the CE model is integrated into the model as a set of constraints. To design the CE, a logistic regression model is developed. The CE helps to plan the transportation schedule to avoid any transportation delay risk. The integrated CE and SCM model is used as a recommender system for risk managers to mitigate risk to the system. Finally, to validate the recommender system, a real case study is analyzed and the solutions of the model with and without the CE are compared and it is shown that the CE-added constraints increase the resiliency of the system significantly while the increase in financial cost is less than 1%. Therefore, the model is validated for use for different risks and disruptions.

Keywords: Supply chain management · Counterfactual explanation · Resiliency · Recommender system

1 Introduction

Artificial intelligence (AI) and machine learning (ML) have achieved many successful results in different applications such as financial, accounting, health care, military, photo OCR, autonomous driving etc. There are many powerful ML

© Springer Nature Switzerland AG 2022
H. Hacid et al. (Eds.): ICSOC 2021 Workshops, LNCS 13236, pp. 103–116, 2022.
https://doi.org/10.1007/978-3-031-14135-5_8

models in which real-world problems may be considered and the outputs of the model are reliable.

One of the recent challenges in AI is the need to increase the transparency and in-terpretability of ML models. Since the outputs of the ML models may be difficult to understand even for data scientists who have designed the model, therefore, it is necessary that model designers clarify the results of the models for stakeholders such as managers, legal organizations, end users etc. In this regard, many methods named post-hoc algorithms have been developed to explain the result of the models.

One of the most commonly used post-hoc methods is the counterfactual explana-tion (CE) method. A large volume of research has been conducted in this area [1–3]. In this method, we assume that the features of each record in a database is assumed to be a vector of amounts (e.g. given a n-dimensional vector) and the result of each record according to the context of the problem may contain differ-ent amounts (which will be the cell number n+1 in the vector). The CE methods will be used in situations where the algorithm predicts an undesirable result for a given input n-dimensional vector. In such a case, the CE model will try to find a perturbation vector to alter the n-dimensional input vector in such a way that the undesirable result of the given record changes to a desirable result.

The CE algorithm must try to find a perturbation vector where the ML model pre-dicts a desirable output but it should not be such a significant change to not be appli-cable. In addition, the altered vector should not be an outlier and the interdependen-cy of different features must be taken into account. This process is done by means of an optimization model which will be discussed more in Sect. 3.1.

One of the most interesting fields of study is supply chain management (SCM). It has many applications in different areas like healthcare, food supply, manufacturing etc. Many researchers are working to design applicable models to manage SCM and make the management of the system more practical.

There are different methods to deal with SCM applications, like optimization models, simulation and ML models. Mathematical modelling is widely used by re-searchers to find the optimal solution for the system as it has the ability to take different constraints into account simultaneously and find optimal solutions for differ-ent kinds of purposes, like financial, environmental and social objective functions.

One of the most important aspects of each supply chain network (SCN) is risk management. It is important to identify the disruptions and the risks and their impact on the network. In addition, the mitigation strategies to reduce their impact are important. Therefore, many different approaches like stochastic programming, robust programming and fuzzy methods have been developed to consider different risks like supply disruptions, demand fluctuations, transporta-tion delay etc. [4,5].

1.1   Problem Statement

In this research, we consider a two-echelon SCN. There are many aspects of SCN to manage and control. In a two-echelon SCN, there are two parts, namely suppliers and consumers. The suppliers provide products for consumers to satisfy their demand, but it is important that other requirements of the system are taken into account, for instance, the maximum capacity of suppliers and vehicles or delivery lead time etc.

One the most important sections of the SCM is managing disruptions and their risks related to the model. There are many different categorizations for SCM disruptions, for instance supply risks, demand risks, environmental risks, process and control risks [6]. The disruptions of any echelon may cause problems for other echelons, for instance transportation delay may cause problems for consumers.

In this research, the target is to propose a recommender system to help risk managers schedule a risk-averse SCN to prevent pre-identified risks. It is easier to avoid risks from occurring and their damage will be reduced and the system's resiliency will be increased.

1.2 Solution Overview

In this paper, we propose an optimization mathematical programming model to manage different aspects of the two-echelon SCN where different constraints such as demand satisfaction, capacity constraints etc. are considered. In addition, the financial costs are considered to be reduced as an objective function of the system. The optimization model id is proposed in Sect. 3.2.

To manage disruptions and risks to the SCN, we use a CE model embedded to the mathematical model which helps the optimization model to plan a transportation schedule in such a way to avoid any future possible risks occurring beforehand. The risk considered in this paper is transportation delay which may cause many problems in the system.

First, a developed ML model predicts the probability of delay in transportation, then the optimization model for CE is developed by considering the ML model. Lastly, the CE optimization model is integrated into the SCM optimization model. In fact, the CE and ML models are used to plan the transportation schedule of the SCN. It predicts the probability of delay in the transportation and tries to avoid it. It is a recommender system that helps risk managers deal with risk before they occur.

1.3 Research Contributions

This research contributes to the research on risk management in SCNs. Although many methods have been developed to reduce risks to the system, most are probabilistic. In our proposed method, the risk is taken into account by the ML model, which predicts the delay before it occurs, based on the analyzed database.

Another significance of this research is a contribution to CE methods to be used as a recommender system to help in planning to avoid risk. This is done by integrating CE into the optimization models of SCM. In addition, by integrating these two optimization models, the context of the network is used to help CE to find solutions in which the interdisciplinary features are taken into account and no outlier solutions will be proposed. As previously mentioned, it is important to find valid solutions in CE. It is difficult to find a solution in which all the relations among the different features are considered, therefore we have the advantage of the SCM optimization model to help the CE model to produce valid solutions. In addition, after planning the transportation schedule, it is easier for risk managers to be aware of the reasons for the decisions since it is based on a ML model and the risk manager can analyze the model.

2   Related Work

As previously mentioned, SCM is one of the most studied problems in the research. Different aspects of the system have been considered. Many research studies have employed probabilistic approaches to design robust SCM systems [7]. In addition, ML has many applications in different aspects of the industry and a large body of research has been conducted in this regard. Recently, many researchers have been keen to harness the advantageous of ML methods in SCM and predict different parameters. It may be useful to use ML to predict disruptions to reduce risks to the system [8]. Therefore, it seems reasonable to utilize the advantages of ML to predicting disruptions and risks and use them to plan the SCN.

The optimization of SCM has some deficiencies, such as expensive computational complexities [9]. But many methods, such meta-heuristic algorithms, have been developed to deal with this problem [10]. In this paper, to minimise cost, a small-size problem was analyzed.

As previously mentioned, CE methods are widely used by researchers to increase the explainability of ML models. These methods provide a solution to counter ML predictions. For instance, the researchers in [3], tried to find solutions to change the ML model's predictions on which individuals are likely to pay their home equity line of credit loan. This would help institutions accept potential customers instead of rejecting them because of an ML prediction. Other researchers used CE in health care and credit prediction cases [1,11]. There are many applications where CE can be helpful, therefore we decided to use it in SCM.

In addition, as previously discussed, one of the challenges in CE methods is outlier error and the interdependency of the model, where CE proposes a perturbation vector which gives the desirable output but it may be unrealistic or not applicable. In addition, the interdependency of parameters is an important factor that should be considered. These two factors usually are considered by calculating the distance of the CE perturbed vector and other records in the database (DB). Many modern approaches have been used to calculate outlier risk and consider interdisciplinary parameters [3,12]. For example, in [3], the researchers designed Mahalanobis's distance and a local outlier factor to analyze outlier risk and feature correlation. These methods are based on the database and not the context of the problem.

In this research, we use the SCM optimization model to manage the SCN and to increase the resiliency of the system. Instead of using common probabilistic approaches in optimization models, we use ML and CE, and a recommender system to mitigate system risk which is one of the important contributions of this paper. In addition, we use CE in SCM to increase the explainability of decisions made by the system for the risk manager. CE is used to predict the delay disruption in transportation and take it into account while the optimization model solves the problem. It is added to the model as a constraint. The nature of CE, which is an optimization problem, helps to integrate it into our SCM model and use these new integrated model capabilities to increase the quality of both the SCM model and CE solutions.

In each iteration, the CE constraints force the model to propose schedules in which no delay will happen. This is a contribution to the SCM model. Furthermore, the contribution of our recommender system to CE is to consider the interdependency of the parameters and to prevent outlier solutions. In our proposed method, there is no need to investigate the DB to reduce the probability of outliers and non-correlated outputs by designing distance functions and outlier metrics. The integrated model gives 100% applicable and correlated output as long as the factors are considered in the mathematical model. For instance, if a supplier provides some specific products, the CE doesn't propose an output in which that supplier provides non-assigned products, since its constraint is considered in the optimization model.

In the remainder of this paper, the problem statement is explained in detail in Sect. 3. The numerical results are detailed in Sect. 4 and the conclusion and suggestion for future studies is given in Sect. 5.

3   Problem Statement

This section explains the proposed integrated counterfactual explanation (CE) and mathematical modelling to consider risk management of a two-echelon SCN. The considered SCN has two echelons, namely suppliers and consumers, where the consumers may be the final clients or warehouses for the final products or they may be factory plants for semi-final products or raw materials. As previously mentioned, one of the most common disruptions in such systems is transportation delays, which may cause problems in the production plan resulting in the need for back orders and possibly lost sales.

In this paper, we review a two-echelon SCN with several suppliers and consumers, and we try to mitigate the risk of delay disruptions in receiving products from suppliers. To do this, first the ML model is trained to predict the delay occurrence based on the features of a transportation (e.g. material shipped, customer, supplier, vehicle type, transportation distance and planned lead time). Then, in the case of any undesirable output, the features of a trip with delay will be changed by the CE algorithm to make it a non-delayed trip. This will reduce the risk of delayed transportation. Finally, as a contribution of this paper, we integrate it into the mathematical modelling of the considered SCN to mitigate the transportation delay risk of the whole network for the future schedule. As the CE is used to plan a SCN, the greatest advantage of the integration of CE and the optimization model is the mitigation of delay risk.

3.1   Counterfactual Explanation

As previously discussed, the CE algorithms are used after the ML algorithms have predicted some targets. For instance, in this paper, we want to predict the probability of transportation delay for a particular trip. Several features will be considered and different algorithms may be developed. In the case of undesirable output, the CE algorithm will be designed. The algorithm consists of an optimization problem in which the main constraint is finding a set of actions that leads to a desirable output of the ML algorithm. The objective function is the effort required for the actions. The CE tries to reduce the effort needed to change the input.

Assume that $H(x) = y$ is the classifier of the input vector x which consists of n features and y is the output of the classifier. In our paper, the classifier tries to predict the delay occurrence using Eq. 1.

$$y = \begin{cases} 1 & \text{if the transportation has delay} \\ 0 & \text{if the transportation is on-time} \end{cases} \tag{1}$$

The optimization model of the counterfactual explanation model is given in Eq. 2–4:

$$minimize \quad Cost(p) \tag{2}$$
$$s.t.$$

$$x\prime = x + p \tag{3}$$

$$H(x\prime) = 0 \tag{4}$$

As previously mentioned, in the case of a delay in transportation, we use a CE algorithm to change the input vector x to $x\prime$, the features to be changed and the degree of change are calculated as vector p, which is a n-dimensional vector with zero and non-zero amounts. Those features which are not changed have zero amounts in vector p, and the features which have been changed by the algorithm have non-zero amounts (e.g. positive and negative amounts) which are added to the original feature amount. The calculation of vector p is shown in Eq. 3.

The constraint in Eq. 4 is used to make sure that the perturbed vector $x\prime$ has the desirable output, that being non-delayed transportation, which is the main goal of the CE algorithm. The output based on our ML model must be desirable. The objective function $Cost(p)$ in Eq. 2 is the effort required for changing the features of the input. The cost of perturbation vector p is calculated based on the nature of the considered system.

For instance, a trip may be predicted to have delay. In this case, the CE algorithm tries to find vector p, and we assume that the CE finds that the trip will be on time if vehicle type 1 is changed to vehicle type 2 or supplier A must is replaced by supplier B. Therefore, the CE suggests that the trip can be continued with vehicle type 2 or supplier B. In this case, the difference in the transportation cost of changing vehicle type will be calculated as Cost(p) and the difference of ordering from supplier B will be calculated as Cost (p/). The question to be answered is which change should be accepted. The CE mathematical model decides to choose the perturbation with the lower cost function.

In our paper, we use logistic regression to predict transportation delay. Although there are many advanced ML models, logistic regression is appropriate for our purpose because it is easier to integrate into an optimization model. The parameters of the logistic regression classifier are coefficients of the features and a constant intercept as shown in Eq. 5.

$$H(x) = sigmoid(Co_i * x_i + IC) \tag{5}$$

where the sigmoid function is defined as Eq. 6:

$$\left(\begin{matrix} sigmoid(z) \\ = \end{matrix} \right) \frac{1}{1 + e^{-z}} \tag{6}$$

where Co_i is the coefficient for x_i which is feature i and IC is the constant intercept of the model. In this paper, the delay probability of the input record is calculated, and there is a threshold amount as ts that helps us to decide the output of the ML predictor as shown in Eq. 7.

$$y = \begin{cases} 1 & H(x) > ts \\ 0 & H(x) \le ts \end{cases} \tag{7}$$

It is noteworthy that in the sigmoid function, Eq. 8 is true and will be used in our integrated optimization model to linearize the model.

$$\begin{cases} z > 0 \Rightarrow & 0.5 < sigmoid(z) < 1 \\ z = 0 \Rightarrow & sigmoid(z) = 0.5 \\ z < 0 \Rightarrow & 0 < sigmoid(z) < 0.5 \end{cases} \tag{8}$$

3.2   Integrated Supply Chain Network Optimization Model

This section describes the parameters, variables and mathematical model of the SCN optimization model. The two-echelon SCN model is proposed for multi-product distribution from various suppliers to different customers to plan the daily scheduling of the transportation in a specified time horizon.

Different aspects of a SCN will be considered and the main constraints are de-mand satisfaction, inventory, and transportation plan. This section details the sets, parameters and variables of the model and then the objective functions and con-straints of the integrated CE and SCN mixed integer linear programming (MILP) are discussed. Table 1 defines the sets of the integrated model.

Table 1. Set definition of the integrated model tables.

Index set	Description
I	Set of suppliers i $\in \{1, 2, ..., |I|\}$
J	Set of consumers j $\in \{1, 2, ..., |J|\}$
T	Set of time periods t $\in \{1, 2, ..., |T|\}$
P	Set of products p $\in \{1, 2, ..., |P|\}$
V	Set of vehicles v $\in \{1, 2, ..., |V|\}$

The parameters of the model are described in Table 2.

Table 2. The parameters of the integrated model tables.

Parameters	Description
$D_{j,p,t}$	Demand of consumer j for product p in period t
$SS_{j,p}$	Safety stock of consumer j for product p
$IS_{j,p}$	Initial stock of consumer j for product p
$CP_{i,p}$	Capacity of supplier i for providing product p
$SC_{i,p}$	If supplier i can provide product p 1, otherwise 0
$LT_{i,j}$	Planned lead time for transportation from supplier i to consumer j
$Dis_{i,j}$	Distance between supplier i to consumer j
$CInv_{j,p}$	The inventory cost of consumer j for each unit of product p
$CT_{i,j,v}$	The transportation cost of trip from supplier i to consumer j for each unit of vehicle v
IC	The intercept of ML model
ts	Threshold of regression model
$CO_{i,j,p,v}$	The combined coefficient of ML model for any transporting trip
M	A large number

The variables of the mathematical model are shown in Table 3.

Table 3. The variables of the SCN model tables.

Variables	Description
$Inv_{j,p,t}$	Integer variable for inventory of consumer j for product p in period t
$X_{i,j,p,v,t}$	Integer variable for the amount of transportation from supplier i to consumer j for product p in time period t by vehicle v
$Z_{i,j,p,v,t}$	Binary variable if supplier i transports product p to consumer j in time period t by vehicle v 1, otherwise 0

The objective function is shown in Eq. 9.

$$\text{minimize Obj} = \sum_{j,p,t} CInv_{j,p} * Inv_{j,p,t} + \sum_{i,j,p,v,t} CT_{i,j,v} * Z_{i,j,p,v,t} \quad (9)$$

The constraints of the model are given in Eq. (10–17).

$$Inv_{j,p,t} = Inv_{j,p,t-1} + \sum_j X_{i,j,p,v,t-LT_{i,j}} - D_{j,p,t} \forall j, p, v, t \in \{2, ..., |T|\} \quad (10)$$

$$Inv_{j,p,1} = IS_{j,p} - D_{j,p,1} \qquad \forall j, p \quad (11)$$

$$Inv_{j,p,t} \geq SS_{j,p} \qquad \forall j, p, t \quad (12)$$

$$\sum_{j,v} X_{i,j,p,v,t} \leq CP_{i,p} \qquad \forall i, p, t \quad (13)$$

$$\sum_{j,v,t} X_{i,j,p,v,t} \leq M * SC_{i,p} \qquad \forall i, p \quad (14)$$

$$X_{i,j,p,v,t} \leq M * Z_{i,j,p,v,t} \qquad \forall i, j, p, v, t \quad (15)$$

$$Z_{i,j,p,v,t} \leq X_{i,j,p,v,t} \qquad \forall i, j, p, v, t \quad (16)$$

$$(Co_{i,j,p,v} + IC) * Z_{i,j,p,v,t} \leq ts \qquad \forall i, j, p, v, t \quad (17)$$

In Eq. 9, the objective function for the financial costs of transportation and the inventory are considered. Equation 10 gives the inventory constraint, where the inventory in the current period is calculated by the inventory of the previous period and the transportation which will arrive in the current period and the demand must be satisfied. Equation 11 calculates the inventory in first period (there are no transportation deliveries in this period). Equation 12 calculates the safety stock constraint. Equation 13 calculates the supply capacity. Equation 14 calculates the products which the supplier can provide. The logical constraints for connecting the variables are shown in Eqs. 15 and 16. The CE constraint is calculated in Eq. 17. For this constraint, in the case of a trip be-tween two nodes, the delay probability is calculated, then the model tries to find a solution where no delay occurs. To avoid making the model nonlinear, the sigmoid function is not used and the prediction function must be lower than zero.

Based on the literature review, several papers on the applications of CE relate to loans and financial matters or health care issues, and they aim to find a perturbation vector to change the applicant's features in which the ML predicts the desirable output. In this paper, CE has been used as a recommender system to avoid delay risk in SCM before it occurs. It is noteworthy that in our paper, CE can be used to plan a transportation resilient schedule. On the other hand, using CE in an optimization SCM model may help to reach better outputs in CE, since interdisciplinary features are considered and outlier output risks are eliminated. Since the CE model is an optimization mathematical model, it is possible to integrate it into the optimization SCM model. In summary, the following summarize the contributions of the integration of CE and the SCM model:

- No delay occurs in the network and all trips have on-time delivery.
- It considers the interdisciplinary parameters in the mathematical model.
- No outlier solution is proposed because the context of the network is considered.

4   Experiments and Results

This section describes the numerical experiments and results. First, in Sect. 4.1, the DB is introduced and the experiments are described. Then, in Sect. 4.2. The numerical outputs are given.

4.1   The Database and ML Development

The DB which is used for the numerical experiments is delivery truck trip data, which is an open source DB on the Internet [13] and contains information on the transportation of various products from suppliers to customers. The target value shows whether a trip is delayed or on time. It is important that we can plan a transportation schedule in which all trips are predicted to be on time.

In this DB, there are 32 features, some of which are irrelevant and do not contribute to predict delay, hence they are omitted. The first group comprises duplicate features, for instance customer ID and customer name code. The second group comprises features which make no contribution to the ML model such as driver details. Two features, namely on time and delay, since both of them are the same the on time feature was deleted and delay feature was set as target. In addition, several features like planned lead time are created by subtraction the booking date from the planned received date and the date data is transformed into integer values, which are more usable in prediction and optimization models. Lastly, the features listed in Table 4 are the features of the system and delay remains as the target for the logistic regression model.

In the real database, about 500 records have missed values so these were deleted first. Then, the DB was analyzed. There are 268 suppliers, 29 customers, 1244 products and 44 vehicles. The original DB does not consider specific suppliers and consumers for limited products. Of the 1244 products, 7 comprise more than 60% of the DB. Therefore, these 7 products were chosen. Using this method 6 of the most frequent customers were chosen, and 16 and 11 of the most frequent suppliers and vehicles were chosen, respectively. Lastly, about 1500 out of 6500 records were chosen. There are two important reasons for this sampling: 1) a large number of suppliers, customers and products may increase the computational complexity of the optimization model. For instance, for the 10 time periods using the original number of records in the DB, Eq. 15 includes $4.2 * 10^9$ constraints. In addition, there is the same number of variables $X_{i,j,p,v,t}$, and less than 1% of them are non-zero, although it is known that most of them are zero but it is not clear which ones, therefore a lot of time is needed to solve the model. 2) It is not efficient for the purpose of this paper to run prediction models for the DB, which has many records related to products, suppliers, consumers, vehicles that are not frequent, and the output of the model is not as useable for integration into the SCM model. In this paper, we consider a limited number of suppliers, customers, products and vehicles and design a schedule for them, as it is more realistic and more applicable for a company which has a couple of factory plants to receive some limited raw materials and semi-final products from some suppliers. Even for large-scale cases, this method can be used but maybe some meta-heuristic algorithms must be developed to solve the optimization problem in a reasonable time. In this paper, deterministic algorithms are used, therefore the small size of the problem is assumed to validate the proposed recommender system.

The DB was selected to predict the transportation delay using logistic regression. The average F1-score is 0.91 and the average recall and precision scores are both 0.91 in the test set, which consists of 30% of the DB records. The test set is chosen randomly to increase the quality of the prediction. Therefore, by considering the scores of the prediction, it can be seen that the prediction model is able to predict the delay probability properly. As discussed in Sect. 3.1, there is one coefficient for each feature, moreover there is an intercept which is constant. These coefficients and the intercept are used in Eq. 5 and 17. The variables' names and their related coefficients and the intercept of the model are shown in Table 4.

Table 4. The calculated coefficients and intercept of the DB

Intercept	Material shipped	Customer ID	Supplier ID	Vehicle type	Distance (km)	Planned LT
-0.97	$-1.99 * 10^{-3}$	$3.31 * 10^{-1}$	$1.95 * 10^{-2}$	$8.86 * 10^{-2}$	$2.24 * 10^{-3}$	-2.08

By considering these parameters of the logistic regression model, and considering 7 products, 6 customers, 16 suppliers, 11 vehicle types and a transportation distance in a range between 9 to 2681 km and lead time in a range between 0 and 21 days, other parameters such as transportation cost, demand etc. are generated. Lastly, the integrated optimization model is solved for 10 time periods and the results are dis-cussed in Sect. 3.2.

4.2 Using the Integrated Optimization Model for the Database

This section presents the numerical experiments of the integrated model, which were described in Sect. 2. The model is solved by GAMS 24.1.2 and the computations are done by CPLEX Solver, and ML model is designed in Python 3.6.3. The model is solved on a laptop with the following configurations: Intel®Core™i7-4510U CPU @ 2.00 GHz. The processor speed is 2594 MHz and there is 8 GB of RAM. The optimization model for this problem size is solved in a reasonable time, less than 5 min.

The problem was solved by considering two different scenarios. In the first one, the scheduling of SCN was planned without considering the CE constraints, which means that there was no control on delay. Lastly, the optimal SCN schedule was planned, taking into consideration the inventory constraints, demand satisfaction constraints and other related requirements. The output schedule was 3179 transportation trips between suppliers and customers, as shown in Fig. 1. The more deeply colored cells show that transportation is more frequent between those nodes. No constraints were considered to control delay. After calculating the delay probability, according to the ML prediction, it is found that 635 transportation trips may cause delay, which is about 20% of all trips. Figure 2 shows the percentage of on-time transportation between each supplier and consumer. The green cells are those pairs of suppliers-consumers whose trips are on time. The yellow cells and red cells are those pairs of suppliers-consumers who experience delay. As shown in Fig. 2, no transportation occurred between several pairs of suppliers-consumers, therefore they are not applicable (N.A.).

Supplier No.

Customer No.	1	2	3	4	5	6	7	8	9	10	11	12	13	14	15	16
1	0	18	45	2	77	0	26	29	38	90	21	0	45	0	59	45
2	0	0	23	52	52	0	0	81	0	9	156	0	0	138	0	61
3	22	111	8	0	0	48	96	48	0	48	0	48	51	0	48	0
4	60	0	96	48	0	0	36	0	0	0	48	48	0	48	96	48
5	6	0	8	0	79	77	48	48	9	28	96	0	48	0	81	0
6	11	0	48	48	83	0	96	0	0	80	0	13	96	0	42	11

Fig. 1. Number of transportation trips between suppliers and consumers without considering CE

Supplier No.

Customer No.	1	2	3	4	5	6	7	8	9	10	11	12	13	14	15	16
1	N.A.	100%	100%	0%	58%	N.A.	0%	100%	100%	100%	100%	N.A.	100%	N.A.	100%	100%
2	N.A.	N.A.	0%	100%	100%	N.A.	N.A.	100%	N.A.	100%	N.A.	100%	100%	N.A.	N.A.	100%
3	100%	100%	100%	N.A.	N.A.	0%	0%	100%	N.A.	100%	N.A.	100%	100%	N.A.	0%	N.A.
4	0%	N.A.	0%	0%	N.A.	N.A.	100%	N.A.	N.A.	N.A.	100%	100%	N.A.	100%	100%	100%
5	100%	N.A.	100%	N.A.	100%	100%	0%	100%	100%	0%	100%	N.A.	100%	N.A.	100%	N.A.
6	100%	N.A.	100%	100%	100%	N.A.	100%	N.A.	N.A.	0%	N.A.	100%	100%	N.A.	100%	100%

Fig. 2. Percentage of on-time transportation trips between suppliers and consumers without considering CE (Color figure online)

In the second scenario, the CE constraints are included in the model. These constraints ensure that none of the trips in the planned transportation schedule will experience delay. Therefore, of the 3245 planned transportation trips which is the out-put of the model, none had any probability of delay according to the ML model. Figure 3 shows the number of trips in the second scenario. All are predicted to be on-time delivery by the ML model. This helps to prevent delay disruption and its related risks before they occur. Therefore, the risks will be reduced. The financial objective function of the optimization model increases about 0.13% in the second scenario, which is acceptable.

Supplier No.

	1	2	3	4	5	6	7	8	9	10	11	12	13	14	15	16
1	0	97	44	0	0	0	10	0	0	79	0	135	46	39	37	19
2	0	7	159	0	0	145	0	60	0	30	53	53	0	0	76	0
3	19	50	0	50	0	50	50	0	118	13	0	0	0	100	50	50
4	61	0	70	48	0	48	0	48	28	0	106	26	0	45	0	48
5	10	22	98	0	49	0	88	0	0	29	0	69	76	49	49	0
6	0	98	0	11	98	0	49	0	0	49	49	87	0	98	0	0

(Customer No. labels the rows 1–6)

Fig. 3. Number of transportation trips between suppliers and consumers considering CE

5 Conclusion and Future Contributions

In this paper, the transportation scheduling of the SCN was considered. The risks caused by delay disruptions were taken into account using a new method. In this method, the prediction model is integrated into the optimization SCN model and ensures that all trips are planned in such a way as to reduce the probability of delay in receiving products by customer. Delay is one of the most probable disruptions which is a risk in SCN. Therefore, by avoiding delay before it occurs, the SCN is more resilient.

Another advantage of the proposed model is that the CE function, which is the ML function constraint, changes the number of trip features if it causes delay simultaneously using the SCN optimization model. Therefore, the output schedule won't experience any delay. The most important contribution to the research on CE is that the interdependency of ML features are considered thoroughly in the integrated model. In addition, no outlier output is proposed by the CE because the optimization model takes the context of the real system into account. For example, a given supplier may only provide certain specific products, it is considered in the model and the CE won't propose a solution in which suppliers are assigned to provide products that they are not capable of providing, or the transportation will be postponed to another time period and it is controlled by inventory constraints.

For future research, we will investigate the use of other ML models and other disruptions and risk. In addition, three-echelon or four-echelon SCNs may be considered. The ripple effect of disruption in the first echelons on the last echelon may be effectively reduced using this method, and the ripple effect may be considered in more detail in further research.

References

1. Cheng, F., Ming, Y., Qu, H.: DECE: decision explorer with counterfactual explanations for machine learning models. IEEE Trans. Visual Comput. Graphics **27**(2), 1438–1447 (2021)
2. Stepin, I., et al.: A survey of contrastive and counterfactual explanation generation methods for explainable artificial intelligence. IEEE Access **9**, 11974–12001 (2021)

3. Kanamori, K., et al.: DACE: distribution-aware counterfactual explanation by mixed-integer linear optimization. In: IJCAI (2020)
4. Rabbani, M., et al.: A hybrid robust possibilistic approach for a sustainable supply chain location-allocation network design. Int. J. Syst. Sci. Oper. Logist. 7(1), 60–75 (2020)
5. Budiman, S.D., Rau, H.: A stochastic model for developing speculation-postponement strategies and modularization concepts in the global supply chain with demand uncertainty. Comput. Ind. Eng. 158, 107392 (2021)
6. Christopher, M., Peck, H.: Building the resilient supply chain (2004)
7. Manupati, V.K., Akash, S., Illaiah, K., Suresh Babu, E., Varela, M.L.R.: Robust supply chain network design under facility disruption by consideration of risk propagation. In: Machado, J., Soares, F., Trojanowska, J., Ivanov, V. (eds.) icieng 2021. LNME, pp. 97–107. Springer, Cham (2022). https://doi.org/10.1007/978-3-030-78170-5_10
8. Lorenc, A., Kuźnar, M.: The most common type of disruption in the supply chain - evaluation based on the method using artificial neural networks. Int. J. Shipp. Transp. Logist. 13(1–2), 1–24 (2021)
9. Fattahi, M., Govindan, K., Maihami, R.: Stochastic optimization of disruption-driven supply chain network design with a new resilience metric. Int. J. Prod. Econ. 230, 107755 (2020)
10. Nikolopoulos, K., et al.: Forecasting and planning during a pandemic: COVID-19 growth rates, supply chain disruptions, and governmental decisions. Eur. J. Oper. Res. 290(1), 99–115 (2021)
11. Holzinger, A.: Explainable AI and multi-modal causability in medicine. i-com 19(3), 171–179 (2021)
12. Mothilal, R.K., Sharma, A., Tan, C.: Explaining machine learning classifiers through diverse counterfactual explanations. In: 3rd ACM Conference on Fairness, Accountability, and Transparency, FAT* 2020. Association for Computing Machinery Inc. (2020)
13. Thiagu, R.: Delivery truck trip data. http://www.kaggle.com/ramakrishnanthiyagu/delivery-truck-trips-data

Improving Multi-label Text Classification Models with Knowledge Graphs

Divya Prabhu$^{(\boxtimes)}$ [ID], Enayat Rajabi [ID], Mohan Kumar Ganta [ID], and Tressy Thomas [ID]

Cape Breton University, Sydney, NS, Canada
{cbu19jxd,enayat_rajabi,cbu19fxf,cbu17pmz}@cbu.ca

Abstract. Multi-label Text Classification (MLTC) is a variant of classification problem where multiple labels are assigned to each instance. Most existing MLTC methods ignore the relationship between the target labels. Since the hierarchical relationship for addressing these problems is significant, a semantic network approach with the help of knowledge graphs can be used. This paper proposes a knowledge graph-based approach together with GRU (Gated Recurrent Unit) neural network model to solve an MLTC problem on a research text dataset. In particular, we leverage the Tax2Vec approach to extract hypernyms from the Word-Net knowledge graph and enrich the dataset. The enrichment results in following a tree-like structure to identify the relationship between the semantic concepts. The result shows that the enriched dataset outperforms the traditional GRU neural network-based model based on different evaluation metrics.

Keywords: Multi-label Text Classification · Knowledge graph · Neural network model · Tax2Vec

1 Introduction

Classification problem in Machine Learning is a technique that categorizes a dataset into classes. Multi-Label Text Classification (MLTC) is a variant of classification problem that multiple labels are assigned to each instance in a dataset. The problem of MLTC is challenging but fundamental to addressing several real-world problems ranging across different fields such as recommendation systems and sentiment analysis and various domains like healthcare. Identification of diseases based on the onset of symptoms in the patients is one of the MLTC applications, as symptoms of multiple diseases might be present in the patients simultaneously. In E-Commerce, the classification system of Amazon, for example, comprises a large hierarchy used for the organization of products. Several techniques are used for solving an MLTC problem through Flat-Based and Hierarchical methods. The Flat-Based methods are commonly used by Naive Bayes classifiers and do not consider the hierarchy of relationship between the labels.

H. Hacid et al. (Eds.): ICSOC 2021 Workshops, LNCS 13236, pp. 117–124, 2022.
https://doi.org/10.1007/978-3-031-14135-5_9

In contrast, a single instance in hierarchical MLTC includes multiple labels simultaneously, wherein these labels are stored in a hierarchical manner [1]. Existing methods of MLTC usually ignore the relationship between the labels [8]. As the hierarchical relationship for addressing these problems is essential, a semantic network approach with the help of Knowledge Graphs could prove useful [2,3,13]. A knowledge graph illustrates the relationship between entities in the form of a triple comprising of Subject, Predicate, and Object. They add meaning or domain knowledge to a dataset. There are multi-fold applications of the knowledge graph. For example, knowledge graphs can be used for text analysis to extract the semantic relationship between entities in a sentence or paragraph. Knowledge graphs as graphs have been proved to be more effective for label structure modeling, ontological knowledge and machine learning [11]. In this study, we are seeking to answer the following research question: "How can knowledge graphs be used in an MLTC problem to enrich a dataset and improve the existing emerging machine learning models?". To answer this question, we developed the neural network-based models to identify the best-performing model for this study and followed a knowledge graph-based approach to enrich a dataset and improve the selected model's performance.

The paper has been structured as follows: Sect. 2 outlines the emerging approaches to the MLTC problem. Section 3 describes the proposed approach. Section 4 explains the experimentation and results. The paper is concluded in Sect. 5.

2   Related Works

Though works have been carried out for the MLTC problems, there still lies many complexities for effectively and efficiently leveraging dependencies amongst the labels for improving classification. Authors in [11] proposed a novel hierarchical taxonomy aware framework for a large-scale MLTC problem. This study proposed an approach to identify the hierarchical relationships among labels in documents for classification purposes. They used a Recurrent attention-based Convolutional Neural Network approach to model the documents and encode each document as a 3-D feature map. The authors used a word2vec model to train 50-dimensional word embedding over the 100 billion words from the Wikipedia corpus. An attention-based graph neural network model was proposed by [10] to capture the attentive dependency structure among the labels of documents. The authors used a feature matrix and a correlation matrix to capture the dependencies and implemented Bidirectional Encoder Representations from Transformers (BERT) embedding to encode the sentences. The model employs Graph Attention Network (GAT) to find the correlation between labels, bidirectional LSTM to obtain the feature vectors, BERT for embedding the words, and then feed it to BiLSTM. They showed that the combination of GAT (Graph Attention Network) with bi-directional LSTM network achieved consistently higher accuracy than those obtained by conventional approaches.

Another attention-based Recurrent Neural Network approach was presented by [5] where the authors proposed a novel framework called Hierarchical

Attention-based Recurrent Neural Network (HARNN) for classifying documents into the most relevant categories level by level via integrating texts and the hierarchical category structure. They modeled the dependencies among different levels by leveraging the hierarchical structure gradually in a top-down fashion. After comparing their model with several state-of-the-art models including Clus-HMC, and HMC-LMLP, they showed that HARNN is more capable for hierarchical multi-text classification tasks with the advantage of tackling hierarchical category structure effectively and accurately.

In terms of using semantic networks in text analytics, [6] proposed using a background knowledge graph-based method, called BaKGraSTeC, to utilize explicit external knowledge and their structure information in a knowledge graph for a short text classification problem. They used Tax2Vec [14] to tokenize each noun with hypernyms from a knowledge graph and extract corpus-relevant semantic information from a knowledge graph.

Our paper extends the [6] approach by extending the Tax2Vec approach to the MLTC problem. The Tax2Vec algorithm utilizes the WordNet hypernyms[1] and also has the capability to include semantic relationships from Microsoft Knowledge Graph. The approach tokenizes the title and abstract of a scholar dataset and enriches it by the hypernyms extracted from the WordNet knowledge graph and the semantic feature vectors from GloVe [12]. While other methods discussed above evaluate the hierarchical nature of relationships in MLTC problems, our objective is to understand to what extent a dataset enrichment with knowledge graphs can improve the performance of a neural network algorithm.

3 Approach

MLTC problems are complex in nature. Our objective is to use an approach that is comparatively simple in application across similar problems in the industry. In the proposed approach, we enhance a dataset with domain knowledge enriched by Tax2Vec and GloVe to improve the performance of a neural network model for the MLTC problem.

We use the Tax2Vec model for semantic feature construction and vectorization. The model has an in-built ability to include Microsoft Knowledge Graph in the form of "Is-a" relationships. Since this model had proven successful in text classification problems with single labels on SVM [7], our approach is to analyze its performance on an MLTC problem by utilizing some state-of-the-art Neural Network models to understand the effect.

The Tax2Vec model embeds the domain knowledge by the addition of Word-Net hypernyms by including a knowledge graph. The model helps create new semantic features that are taxonomy-based and can extract corpus-specific semantic keywords. This is in continuation with the studies that were done by the original authors, where they used Tax2Vec to analyze the performance of the linear models on single-label text classification tasks using linear classifiers.

---

[1] https://wordnet.princeton.edu/.

To develop an MLTC problem using Neural Network models, we use LSTM, RNN, and GRU to compare their results on the dataset and choose the best fit model. We shortlist a set of machine learning models based on relevancy to our MLTC problem and the literature review in similar MLTC problems. We first evaluate the performance of existing emerging Neural Network models such as Recurrent Neural Networks (RNN), Long Short-Term Memory (LSTM), and Gated Recurrent Units (GRU). The evaluation metrics we use are F1-score, Precision, and Recall. The metrics from these baseline data models are used to identify the better-performing model for the experimentation.

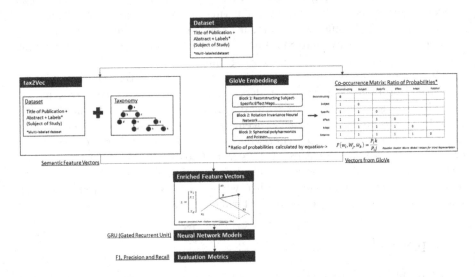

Fig. 1. Using Tax2Vec on a MLTC problem

3.1 Problem Transformation

In an MLTC problem, we have a set of document text (D) in the form of abstract and titles in the research text dataset. Each unique D has a set of labels $(y1, y2, y3.....yi)$. We enhance a dataset using Tax2Vec and GloVe methods. In the Tax2Vec approach, each word t in document D is counted in terms of the number of times the word or its hypernym appears. For each term t in document D,the frequency of term and hypernym (from WordNet taxonomy) in the specific document is counted for a specific selected normalization factor (K). Feature value is calculated, and features are selected by term counts by frequency of occurrence. Term counts are aggregated into n vectors, wherein n is the number of documents. A real-valued, sparse matrix-vector space, where columns represent the terms from the corpus of documents, a matrix of selected features is returned. We vectorize the data using the GloVe method. This method creates a co-occurrence matrix from word blocks, the ratio of probabilities later determines the vectors from GloVe from a 6 billion corpus of words [12]. The vectors from GloVe F are generated by co-occurrence matrix from D in the form of feature vectors. Vectors from Tax2Vec and GloVe form enriched feature vectors X. A neural network model like GRU can be utilized, wherein X is embedded, the neural network model operations learn about the embedding vector dependencies, a max-pooling operation reduces the sequence of feature vectors into a single feature vector. A connected layer map features to binary outputs. Sigmoid operation is later utilized for learning the binary cross-entropy loss between the target and the output labels.

4   Experimentation

We implemented RNN, LSTM, and GRU machine learning models on a research text dataset. Based on the performance of these models, we shortlisted GRU for further evaluation. The next step was to understand how a knowledge graph improves the performance of GRU. The code and the dataset are available for re-use on GitHub[2].

4.1   Dataset

The dataset, which was taken from Kaggle website[3], comprises text data including titles and abstracts of various research papers along with their labels. The research papers are classified into six categories or classes: Computer Science, Physics, Mathematics, Statistics, Quantitative Biology, and Quantitative Finance. There are 20,971 unique records, two input variables (abstracts and titles), and six target variables (categories) in the dataset. The target variables (classes) were encoded as binary values denoting the six categories, while the title and abstract are String.

4.2   Pre-processing and Parameters

As a first step, we cleaned the dataset by removing null values, punctuation, and special characters, and stop words with the help of corpus. As the last part of pre-processing, we stem the text in the dataset before creating tokens from it. We used a single input (combination of title and abstract) and six dense layers as output in the models. Eighty percent of data was used to train the model and ran the models with three different epochs. The activation layer was set to "sigmoid" from which we got the better results based on experimentation. A Neural Network for text classification needs to accept queries with variable lengths and predict their labels. In our experiment, words are transformed into word embeddings. After which, we applied Bidirectional layers and Max-pooling over time to extract fixed-length feature vectors and feed them into the output layer to predict the label for the query. We shortlisted binary cross-entropy as

---

[2] https://github.com/MohanKumarGanta/MLTC-with-KG.
[3] https://www.kaggle.com/shivanandmn/multilabel-classification-dataset.

a loss function, as it outperforms the pairwise ranking loss. We considered the "ReLU" activation function in the dense layers. We also used the optimizer as "Adam", maintained the default learning rate of 0.001, and used the node values as 64 and 128 for all the algorithms. The output of the final "ReLU" layer was used to get the result in binary format (predicting the label as 1 or 0) with the "Sigmoid" activation function. The next stage of our experimentation involved using Tax2Vec to augment the data before utilizing the model on the enhanced data set. In the case of Tax2Vec, we found that the "rarest_terms" heuristic performs better and max features were set to 2 to get the hypernyms from a subset of the Microsoft knowledge graph created with the Is-A relationship between the words (so-called the refined text file). The hypernyms from the Tax2Vec are then appended to the original word tokens and passed into the model.

4.3   Performance Metric

To evaluate the performance of the models, we used the metrics like F1-score, recall, and precision which according to the authors in the [4]. Since the dataset is multi-labeled and the records are classified into different categories, we have considered the micro average of the metrics mentioned above, as it is obtained by summing over all True Positives (TPs), True Negatives (TNs), False Positives (FPs) and False Negatives (FNs) for each class and then the average is taken using the formulas given below, as [9] mentioned it as the best way to assess the models.

$$Microaveraging\ Precision\ Prc^{micro}(D) = \frac{\sum_{c_i \epsilon C} TPs(c_i)}{\sum_{c_i \epsilon C} TPs(c_i) + FPs(c_i)}$$

$$Microaveraging\ Recall\ Rcl^{micro}(D) = \frac{\sum_{c_i \epsilon C} TPs(c_i)}{\sum_{c_i \epsilon C} TPs(c_i) + FNs(c_i)}$$

4.4   Results

We evaluated the performance of the Neural Network models based on the metrics discussed in the above section and have included 50% of data for validation. We experimented with varying levels of Epochs and compared the output for various models. Table 1 shows the results before implementing the Tax2Vec to the dataset for various machine learning algorithms:

Since the performance of GRU was best amongst the other models used, with a Precision of 80%, Recall of 72% and F1-score of 76%, we considered GRU for the next step of our experimentation. We enriched the input data by adding three new hypernyms extracted using Is-A relationship from the Tax2Vec model. After using the Tax2Vec approach along with the GRU model, we found that the enriched dataset by Tax2Vec led to an improvement of the Recall from 72%

Table 1. Performance metrics before applying knowledge graph

Accuracy for algorithms before KG embedding		
Algorithm	f1 score	Precision	Recall
RNN	74.00	75.00	74.00
LSTM	72.00	78.00	67.00
GRU	76.00	80.00	72.00

to 74% and F1-score from 76% to 77%. An increase in Recall means that the model was able to find relevant cases within the data set. However, we noticed that there was not any noticeable improvement in the Precision of the GRU model that was enhanced with the Tax2Vec embeddings from the knowledge graph. The readings in both without the embeddings from Tax2Vec and after the embeddings remained constant at 80%.

5 Conclusions and Future Works

In this paper, we applied a knowledge graph-based method to enrich an MLTC dataset and improve the performance of a neural network model to solve the problem. To this end, we showed that the GRU machine learning model generally performs better compared to LSTM and RNN algorithms on MLTC problems. Then, we improved the GRU model by leveraging Tax2Vec in terms of Recall and F1-score. Though the Recall improved, precision remained constant at 80% before and after the Tax2Vec implementation. Our next step entails an investigation on improving the precision score and applying the same approach to the other MLTC datasets. We will also test the approach on a variety of balanced and unbalanced datasets, given that the dataset under consideration for this study was not perfectly unbalanced. Additionally, there are more semantic relationships such as sameAs and sub-class-of that we plan to include in our future works.

Acknowledgement. The work presented in this paper was funded by Cape Breton University (RISE grant).

References

1. Aljedani, N., Alotaibi, R., Taileb, M.: HMATC: hierarchical multi-label Arabic text classification model using machine learning. Egypt. Inform. J. **22**, 225–237 (2020)
2. Beheshti, A., Benatallah, B., Sheng, Q.Z., Schiliro, F.: Intelligent knowledge lakes: the age of artificial intelligence and big data. In: U, L.H., Yang, J., Cai, Y., Karlapalem, K., Liu, A., Huang, X. (eds.) WISE 2020. CCIS, vol. 1155, pp. 24–34. Springer, Singapore (2020). https://doi.org/10.1007/978-981-15-3281-8_3
3. Beheshti, S.-M.-R., Benatallah, B., Motahari-Nezhad, H.R.: Scalable graph-based OLAP analytics over process execution data. Distrib. Parallel Databases **34**(3), 379–423 (2014). https://doi.org/10.1007/s10619-014-7171-9

4. Gargiulo, F., Silvestri, S., Ciampi, M.: Deep convolution neural network for extreme multi-label text classification. In: Proceedings of the 11th International Joint Conference on Biomedical Engineering Systems and Technologies (2018)
5. Huang, W., et al.: Hierarchical multi-label text classification: an attention-based recurrent network approach. In Proceedings of the 28th ACM International Conference on Information and Knowledge Management, pp. 1051–1060 (2019)
6. Jiang, X., Shen, Y., Wang, Y., Jin, X., Cheng, X.: Bakgrastec: a background knowledge graph based method for short text classification. In: 2020 IEEE International Conference on Knowledge Graph (ICKG) (2020)
7. Škrlj, B., Martinc, M., Kralj, J., Lavrač, N., Pollak, S.: Tax2vec: constructing interpretable features from taxonomies for short text classification. Comput. Speech Lang. **65**, 101104 (2021)
8. Pal, A., Selvakumar, M., Sankarasubbu, M.: Magnet: multi-label text classification using attention-based graph neural network. In: Proceedings of the 12th International Conference on Agents and Artificial Intelligence - vol. 2: ICAART, pp. 494–505. INSTICC, SciTePress (2020)
9. Pal, A., Selvakumar, M., Sankarasubbu, M.: Magnet: multi-label text classification using attention-based graph neural network. In: Proceedings of the 12th International Conference on Agents and Artificial Intelligence (2020)
10. Pal, A., Selvakumar, M., Sankarasubbu, M.: Multi-label text classification using attention-based graph neural network. arXiv preprint arXiv:2003.11644 (2020)
11. Peng, H., et al.: Hierarchical taxonomy-aware and attentional graph capsule RCNNs for large-scale multi-label text classification. IEEE Trans. Knowl. Data Eng. **33**, 2505–2519 (2019)
12. Pennington, J., Socher, R., Manning, C.D.: Glove: global vectors for word representation. In: Empirical Methods in Natural Language Processing (EMNLP), pp. 1532–1543 (2014)
13. Sharifirad, S., Jafarpour, B., Matwin, S.: Boosting text classification performance on sexist tweets by text augmentation and text generation using a combination of knowledge graphs. In: Proceedings of the 2nd Workshop on Abusive Language Online (ALW2), pp. 107–114 (2018)
14. Škrlj, B., Martinc, M., Kralj, J., Lavrač, N., Pollak, S.: Tax2vec: constructing interpretable features from taxonomies for short text classification. Comput. Speech Lang. **65**, 101104 (2021)

An Iterative Model for Quality Assessment in Collaborative Content Generation Systems

Fariba Abedinzadeh, Haleh Amintoosi$^{(\boxtimes)}$, and Mohammad Allahbakhsh

Computer Engineering Department, Faculty of Engineering,
Ferdowsi University of Mashhad, Mashhad, Iran
fariba.abedinzadehzare@mail.um.ac.ir, {amintoosi,allahbakhsh}@um.ac.ir

Abstract. In online collaborative content generation systems, a group of contributors collaboratively generate artifacts. The main concern in these systems is the quality because of varying quality of human-generated contents. Several techniques and methods have been proposed for quality assessments in these systems. However, almost all of them are either based on prone to error techniques such as simple or weighted averaging, or they ignore the interrelation between the quality factors such as quality of artifacts and quality of contributors.

In this paper, we present a novel iterative model for quality in a collaborative content generation system. We then present an algorithm, based on our proposed quality model, that takes into account several factors such as popularity, community attention, and relationships between artifacts and contributors, and computes meaningful accurate quality scores. We compare the performance of our model with a well-known selected work. The comparison results show the superiority of our model.

Keywords: Quality score · Iterative model · Collaborative content generation system · Collusion

1 Introduction

Web technologies have enabled people to easily generate content and make it available to others through crowdsourcing/social platforms [1]. Sometimes these contents are generated individually, such as the tweets posted on Twitter[1] or posts on Facebook[2]. People also can generate content in collaboration with others. In this case, a group of people, called contributors, contribute to content, also called an artifact. This contribution can be either synchronous or asynchronous. In synchronous collaboration, contributors collaborate explicitly, willingly, and at the same time, usually upon the creation of the content. Writing research papers is an example of such a collaboration. In the asynchronous form of collaboration, the artifact is generated by one contributor and other contributors

---

[1] http://www.twitter.com.
[2] http://www.facebook.com.

© Springer Nature Switzerland AG 2022
H. Hacid et al. (Eds.): ICSOC 2021 Workshops, LNCS 13236, pp. 125–138, 2022.
https://doi.org/10.1007/978-3-031-14135-5_10

contribute to the artifact over time and, often, without any intention of collaboration. Asynchronous collaborative systems fall into two main categories. In one category artifacts are created from scratch, while in another group artifacts can be forked/derived from an existing artifact and, hence, there is a parent-child relationship between some artifacts. Wikipedia (wikipedia.com) represents the former category and Github (github.com) is a good represent for the latter. In what follows, by ACCG, we mean the asynchrony collaborative content generation systems of the second category, the category on which this paper focuses.

Quality is always a challenge when dealing with human-generated content, especially in collaborative systems, and a large body of research has been proposed for addressing this issue [2–4]. The quality control even becomes more challenging in asynchronous collaborations, because in these systems the collaboration takes place over time and the quality factors of involved entities, specifically contributors trustworthiness, the share of contributors in the creation of the existing version of the artifact and many other factors changes over time. Deciding on how to take into account these factors and dynamics is a serious challenge that needs investigations and research.

This problem has been investigated in several research efforts [3–8]. The major group of related investigations has adopted simple aggregation or weighted aggregation techniques such as expert review, majority consensus (e.g. IMDb, Amazon, Waze), contributor evaluation, and ground truth in order to assess the quality of human-generated contents [3,9]. These techniques are simple to understand and implement. However, they are prone to attacks and malicious behaviours such as collusion [3,5,10]. To address this problem, iterative techniques have been proposed. In these approaches, quality metrics are computed iteratively and in an interdependent manner. They also take into account the overall architecture of the system and the relationships between the contributors and artifacts [4,5,7]. This makes them more robust against collusive attacks [4,5].

Specifically, for the ACCG systems, a few iterative works have been proposed [11,12]. These techniques aim at computing robust quality scores for artifacts and contributors. However, to the best of our knowledge, in almost all of them, quality and popularity have been considered the same. While popularity contributes to the quality, they are different as popularity is one of the aspects of quality that reflects the community's attention to the artifact.

In this paper, we propose an iterative model to calculate the quality of artifacts in ACCGs. In our model, other factors such as the quality of the content and the collaboration of contributors have been considered besides popularity. More precisely, we consider the popularity of an artifact, number of community feedbacks, number of forked children, the quality of the contributors, and the relationships between the contributors (following each other) in order to compute meaningful and robust quality scores for both artifacts and contributors. In summary, the main contributions of this paper are as follows:

– We propose a novel model for quality in ACCGs in which, popularity, community attention, and relationships between involving entities are taken into account.

- Based on the proposed model, we present an iterative algorithm to compute quality and trust scores for artifacts and contributors in ACCGs.
- We compare the performance of our model with other well-known related models and show the efficiency of our model.

The remainder of the paper is organized as follows. In Sect. 2 we introduce the basic notations and also an example application scenario. In Sect. 3 we propose our quality model, formulations, and our algorithm. We evaluate the performance of the proposed model in Sect. 4. We study related literature in Sect. 5, and finally, conclude in Sect. 6.

2 Data Model and Notation

Assume that in an ACCG system, a set of n_C contributors, denoted by $C = \{c_i | 1 \le i \le n_C\}$, collaboratively contribute to a set of n_A artifacts, denoted by $A = \{a_j | 1 \le j \le n_A\}$.

An artifact is the content generated collaboratively. Each artifact a is identified by a unique Id and has an associated quality score, denoted by Q_a, which is the quality score of a. Moreover, an artifact might also have other application-specific attributes, such as name, title, and URI.

A contributor is a human who contributes to the process of generating content. Each contributor c is identified by a unique Id and has an associated quality metric denoted by T_c, which is the trust score of c. A contributor might also have other application-dependent attributes, such as name and address.

GitHub, as an open-source software community, is a motivating scenario for ACCG systems in which, software developers from different areas and with different technical backgrounds collaborate on open-source software projects, called repositories. Developers can contribute to a repository by committing codes, fork/star a repository, and following other developers. In Github, developers are contributors and repositories are artifacts. Therefore, in this application scenario repositories are artifacts and developers are contributors.

3 Proposed Method

In this section, we propose our model to calculate quality metrics for artifacts and contributors. The main idea behind our model is to propose an iterative definition for the quality of artifacts and contributors in a ACCG system. In what follows, we first propose the notations that we will use in the rest of the paper. Then, we show how we compute the quality metrics, and finally, we propose an iterative algorithm.

3.1   Quality of Artifact

The quality of an artifact depends on the quality of the content of the artifact, community attention, and quality of its parent (e.g., the repository that forked from). In what follows, we explain how we compute these parameters when assessing the quality of an artifact.

Quality of the Content. The content of an artifact can be text, graphics, video, audio, etc. So, it is not possible to define the same set of attributes for all types of content. Also, in many cases, we do not have the quality of the content directly, unless they are evaluated by domain experts, which is not common. Therefore, we use other parameters that indirectly reflect the quality of the content of an artifact. The first parameter is the number of times that an artifact has been liked/promoted by users. A high number of likes somehow shows the high quality of the artifact.

The second parameter is the quality of changes applied to the artifact. Each artifact is subject to contributions/changes during its lifetime. Let's call each change a commit. The quality of a commit directly impacts the quality of the corresponding artifact. On the other hand, the quality of a commit is directly related to the quality of its contributor. Moreover, the changes in the quality of an artifact when made by a large number of commits are more credible than a quality change that comes from just a few commits. So, the gain of an artifact from the commits comes from the combination of two parameters: quality of commits, denoted by Q_a^K, and the number of commits, denoted by K_a. More precisely, assume that a group of contributors has committed changes to artifact a, denoted by C_a. Let K_a be the number of merged commits to the artifact a, and K_a^c be the number of merged commits of contributor c in artifact a. We compute the quality of commits as follows and denote it by Q_a^K:

$$Q_a^k = \sum_{c:c \to a} T_c \times K_a^c \tag{1}$$

In Eq. 1, $:c \to a$ means that c has contributed to a. We use a fuzzy inference engine to combine the quality of commits and number of commits to produce the gain of artifact from the commits, i.e.,

$$Q_a^e = Fuzzy(Q_a^k, K_a) \tag{2}$$

Our proposed framework employs fuzzy logic to calculate Q_a^e. The use of fuzzy logic allows us to achieve a meaningful balance between Q_a^K and K_a. The inputs to the fuzzy inference system are the crisp values of Q_a^K and K_a. The fuzzifier converts the crisp values of input parameters into a fuzzy value according to their membership functions. In other words, it determines the degree to which

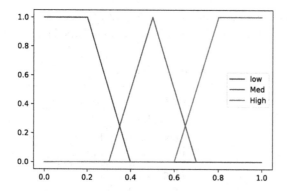

(a) Membership function for the inputs

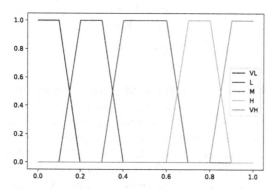

(b) Membership function for the output

Fig. 1. Membership functions

these inputs belong to each of the corresponding fuzzy sets. The fuzzy sets for Q_a^K, K_a and Q_a^e are defined as:

$$S(Q_a^K) = S(K_a) = \{Low, Med, High\}$$
$$S(Q_a^e) = \{VL, L, M, H, VH\}$$

The inputs of fuzzy (Q_a^K and K_a) are then converted to the fuzzy output by using If-Then type fuzzy rules. The mixture of the above-mentioned fuzzy sets creates nine different states, which have been represented by nine fuzzy rules, as shown in Table 1. The rule base design has been done manually, based on the experience and beliefs on how the system should work [10]. To define the output zone, we used the max-min composition method. The result is Q_a^e which is a linguistic fuzzy value. Finally, to convert the Q_a^e fuzzy value to a crisp value in the range of [0, 1], we employ the Centre of Gravity (COG) [13] Defuzzification method, which computes the COG of the area under the content quality membership function.

(Fig. 1) represents the membership function of Q_a^K and K_a, (Fig. 2) represents the Q_a^e membership function. We used trapezoidal-shaped membership functions.

Table 1. Fuzzy rules set

Rule No.	if Q_a^k	and K_a	Then Q_a^e
1	Low	Low	VL
2	Low	Med	L
3	Low	High	M
4	Med	Low	L
5	Med	Med	M
6	Med	High	H
7	High	Low	M
8	High	Med	H
9	High	High	VH

Gain of Artifact from Community Attention. Community attention shows another aspect of quality that shows specifically the popularity of the artifact. In our proposed approach, community attention is represented by two parameters. The first parameter is the number of direct feedbacks given on the artifact in the form of likes, promotions, starts, etc. An artifact with a great number of likes is deemed to be more attractive for the community, especially when trustworthy users have liked it. Let's denote the number of likes by n_l and the quality gain from likes by Q_a^l. Q_a^l is calculated as follows:

$$Q_a^l = \frac{\sum_{c:a \mapsto c} T_c}{n_l} \tag{3}$$

In Eq. 3: $a \mapsto c$ means that c has liked a. The second parameter of community attention is the number of children. In some systems, users can take a copy of an artifact, edit it and post it as a new artifact. In such a case we say the new artifact is a child of the original one. An artifact with a great number of children seems to be deemed as a high-quality artifact from the community point of view Assume that a has n_h children. The gain of artifact a for its children, denoted by Q_a^h, is calculated as follows:

$$Q_a^h = \frac{\sum_{c:c \Rightarrow a} T_c}{n_h} \tag{4}$$

In Eq. 4, $:c \Rightarrow a$ means that c has created a child from a. The gain from likes and children might have different weights while combined. Let W_a^l be the weight of Q_a^l, and W_a^h be the weight of Q_a^h. The community attention for the artifact a is denoted by Q_a^k and is computed as follows:

$$Q_a^k = W_a^l \times Q_a^l + W_a^h \times Q_a^h \tag{5}$$

In Eq. 5, W_a^l and W_a^h are two numbers that are in the range of $[0,1]$ and reflect the level of importance of two parameters. These weights are selected so that $W_a^l + W_a^h = 1$.

Gain of Artifact from its Parent. As stated in the previous section, a user can create a child from an artifact. In such a case, when an artifact is forked from another artifact, it gains a part of its quality from its parent. We denote the quality of the parent of artifact a by Q_a^p.

Computing Quality of Artifact. In our proposed model, the quality of the artifact a is composed of three elements: Q_a^e, Q_a^k and Q_a^p. In different application domains, each of these elements might have a different level of importance. In some systems, the content might be the most important one, while in some others, community attention or parent. To represent different levels of importance, we assume that the importance of these elements is reflected in their weights, which are W_a^e, W_a^k, and W_a^p, for Q_a^e, Q_a^k and Q_a^p, correspondingly. Based on these weights, we calculate the quality of the artifact a as follows:

$$Q_a = W_a^e \times Q_a^e + W_a^k \times Q_a^k + W_a^p \times Q_a^p \tag{6}$$

In Eq. 6, weights are positive numbers in the range $[0,1]$ and are selected so that $W_a^e + W_a^k + W_a^p = 1$.

3.2 Quality of Contributor

The quality of a contributor c depends on the quality of artifacts to which c has contributed, and the number of his followers. A contributor receives gain from each of these parameters. Let's T_c^a, be the gain of the contributor from his contribution to artifacts, and T_c^f be the gain of the contributor from his followers. In what follows, we explain how to compute these gains.

Gain of Contributor from Artifacts. The gain of c from all the artifacts to which he has contributed is denoted by T_c^a and is calculated as follows:

$$T_c^a = \frac{\sum_{a|c \to a} W_c^\alpha \times Q_a}{\sum_{a|c \to a} W_c^\alpha} \tag{7}$$

In Eq. 7, W_c^a is a weight that specifies the impact of an artifact on T_c^a, and is computed as follows:

$$W_c^\alpha = \begin{cases} W_o & \text{if } c \text{ is the creator of a} \\ W_{no} & \text{otherwise} \end{cases} \tag{8}$$

Gain of Contributor from his Followers. In ACCGs, users can follow other contributors. A contributor with a great number of followers is deemed to be a high quality contributor, especially when he has been followed by trustworthy users. We denote gain of contributor from his followers by T_c^f, and compute it as follows:

$$T_c^f = \frac{\sum_{f:f - \to c} T_f}{|f : f - \to c|} \tag{9}$$

In Eq. 9, $f - \to c$ means that contributor f has followed c, and $|f : f - \to c|$ means the number of followers.

Computing Quality of Contributor. We use two gains to compute the quality of a contributor, T_C as follows:

$$T_c = W_c^a \times T_c^a + W_c^f \times T_c^f \tag{10}$$

In Eq. 10, W_c^a, W_c^f and are the weights for T_c^a and T_c^f, correspondingly. These weights are positive numbers in the range $[0,1]$ and selected so that $W_c^a + W_c^f = 1$.

3.3   Proposed Algorithm

We propose an iterative algorithm, inspired by SciMet [4], to calculate the quality of artifacts and contributors. The intuition behind the algorithm is based on the bidirectional relationship between the quality of contributors and artifacts. In other words, a high-quality artifact is an artifact that is created/modified by high-quality contributors, and a high-quality contributor is a contributor who has contributed to high-quality artifacts. Looking at this definition reveals that the proposed method should be an iterative one since the notion of quality of artifacts and contributors are interdependent.

The proposed algorithm is shown in Algorithm 1. The algorithm starts with a set of initializations for weights and the convergence condition. Then it starts computing quality scores and trust ranks iteratively. In the $(p + 1)^{th}$ step, the quality scores are computed based on the quality factors computed in the $(p)^{th}$ step. Then the new trust scores are computed based on the new quality scores.

This iteration will continue, until reaching a convergence point for computed quality scores of artifacts. The convergence of the iterative algorithm is reached when the values of Q in two consecutive iterations are close enough to say they have not changed. This is checked using the root mean square error (RMSE) between $Q^{(p+1)}$ and $Q^{(p)}$. If the RMSE is smaller than a very small threshold (ϵ), the model has converged, and iteration stops.

4   Experimentation and Evaluation

In this section, we evaluate the performance of our proposed model. We first introduce the dataset that we use for the purpose of performance evaluation.

Algorithm 1. Our proposed iterative algorithm

Initialization:

Let:

$\varepsilon > 0$ be the precision threshold

D the set of quality ranks of artifacts

T the set of trust scores of contributors

$$T_c^{(0)} = 1$$
$$Q_a^{(0)} = 1$$

Repeat:

$$Q_a^m = \sum_{c:c \to a} T_c^{(p)} \times M_a^c;$$

$$Q_a^k = W_a^l \times \frac{\sum_{c:c \to a} T_c^{(p)}}{n_l} + W_a^l \times \frac{\sum_{c:c \to a} T_c^{(p)}}{n_h};$$

$$Q_a^{(p+1)} = W_a^e \times \left(Fuzzy(Q_a^k, K_a)\right) + W_a^k \times Q_a^k + W_a^p \times Q_a^p;$$

$$T_e^{(p+1)} = W_c^a \times \left(\frac{\sum_{c \to c} W_c^\alpha \times Q_a^{(p+1)}}{\sum_{c \to c} W_c^\alpha}\right) + W_c^f \times T_c^f;$$

until: $\|\vec{Q}^{(p+1)} - \vec{Q}^{(p)}\|_2 < \varepsilon.$

Then, we introduce the evaluation metrics that are used throughout the evaluation process.

4.1 Dataset

In order to evaluate the performance of the proposed algorithm, we set up some experiments on a real-world dataset. We use the MSR 2014 challenge dataset[3], which is a (very) trimmed down version of the original GHTorrent dataset. It includes data from the top 10 starred software projects for the top programming languages on GitHub, i.e., 70 projects and their forks. For each project, it contains data including issues, pulls requests organizations, followers, stars, and labels.

Before using the dataset, we preprocess its data and normalize the outliers. More precisely, for numeric attributes such as number of likes, we set a threshold and replace the values greater than $\mu + 2 \times \sigma$ with this threshold, i.e., $\mu + 2 \times \sigma$. In this threshold, μ is the mean and σ is the standard deviation of the field values.

4.2 Experimentation Setup

We conduct a set of experiments on a laptop running Windows 7.0 and having 4 GB of RAM. We used Spyder (python 3.8) for developing the model. As there

---

[3] https://ghtorrent.org/msr14.html.

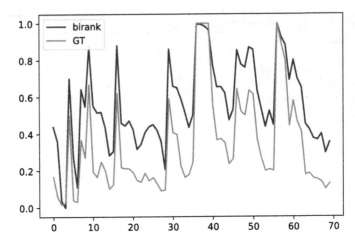

Fig. 2. Conformance of the proposed model results with ground truth

are some wights and constants that need initialization, we set their initial values, as reflected in Table 2, based on our intuition from the performance of a ACCG.

Table 2. Parameter initialization

W_a^e	W_a^e	W_a^e	W_a^e	W_a^e	W_a^e	ϵ
0.3	0.2	0.4	0.1	0.6	0.4	0.01

In a normal system, assuming that there are few or no misbehavior, the number of likes can partially reflect the quality of artifacts. Therefore, we use it as a ground truth (GT). For performance comparison, we selected an iterative technique, called BiRank [14], which has been proposed for popularity computation. This is one of the most related works to our proposed model, it is published in an IEEE TKDE which is a high ranked journal, and it has been cited more than a hundred times, based on Google Scholar.

We apply our algorithm as well as the BiRank to the selected dataset. Then, we use the difference between the computed rank and ground truth as the error rate. We use two error rates for the purpose of performance comparison: Mean-Absolute-Error (MAE) and Mean-Squared-Error (MSE).

4.3   Performance Evaluation

First, we show that our proposed algorithm computes decent meaningful scores. To do this, we check the conformance of the results proposed with our system with the selected ground truth. The results of this conformance check are depicted in Fig. 2. As the figure shows, the performance of our model highly

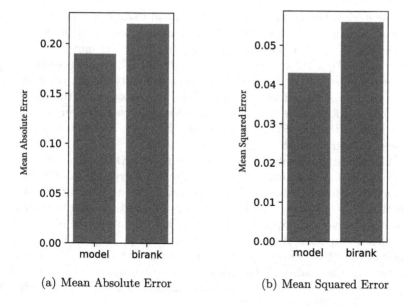

(a) Mean Absolute Error   (b) Mean Squared Error

Fig. 3. Performance comparison

conforms with the ground truth. Just in few points, our model has moved in the opposite direction of ground truth, and that is because we use the effect of contributors' trust as well.

Therefore, in a collusion-free environment, the ranks computed by our model can reflect the true quality of artifacts. Moreover, as our proposed model is iterative, it is more robust against manipulations, and as there are more quality factors involved in our formulations, the scores are more meaningful than the simple form of number of likes.

4.4 Performance Comparison

We compare the performance of our proposed model with Birank in terms of accuracy, using MAE and the MSE metrics. Figure 3a shows the mean absolute error of our model compared with the BiRank. The mean-squared errors are compared in Fig. 3b. As depicted in both parts of Fig. 3, both the MAE and MSE of our model are smaller than the BiRank model. This shows that our proposed model outperforms the BiRanks as a well-known related work.

5   Related Work

Assessing the quality of online artifacts specifically when generated collaboratively, is always a challenge.

In synchronous collaboration mode, the work presented in [6] shows that the number of words in a text document is a good predictor for article quality. In [15], authors consider the authority of the authors and the reviewers to compute the quality of each word in the article. Several models such as Peer Review and Prob Review have been proposed that are based on the mutual dependency between articles' quality and contributors' authority. The paper presented in [11] formulates the interdependent relation between quality factors of articles and authors with an Article-Editor network. The model computes the quality of a document according to the editing relationships between article and editor nodes using a form of the PageRank [7] algorithm. However, it does not consider the relations between the authors.

In [8], the authors implicitly use the co-edit network structure with h-index measure to calculate the authority of the editors. However, the quality of an article is directly computed using only the derived authorities. In [12], the assumption is that high-quality articles are written by good editors and vice versa. The relationship between the quality of the text and the authority is presented in a way that the quality of an article is determined by the quality of text and the level of expertise of a user on each part of the text that he/she has written and approved. The quality of the articles is calculated by considering the editing history of articles and also extracting the amount of content that each editor has left in each edition.

Github, as an example of asynchronous collaboration, and the precursor storeroom in the field of code hosting [16], not only allows users to modify, and comment on repositories, but also provides users to star, fork, and clone repositories. Users could show their satisfaction with a repository through their starring or forking. Therefore, a star or fork is regarded as an appearance of popularity [17,18]. Some studies have tried to predict the popularity of GitHub repositories. Former studies used the number of stars or the number of forks as delegates for the popularity of a GitHub project [19]. Since GitHub is the largest source code repository in the world, researchers started to study the popularity of GitHub repositories. Later studies found that the number of users, fork depth, number of followers, pull requests, and reported issues share statistically considerable relationships with repository stars based on software families comprised of software repositories [20]. To solve the ranking problem on bipartite networks in [14], authors proposed BiRank, which is a ranking method on bipartite networks that normalizes the dataset in an iterative form and ranks the products by their popularity. The work presented in [21] proposed a stargazer-influence-based approach, called StarIn, to predict the GitHub repository's popularity. It uses the followers in GitHub as the main dataset.

As a summary, the existing related works, mostly rely on simple or weighted aggregation techniques, i.e., they compute the quality ranks in a one-way manner. For instance, they compute the quality of artifacts based on the quality of contributors, or vice-versa. They ignore the interdependency between them and the iterative nature of quality in ACCGs [4]. The few iterative techniques consider just a few popularity factors and ignore the relationships between artifacts and contributions. This is exactly where the contributions of this paper fit.

6   Conclusion

In this paper, we proposed an iterative quality score algorithm that calculates the quality of artifacts and contributors based on both their quality factors and the relationships between them. We presented an iterative definition for quality in ACCG systems in which, the quality of artifacts and contributors are computed interdependently. The evaluation results confirm the meaningfulness as well as the credibility of the results of our model, in comparison with a well-known related work.

References

1. Doan, A., Ramakrishnan, R., Halevy, A.Y.: Crowdsourcing systems on the world-wide web. Commun. ACM **54**(4), 86–96 (2011)
2. Allahbakhsh, M., Benatallah, B., Ignjatovic, A., Motahari-Nezhad, H.R., Bertino, E., Dustdar, S.: Quality control in crowdsourcing systems: Issues and directions. IEEE Internet Comput. **17**(2), 76–81 (2013)
3. Daniel, F., Kucherbaev, P., Cappiello, C., Benatallah, B., Allahbakhsh, M.: Quality control in crowdsourcing: a survey of quality attributes, assessment techniques, and assurance actions. ACM Comput. Surv. (CSUR) **51**(1), 1–40 (2018)
4. Allahbakhsh, M., Amintoosi, H., Behkamal, B., Beheshti, A., Bertino, E.: SCiMet: stable, scalable and reliable metric-based framework for quality assessment in collaborative content generation systems. J. Informet. **15**(2), 101127 (2021)
5. Allahbakhsh, M., Ignjatovic, A.: An iterative method for calculating robust rating scores. IEEE Trans. Parallel Distrib. Sys. **26**(2), 340–350 (2014)
6. Thomas Adler, B., Chatterjee, K., De Alfaro, L., Faella, M., Pye, I., Raman, V.: Assigning trust to Wikipedia content. In: Proceedings of the 4th International Symposium on Wikis, pp. 1–12 (2008)
7. Page, L., Brin, S., Motwani, R., Winograd, T.: The PageRank citation ranking: bringing order to the web. Technical report, Stanford InfoLab (1999)
8. Suzuki, Yu.: Quality assessment of Wikipedia articles using h-index. J. Inf. Process. **23**(1), 22–30 (2015)
9. Allahbakhsh, M., et al.: Quality control in crowdsourcing systems: issues and directions. IEEE Internet Comput. **17**(2), 76–81 (2013)
10. Allahbakhsh, M., Amintoosi, H., Ignjatovic, A., Bertino, A.: A trust-based experience-aware framework for integrating fuzzy recommendations. IEEE Trans. Serv. Comput. **15** (2019)
11. Li, X., Tang, J., Wang, T., Luo, Z., de Rijke, M.: Automatically assessing Wikipedia article quality by exploiting article–editor networks. In: Hanbury, A., Kazai, G., Rauber, A., Fuhr, N. (eds.) ECIR 2015. LNCS, vol. 9022, pp. 574–580. Springer, Cham (2015). https://doi.org/10.1007/978-3-319-16354-3_64
12. de La Robertie, B., Pitarch, Y., Teste, O.: Measuring article quality in Wikipedia using the collaboration network. In: 2015 IEEE/ACM International Conference on Advances in Social Networks Analysis and Mining (ASONAM), pp. 464–471. IEEE (2015)
13. Van Leekwijck, W., Kerre, E.E.: Defuzzification: criteria and classification. Fuzzy Sets Syst. **108**(2), 159–178 (1999)
14. He, X., Gao, M., Kan, M.-Y., Wang, D.: BiRank: Towards ranking on bipartite graphs. IEEE Trans. Knowl. Data Eng. **29**(1), 57–71 (2016)

15. bibitemch10r6 Hu, M., Lim, E., Sun, A., Wirawan Lauw, H., Vuong, B.: Measuring article quality in Wikipedia: models and evaluation. In: Proceedings of the Sixteenth ACM Conference on Information and Knowledge Management, pp. 243–252 (2007)

16. Long, Y., Siau, K.: Social network structures in open source software development teams. J. Database Manag. 18(2), 25–40 (2007)

17. Borges, H., Tulio Valente, M.: Hat's in a GitHub star? Understanding repository starring practices in a social coding platform. J. Syst. Softw. 146, 112–129 (2018)

18. Jiang, J., et al.: Why and how developers fork what from whom in GitHub. Emp. Softw. Eng. 22(1), 547–578 (2017)

19. Han, J., Deng, S., Xia, X., Wang, D., Yin, J.: Characterization and prediction of popular projects on GitHub. In: 2019 IEEE 43rd Annual Computer Software And Applications Conference (COMPSAC), vol. 1, pp. 21–26. IEEE (2019)

20. Brisson, S., Noei, E., Lyons, K.: We are family: analyzing communication in GitHub software repositories and their forks. In: 2020 IEEE 27th International Conference on Software Analysis, Evolution and Reengineering (SANER), pp. 59–69. IEEE (2020)

21. Ren, L., Shan, S., Xu, X., Liu, Yu.: StarIn: an approach to predict the popularity of github repository. In: Qin, P., Wang, H., Sun, G., Lu, Z. (eds.) ICPCSEE 2020. CCIS, vol. 1258, pp. 258–273. Springer, Singapore (2020). https://doi.org/10.1007/978-981-15-7984-4_20

Log Attention – Assessing Software Releases with Attention-Based Log Anomaly Detection

Sohail Munir[1]([⊠]) [iD], Hamid Ali[2] [iD], and Jahangeer Qureshi[2] [iD]

[1] Dubai Digital Authority, Dubai, UAE
lsohail.munir@digitaldubai.ae
[2] Xeric.ai, Dubai, UAE

Abstract. A Software Engineering Manager (EM) has to cater to the demand for higher reliability and resilience in Production while simultaneously addressing the evolution of software architecture from monolithic applications to multi-cloud distributed microservices. Pre-release functional testing is no longer sufficient to eliminate faults as more and more issues are generated at runtime, which is challenging to diagnose due to complex inter-service dependencies and dynamic late binding of services. Bugs in Production are known to propagate across software components and become critical as they go undetected.

This paper introduces LogAttention, a methodology based on analysis of runtime logs that provides actionable insights to the EM to identify faults and preempt failure in Production. LogAttention is a Log Anomaly Detection (LAD) technique that uses Attention-based Transformer Models to identify Anomalous Log Messages. LogAttention assigns a quality score to the software release in Production and presents remarkable logs to the EM to analyze, predict, and preempt failure. This paper presents empirical evidence showing that LogAttention outperforms existing LAD techniques to identify anomalous log messages and ensure that the detected log anomalies are reliable indicators of the health of a software release.

Keywords: AIOps · Attention models · Log analysis · Log anomaly detection · Log attention · Software release quality

1 Introduction

Software today has evolved from large monolithic applications deployed on-premises at enterprise data centers to multi-cloud, cloud-native distributed applications based on Microservice Architecture (MSA) [1]. The Software Engineering Manager (EM) in a Software Product Organization (SPO) has the responsibility to build and support software applications that should be scalable, resilient, and robust while adapting to the dynamic and evolving user requirements. However, at the same time, MSA brings the complexity of having hundreds or thousands of interdependent microservices which become cumbersome to manage in production.

An EM can use traces, metrics, or logs to detect bugs in Production [2, 3]. In most cases, logs are the only available data that record software runtime information; however,

© Springer Nature Switzerland AG 2022
H. Hacid et al. (Eds.): ICSOC 2021 Workshops, LNCS 13236, pp. 139–150, 2022.
https://doi.org/10.1007/978-3-031-14135-5_11

a log-based approach must deal with the enormous volume of log data averaging giga-bytes of data per hour for a typical commercial cloud application. Unlike pre-release testing, an EM cannot directly look for bugs in a running production environment. Instead, she relies on advanced techniques to look for the effects of bugs. She could use Log Anomaly Detection (LAD) to find anomalous logs and expect that some bug has caused the anomalies. Anomalous system behavior could mean functional inconsistencies, performance degradation, system compromise, or a change in log patterns and would indicate that some possible bugs exist, and the EM should proactively discover these bugs and address them.

This paper presents a methodology for assessing software release quality and identifying bugs in Production, based only on runtime logs. The proposed model uses LAD based on Attention-based Transformer Models to present remarkable logs to the EM along with a numeric score depicting the health of the software release. The research evaluates the proposed LogAttention model against three log datasets which include two different releases SE-A and SE-C of a software product SE having more than 50,000 daily users and one release DN-C of a software product DN having more than 100,000 monthly active users and more than 500,000 total users. The results demonstrate the superior performance of LogAttention over the current techniques.

2   Methodology

The paper proposes LogAttention, a methodology for assessing the quality of a software release in Production based on runtime logs. The proposed methodology uses Attention-based transformer models to classify log messages as Remarkable. A log message classified as remarkable would mean that the model predicts it as anomalous and representing a bug in Production.

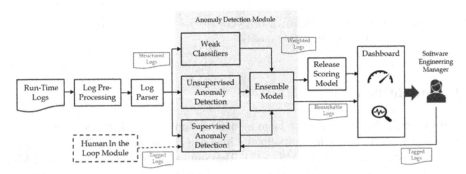

Fig. 1. LogAttention model

The EM can view the remarkable log messages, analyze the log's sequence, and determine if it represents a bug that she should address. The objective is for LogAttention to provide a consistently accurate prediction and a very low detection prevalence for the recommendations to be meaningful for the EM. Figure 1 illustrates the LogAttention methodology explained in the subsequent sections.

The LogAttention model presented in this paper comprises multiple modules. These modules can be categorized based on their role in data understanding, preparation, and analysis. This order also reflects the progression of our methodology. Each section explains our approach to data understanding, preparation, and analysis, along with the description of corresponding models and functions.

2.1 Log Preprocessing

In LogAttention, there is a Log Pre-processing module that cleanses the logs to parse them efficiently. The preprocessing function is fine-tuned to each new production environment to cleanse the log files and automatically extract a log format as an input to log parsers. This results in a "burn-in" period which is essential to determine the most appropriate log parser by testing and evaluating results on the preprocessed log data.

Logs contain an abundant amount of information regarding software activity, e.g., events, parameters, execution details. While there is no fixed structure, a log message typically consists of a variable part (log parameter) and a constant part (log event). We preprocess the log files using the following preprocessing functions:

1. In the scenario where the date is not provided in the logs, a function is used to find the date in the log file name.
2. Verify if a given string is the beginning of a logline.
3. Clean a log file by appending log traces and outputs into a singular line.
4. Remove any log headings for the log file to remove any initial redundant log strings.
5. Extract the log format by identifying Date, Time, Log Level, and Content Token positions.

However, after these preprocessing functions have been applied the logs are still not in a state where they can be processed efficiently. To parse these logs effectively log parsers are essential. After an exhaustive literature review for recent advances in Log Parsing, the Log Parsers that we found of particular interest were DRAIN [4], a Log Structure Heuristics-based Log Parser, and NuLog [5] a neural network-based Log Parser.

These parsers demonstrated impressive performance, yet both seemed to function with varying efficiency based on the amount of data present for parsing. While NuLog outperforms DRAIN on large log files, when insufficient data is available, DRAIN has been observed to be a better choice. Nevertheless, this uncertainty makes a burn-in period essential to parse any new log datasets properly. An implementation of the technique should offer the EM to choose between NuLog, DRAIN, or other viable log parsers.

2.2 Log Anomaly Detection Module

Once the structured logs are extracted, log data is input to the Log Anomaly Detection (LAD) Module. The Machine Learning based LAD Module consists of the Weak Classification (WC) Module, the Supervised Anomaly Detection (SAD) Module, and the Unsupervised Anomaly Detection (USAD) Module.

Weak Classification (WC) Module – WC is based on heuristics from the analysis of different log datasets. Their composition and distribution of anomalies in these datasets help reach immediate approximation. This module classifies each log message as either Normal or Remarkable (i.e., possibly anomalous). This module allows the system to classify logs without any training data despite being less accurate. Based on heuristic measures, this module guides the system into a relevant solution space that further allows the curation of an initial dataset for labeling saving computing overhead. The WC module classified data uses three distinct labeling functions as follows:

Log Level Classifier – A log message contains an associated log level that gives a rough guide to the message's importance and urgency, like INFO, WARNING, or ERROR. Log Level-based classification function utilizes this log level to classify the incoming log messages, and any level other than the "INFO" level is classified as "Remarkable."

Double σ Classifier – A standard log message contains a timestamp that denotes the exact time of the event that has triggered the log. Calculating the differential between the timestamp of the current log and that of the previous log helps estimate the event's execution time that triggered the previous log. This allows the function to calculate the mean execution time for every unique log message. If the execution time of a log message exceeds by two standard deviations, it indicates an abnormal instance of that log message. The EM may manually adjust the standard deviation limits beyond which an event will be classified/labeled as remarkable. This customization helps the EM adjust the system based on historical performance/operational context, thus contributing to early classification accuracy.

Error Lexicon Classifier – Each log entry contains a message that holds information regarding the event that triggered it. This message comprises alphanumeric content that helps understand the nature of the event. By using a lexicon of common error terminologies which occur in log messages, the Error Lexicon-based classification function classifies logs with mentions of erroneous tokens as "Remarkable."

This function's efficiency depends on the accuracy of the lexicon dictionary being used as a reference. To create a thorough and efficient lexicon, this research followed a novel approach using issues from GitHub Activity Data, containing the keywords related to faults like Errors and Exceptions. The Error Lexicon is created by scraping over one million issues of top-rated GitHub repositories, of which around 50,000 issues are related to either an exception or an error.

Unsupervised Anomaly Detection (USAD) Module – USAD refers to the practice of detecting statistical outliers from a dataset with no reliance on labels. This methodology is especially beneficial for the classification of a dataset with no prior training. Additionally, the possible biases in a human-labeled training dataset are also nullified, thus allowing the detection of anomalies that are not explicitly sought out by human users. USAD has two steps, learning representative embeddings from the datasets and classifying the logs as either Normal or Anomalous.

Embeddings – Word Embeddings or Word vectorization is a methodology in Natural Language Processing (NLP) that maps words or phrases from vocabulary to a corresponding vector of real numbers. This mapping allows downstream functions to solve language-related problems such as word predictions, word similarities, and semantics. Log messages contain strings of information in an alphanumeric format, and any analysis of this data, including LAD, requires the conversion of this information into the downstream-functions-ready format.

Attention-based Transformer Models – Attention in neural networks is a mechanism that a model can learn to make predictions by focusing on (attending to) a small sub-set of data. To predict or infer one element (sequence), the model estimates using the attention vector how strongly it is correlated with ("attends to") other elements (sequences) and takes the sum of their values weighted by the attention vector as the approximation of the target. Self-Attention is a category of Attention models that "attends to" different positions in the same input sequence [6].

Transformer is an architecture based on Self-Attention models that transform one sequence into another with the help of the Encoder and Decoder [7]. Transformers do not require sequential data to be processed in order, allowing massive parallelization during training, enabling training on very large datasets very fast.

Language Modeling – Language modeling is a technique that builds language models to help predict a sequence of recognized words and phonemes that are used for real-world problems relating to Natural Language Processing (NLP) using statistical techniques, neural networks, and deep learning methods [8] specially Attention-based Transformer Models. Specifically, word embedding is adopted to use a real vector representing each word in the project vector space based on their usage, allowing words with a similar meaning to have an equal representation while still retaining their contextual information.

Training a Transformer model for use in a particular NLP task is simple. Start with a randomly initialized Transformer model, put together a huge dataset containing text in the language or languages of interest, pre-train the Transformer on the huge dataset, and fine-tune the pre-trained Transformer on the particular task in question, using the task-specific dataset. The advantage of this method is that only labeled data is needed for the final step of fine-tuning. Language models are categorized as:

1. Causal Language Models (CLM): In CLM, the model tries to predict or generate the next word(s) given a sequence of words.
2. Masked Language Modeling (MLM): In MLM, a certain proportion (token) of the text sequence, which may be a complete word or a part of a word, or a text sequence, is masked, and the model then predicts the token that is masked. The masked token may then be replaced with an actual mask or replaced with another random token from the vocabulary.

Recent research has introduced many high-performing Transformer models that could be used for USAD. This research analyzes multiple light-weight transformer models such as ELECTRA [9], DistilBERT [10], and GloVe [11], as well as traditional machine learning models and tested them on datasets SE-A and DN-C against multiple classifiers. The results indicate that the language models allowed far better representations to be learned for the data and provided better results. Figure 2 shows Macro-F_1 scores on SE-A and DN-C datasets, respectively, in a completely unsupervised setting, using different classifiers and embeddings. As evident from Fig. 2, DistilBERT gives the best Macro-F_1 scores on both datasets using Self Organising Maps (SOM) [12] and Isolation Forest (IF) [13] classifiers. DistilBERT is therefore selected as the Transformer model of choice, and SOM and IF as the classifiers of choice for the USAD module.

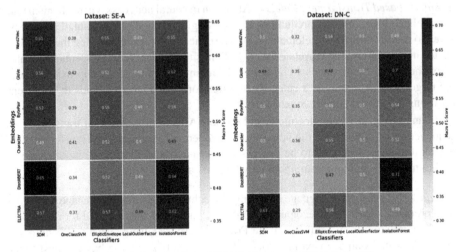

Fig. 2. Model selection of USAD (SE-A and DN-C Datasets)

Supervised Anomaly Detection (SAD) Module – SAD uses active learning to detect anomalies in a supervised paradigm. It uses transfer learning (learning of pre-trained models) and a novel data curation technique for data labeling by the Human-in-the-Loop (HIL), which can be Domain Experts(s), the software developers, or the EM.

The transfer learning and data curation aspects of SAD depicted in Fig. 3 below are as follows:

Pre-trained Model – The pre-trained model is trained on a large dataset comprising real log data, and the weights of this pre-trained model are then applied to the SAD module. When the embedded data is input to the module, it can provide better outputs in lesser iterations based on transfer learning.

Human in the Loop – An important component of training a supervised model is the labeled dataset. This labeling is done by a HIL, who is typically a software engineer from the agile development team who assigns a priority score (i.e., High, Medium, or Low) to each data point (i.e., a log event). While this helps train the supervised model, it also helps the model understand the software engineering team's priority based on these labels. This customization is essential in helping the model focus on specific aspects the developers might find significant and learn the preferences of different developers in the process while still allowing the system to maintain its search for statistically significant logs. Another significant input comes from the EM, which rates the anomaly detection system's output, serving as a labeling feedback system to the SAD module. This labeling serves as an important input for the model because eventually, the model assists the EM in making decisions relating to software release. Since labeling large swathes of data is a tedious task with significant budgetary and computing overheads, the dataset to be labeled needs to be curated smartly. The model leverages the predictions from the Weak Classification (WC) module and the Unsupervised Anomaly Detection (USAD) module to achieve this. In cases of availability of pre-trained models, the Supervised Anomaly Detection (SAD) module predictions are also utilized. It may be noted that the HIL step is needed only once every major release.

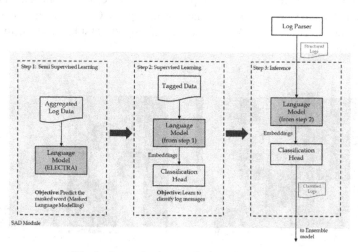

Fig. 3. Supervised anomaly detection module

The paper proposes a novel approach for sampling the data, *MCR Sampling*, to make the size of the dataset tagged by the HIL manageable. The dataset comprises sampling from the following three techniques in equal proportions, Marginal Sampling, Certainty Sampling, and Random Sampling. Kazerouni, et al. [14] have proposed Marginal Sampling (Exploitation) and Random Sampling (Exploration) for Active Learning from a skewed dataset. LogAttention HIL model proposes MCR Sampling, which introduces Certainty Sampling to ensure the model also performs well in a completely new environment.

1. Marginal Sampling – Marginal sampling involves the data samples for which the model is least certain about the classification, i.e., Remarkable or not. Selecting this type of most informative and uncertain data samples for which the model has less confidence in predicting, i.e., prediction on the margins, improves the model's discriminative ability the most. These samples are close to the decision boundary where uncertainty is the highest, and hence Marginal Sampling is used as an Exploitation method.

2. Certainty Sampling – While Marginal sampling helps the model where the model is unsure about the correctness of its labeling, Certainty sampling, on the other hand,

helps the model consolidating its labeling, which the model is fully certain about and, in a way, tests the current state of the model. These samples are usually away from the decision boundary, and the prediction values are over 0.9.

3. Random Sampling – Random sampling involves the selection of random data samples as an exploratory method.

Training the Model – LogAttention firstly trains a Transformer Model with a language modeling head to learn the structure of a log file, referred to as the Self-Supervised learning step. In the next step, the Fine-Tuning step, the language modeling head is replaced with a classification head composed of a simple, fully connected layer and trained using the labels generated from the human experts. Once done, the model infers new logs using the Transformer Model with the fine-tuned classification head to attribute the correct label to each data instance.

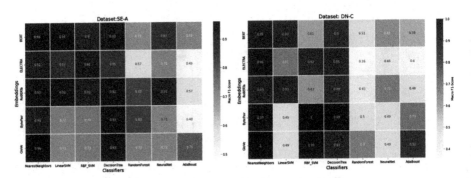

Fig. 4. Model selection of SAD (SE-A and DN-C Datasets)

Model Selection Journey. Like the USAD module, we select a Transformer model for the SAD to benefit from the relatively small amount of data required to train the model and its capacity for transfer learning. As shown in Fig. 4, the transformer models are the most reliable source of embeddings. Additionally, transformer model-based embeddings allow the possibility of zero-shot performance and transfer learning. Given the very slight differences in performance between ELECTRA and RoBERTa [15], we selected ELECTRA as our transformer for classification purposes because it is faster to train than any other model evaluated.

2.3 Ensemble Model

The output from all three ML modules, i.e., WC, USAD, and SAD, is input to the Ensemble model where each classifier is assigned an appropriate weight given its performance on the labeled set of structured log data. The ensemble model outputs a single prediction vector embodying all the best aspects of the previous modules indicating which logs are remarkable. The weights are assigned based on input from tagged data. The EM may adjust the assigned weights based on the production environment as appropriate. In the future, an optimization model may be developed to assign optimal weights that get updated automatically based on the status changes in the production environment. The Ensemble model outputs the weighted logs to the Release Scoring Model and presents Remarkable logs to the dashboard.

2.4 Release Scoring Model

While the Machine Learning module classifies remarkable logs that help identify anomalous instances from the enormous volume of log data of a production release, this information becomes more useful when presented in an aggregated form that helps represent the state of the release with actionable insights for the EM. The EM is then better placed to make an informed decision based on a more actionable and explainable input.

To further simplify it for the EM, the LogAttention methodology assigns a Release Score ($\underline{\bigcirc}$) to the release, which is a numerical representation of the release quality, enabling her to gauge the health of the release by referring to a single parameter. This release scoring model is depicted in Eq. 1 below. The release scoring model uses heuristics to assign the weight to the High Priority (Remarkable), Medium Priority, and Low Priority logs. The model provides an interface for the EM to adjust the weights based on the runtime environment. The EM may also tweak the weights based on her experience and preferences. The release score is calculated as follows:

$$\underline{\bigcirc}_j = 1 - w_p \cdot \left| \min_{0 < i \leq j} \frac{|R_i|}{|L_i|} - \frac{|R_j|}{|L_j|} \right| - \left(w_r \cdot |R_j| + w_m \cdot |M_j| + w_o \cdot |O_j| \right) \quad (1)$$

where, j is the number of the current release in a production environment assessed by LogAttention. $\underline{\bigcirc}_i$ = LogAttention score for the ith release, L_i = {Set of Log Messages for the ith release}, R_i = {Set of Remarkable/High Priority Log messages for the ith release}, M_i = {Set of Medium Priority Log messages for the ith release}, O_i = {Set of Low Priority Log messages for the ith release, w_r = weight assigned to Remarkable/High Priority Logs, w_m = weight assigned to Medium Priority Logs, w_o = weight assigned to Low Priority Logs, and w_p = weight assigned to min detection prevalence.

2.5 Dashboard

LogAttention presents the release score ($\bigcirc$) and the remarkable logs on a dashboard for the EM to gauge the health of the release in production. The LogAttention dashboard allows the EM to conduct detailed analysis by looking at the logs before and after the remarkable logs and fetch further information regarding the issue highlighted in the log message through third-party sources like Stack Overflow. She may also reclassify a certain log as appropriate using the interface provided.

3 Results

The paper evaluates LogAttention against three benchmark models DeepLog, LogAnomaly, and RobustLog. The consolidated results are presented in Table 1.

DeepLog [16] and LogAnomaly [17] are Semi-Supervised models and are two of the best performing LAD techniques; whereas, RobustLog [18] is a Supervised Log Anomaly Detection technique that works very well after training which requires tagging of the entire dataset, but does not work well without training.

Table 1. Model Performance results across datasets.

Model	Logs Processed	Remarkable Logs	Detection Prevalence	TP	FP	TN	FN	Accuracy	Precision	Recall	F₁ Score	Macro F₁ Score	MCC
SE-A dataset													
DeepLog	32,873	3,764	0.1145	558	3,206	28,445	664	0.8823	0.1482	0.4566	0.2238	0.5801	0.2111
LogAnomaly	32,873	4,130	0.1256	457	3,673	27,978	765	0.8650	0.1110	0.3740	0.1750	0.5186	0.1532
LA (USAD)	32,873	2,664	0.0810	846	1,818	29,833	376	0.9333	0.3176	0.6923	0.4354	0.7000	0.4401
LA (USAD+WC)	32,873	2,664	0.0810	846	1,818	29,833	376	0.9333	0.3176	0.6923	0.4354	0.7000	0.4401
RobustLog (SR)	25,971	2,375	0.0914	430	1,945	23,082	514	0.9053	0.1811	0.4555	0.2591	0.6043	0.2453
RobustLog (FT)	23,012	824	0.0358	791	33	22,124	64	0.9958	0.9600	0.9251	0.9422	0.9570	0.9402
Log Attention (SR)	24,027	697	0.0290	694	3	23,151	179	0.9924	0.9957	0.7950	0.8841	0.9401	0.8862
Log Attention (FT)	24,027	721	0.0300	721	0	23,154	152	0.9937	1.0000	0.8259	0.9046	0.9507	0.9058
SE-C dataset													
DeepLog	106,906	20,911	0.1956	1,245	19,666	83,792	2,203	0.7941	0.0595	0.3611	0.1022	0.4834	0.0743
LogAnomaly	106,906	20,492	0.1917	1,338	19,154	84,304	2,110	0.8014	0.0653	0.3881	0.1118	0.4999	0.0931
LA (USAD)	106,906	15,255	0.1427	1,992	13,263	90,195	1,456	0.8625	0.1306	0.5777	0.2130	0.5683	0.2271
LA (USAD+WC)	106,906	8,675	0.0811	2,831	5,844	97,611	617	0.9396	0.3263	0.8211	0.4670	0.7175	0.4917
RobustLog (SR)	63,647	1,595	0.0251	202	1,393	61,005	1,047	0.9617	0.1266	0.1617	0.1421	0.5612	0.1237
RobustLog (FT)	74,835	2,536	0.0339	2,414	122	72,299	0	0.9983	0.9519	1.0000	0.9754	0.9873	0.9748
Log Attention (SR)	75,295	2,376	0.0316	2,092	284	72,595	324	0.9919	0.8805	0.8659	0.8751	0.9345	-0.8690
Log Attention (FT)	75,295	2,414	0.0321	2,414	0	72,879	2	1.0000	1.0000	0.9992	0.9996	0.9998	0.9996
DN-C dataset													
DeepLog	55,290	2,224	0.0402	221	2,003	52,643	423	0.9561	0.0994	0.3432	0.1541	0.5658	0.1654
LogAnomaly	55,290	2,874	0.0520	394	2,480	52,166	250	0.9506	0.1371	0.6118	0.2240	0.5992	0.2738
LA (USAD)	55,290	1,570	0.0284	413	1,157	53,489	231	0.9749	0.2631	0.6413	0.3731	0.6801	0.4006
LA (USAD+WC)	55,290	1,208	0.0218	507	701	53,945	137	0.9848	0.4197	0.7873	0.5475	0.7699	0.5684
RobustLog (SR)	41,809	15,110	0.3614	265	14,845	26,615	84	0.6429	0.0175	0.7593	0.0343	0.3075	0.0550
RobustLog (FT)	38,703	471	0.0122	451	20	38,232	0	0.9995	0.9575	1.0000	0.9783	0.9896	0.9783
Log Attention (SR)	38,235	361	0.0094	358	3	37,724	150	0.9960	0.9917	0.7047	0.8239	0.9110	0.8343
Log Attention (FT)	38,235	481	0.0126	481	0	37,727	27	0.9993	1.0000	0.9469	0.9727	0.9862	0.9727

LogAttention is an ensemble of WC, USAD and SAD. The performance of the LogAttention USAD module alone is labeled as LA (USAD). The performance of the LogAttention USAD module together with the WC module labeled as LA (USAD + WC). The training dataset is then used to evaluate the performance of the LogAttention SAD module. The dataset for SE-A, SE-C, and DN-C is randomly split into 30% training and 70% test data. After training on the training dataset, this fine-tuned LogAttention model was evaluated, which is an ensemble of WC, USAD, and SAD against the test dataset. The results of this fine-tuned model are labeled as LogAttention (FT). To eliminate the possibility of Overfitting, a separate training dataset is curated by dropping the messages (event templates) from the training dataset that were also found in the test dataset. The LogAttention SAD model was retrained on this similarity removed training dataset and evaluated using the test dataset ensembling all three modules of Log Attention, and report the results of this model labeled as LogAttention (SR). The same split dataset was used to train RobustLog. However, since the process of removing similar logs from the data set is random during reassigning labels to original structured logs, RobustLog's manner of preprocessing structured logs both during training and inference yields marginally different amounts of resulting test logs when reassigned to the original datasets, as compared to LogAttention. Just like in the case of log attention, two separate results of RobustLog are reported, fine-tuned, and similarity-removed with the results labeled as RobustLog (FT) and RobustLog (SR). Figure 5 presents the evaluation in terms of Macro F_1, score. It can be observed that even the unsupervised models of

LogAttention perform better than the semi-supervised benchmark models DeepLog and LogAnomaly for all datasets across all metrics.

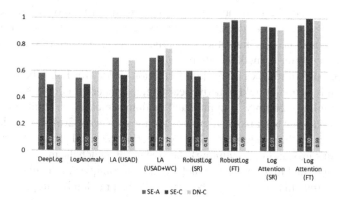

Fig. 5. Model Macro F_1 score across datasets

LogAttention (SR) yields very good results in supervised settings which are almost as good as RobustLog (FT). Without removing the similar templates between the training and test datasets, LogAttention (FT) yields near-perfect Precision, a very high Recall, and high Macro F_1. With the LogAttention (FT) model, one may suspect overfitting of data. However, log data is inherently repetitive and has repetitive patterns, and LogAttention (FT) seeks to benefit from these repetitive patterns to give accurate insights to the EM after analyzing a very small proportion of the dataset.

The Macro F_1 score and Matthews Correlation Coefficient (MCC) of LogAttention are superior to DeepLog, LogAnomaly, and RobustLog. It is apparent that with training, the performance of LogAttention significantly improves and becomes near perfect whereas even in new environments it has a good headstart over other techniques.

4 Conclusion

LogAttention uses Language Models to generate more representative embeddings for the log messages. Instead of working with just the template of the log message, it proposes a mechanism to ingest the complete message and let the model learn from the embedded reply in the message as well. This is a novel approach and has not been seen in literature before. The approaches used in the literature DeepLog and LogAnomaly rely on efficient log parsing because they expect a log message divided into event templates and log parameters. State-of-the-art log parsers are not as efficient in parsing logs from datasets in the wild. This also affects the efficiency of Anomaly detection. The LogAttention approach circumvents this issue by using the whole content of the log message and provides superior results when compared to other approaches in LAD.

LogAttention is pioneering work in using embeddings from Attention-based Transformer Language Models for LAD to analyze run-time logs from a software release in production to assess the software release quality by accurately identifying anomalous logs to predict and preempt failure. To give the models a head start, LogAttention also uses heuristic-based weak classifiers enabling the model to achieve a certain level of even without the availability of labeled data. The model also uses heuristics such as the log levels already built in the messages, the execution time of the log event, and an evolving error lexicon to detect any unusual text in the log message itself. These classifiers, although not as efficient individually, allows the system a better start instead of relying solely on un-trained classifiers.

The paper also introduces a novel data curation system for HIL labeling. Learning from the rapidly evolving field of active learning, the paper introduces MCR Sampling (marginal sampling, certainty sampling, and random sampling) to curate the most pertinent tagged data for the model. It curates the most relevant 20% from a log dataset consisting of one to three days of log data by selecting the equal parts, data about which the model is unsure (marginal sampling), data regarding which the model is certain (certainty sampling), and random data to include random samples in the data as well. The results demonstrate this technique to be superior to just randomly sampling data.

References

1. Di Francesco, P., Malavolta, I., Lago, P.: Research on architecting microservices: trends, focus, and potential for industrial adoption. In: ICSA 2017 (2017)
2. Mariani, L., et al.: Localizing faults in cloud systems. In: IEEE ICST 2018 (2018)
3. Wu, L., Tordsson, J., Elmroth, E., Kao, O.: MicroRCA: root cause localization of performance issues in microservices. In: IEEE/IFIP NOMS 2020 (2020)
4. He, P., Zhu, J., He, S., Li, J., Lyu, M.R.: An evaluation study on log parsing and its use in log mining. In: IEEE/IFIP DSN 2016 (2016)
5. Nedelkoski, S., Bogatinovski, J., Acker, A., Cardoso, J., Kao, O.: Self-supervised log parsing. In: Dong, Y., Mladenić, D., Saunders, C. (eds.) ECML PKDD 2020. LNCS (LNAI), vol. 12460, pp. 122–138. Springer, Cham (2021). https://doi.org/10.1007/978-3-030-67667-4_8
6. Cheng, J., Dong, L., Lapata, M.: Long short-term memory-networks for machine-reading (2017)
7. Vaswani, A., et al.: Attention is all you need (2017)
8. Klosowski, P.: Deep learning for NLP and language modeling. In: SPA 2018 (2018)
9. Clark, K., Luong, M., Le, Q.V., Manning, C.D.: ELECTRA: Pre-training text encoders as discriminators rather than generators (2020)
10. Sanh, V., Debut, L., Chaumond, J., Wolf, T.: DistilBERT, a distilled version of BERT: Smaller, faster, cheaper, and lighter. In: Co-Located with NeurIPS 2019, 5th edn. (2019)
11. Pennington, J., Socher, R., Manning, C.: GloVe: global vectors for word representation. Proceedings of the 2014 Conference on Empirical Methods in NLP (EMNLP) (2014)
12. Vellido, A., Gibert, K., Angulo, C., Martín Guerrero, J.D. (eds.): WSOM 2019. AISC, vol. 976. Springer, Cham (2020). https://doi.org/10.1007/978-3-030-19642-4
13. Liu, F.T., Ting, K.M., Zhou, Z.: Isolation forest. In: Paper presented at the Eighth IEEE International Conference on Data Mining, pp. 413–422 (2008)
14. Kazerouni, A., et al.: Active Learning for Skewed Data Sets (2020)
15. Liu, Y., et al.: RoBERTa: A robustly optimized BERT pretraining approach (2019)
16. Du, M., Li, F., Zheng, G., Srikumar, V.: DeepLog: anomaly detection and diagnosis from System Logs through deep learning. In: CCS 2017 (2017)
17. Meng, W., et al.: LogAnomaly: Unsupervised detection of sequential and quantitative anomalies in unstructured logs (2019)
18. Zhang, X., et al.: Robust log-based anomaly detection on unstable log data (2019)

AIOPS

A Taxonomy of Anomalies in Log Data

Thorsten Wittkopp$^{(\boxtimes)}$, Philipp Wiesner$^{(\boxtimes)}$, Dominik Scheinert$^{(\boxtimes)}$, and Odej Kao$^{(\boxtimes)}$

Technische Universität Berlin, DOS, TU-Berlin, Berlin, Germany
{t.wittkopp,wiesner,dominik.scheinert,o.kao}@tu-berlin.de

Abstract. Log data anomaly detection is a core component in the area of artificial intelligence for IT operations. However, the large amount of existing methods makes it hard to choose the right approach for a specific system. A better understanding of different kinds of anomalies, and which algorithms are suitable for detecting them, would support researchers and IT operators. Although a common taxonomy for anomalies already exists, it has not yet been applied specifically to log data, pointing out the characteristics and peculiarities in this domain.

In this paper, we present a taxonomy for different kinds of log data anomalies and introduce a method for analyzing such anomalies in labeled datasets. We applied our taxonomy to the three common benchmark datasets Thunderbird, Spirit, and BGL, and trained five state-of-the-art unsupervised anomaly detection algorithms to evaluate their performance in detecting different kinds of anomalies. Our results show, that the most common anomaly type is also the easiest to predict. Moreover, deep learning-based approaches outperform data mining-based approaches in all anomaly types, but especially when it comes to detecting contextual anomalies.

Keywords: AIOps · Log analysis · Log anomaly taxonomy

1 Introduction

The operation and maintenance of data centers and corresponding IT infrastructure are becoming increasingly difficult, due to the continuous growth of cloud computing. To cope with this complexity, systems are using more and more levels of abstraction, leading to the creation of large multilayered systems. However, from an IT operator perspective, these layers can even aggravate the problem by adding further technical complexity under the hood. At the same time, unpredictable events such as downtimes can cause severe financial damage, especially in case of service level agreement (SLA) violations [19]. The area of artificial intelligence for IT operations (AIOps) tries to manage this newly introduced complexity, by supporting cloud operators to ensure operational efficiency as well as dependability and stability [6]. A core component of AIOps systems is the detection of anomalies in monitoring data such as metrics, traces, or log data. Especially logs are an important resource for troubleshooting, as

© Springer Nature Switzerland AG 2022
H. Hacid et al. (Eds.): ICSOC 2021 Workshops, LNCS 13236, pp. 153–164, 2022.
https://doi.org/10.1007/978-3-031-14135-5_12

they record events during the execution of IT applications [9]. For this reason, a large number of methods have been proposed in the field of log data anomaly detection, mostly building on data mining [2] or deep learning techniques [18]. While supervised methods mostly perform better in anomaly detection [26], they have the drawback that the anomalies must be known at training time, which is not always the case. Furthermore, is costly and time consuming to create labeled log data, and thus unsupervised methods are of high relevance.

The wide variety of approaches to anomaly detection present IT operators with the challenge of choosing the right methods for their systems. Although there exist some commonly used datasets for evaluating approaches such as HDFS, BGL, Thunderbird, and Spirit [17,24], the characteristics and properties that distinguish these datasets are often not sufficiently clarified. Furthermore, there is no common schema on how to utilize different datasets in performance evaluations. Hence, different anomaly detection methods perform diverse evaluations, e.g. based on time windows or individual log lines [5,7,12,25]. The evaluations of different research papers are therefore not always comparable. It remains hard to estimate the performance of methods on new, unknown datasets based on their performance on benchmark datasets, without having more insights of the anomaly types. We want to address this lack of understanding by making the following contributions:

- We propose a taxonomy for different kinds of log data anomalies based on a well established general categorization for anomalies [4].
- Using this taxonomy, we introduce a method to classify anomalies in labeled datasets and analyze the benchmark datasets BGL, Thunderbird, and Spirit.
- We evaluate the performance of the widely used unsupervised anomaly detection methods DeepLog, A2Log, PCA, Invariants Miner, and Isolation Forrest in detecting the different types of anomalies.

The remainder of this work is structured as follows: Sect. 2 explains the common distinction of point and contextual anomalies and provides examples in the context of log data. Section 3 surveys the related work. Section 4 introduces our taxonomy and presents a method for classifying anomalies. Section 5 analyses three common benchmark datasets by applying our method. Furthermore, we evaluate the performance of five unsupervised anomaly detection approaches on the different types of anomalies. Section 6 concludes the paper.

2   Towards an Anomaly Taxonomy for Log Data

Labeled anomaly detection datasets contain normal samples $\mathcal{N}$ and anomalous samples $\mathcal{A}$. Each sample is described through its feature-set, which varies depending on the domain of the underlying data. For example, for time series data, a sample is usually described by its position in a multidimensional space and a temporal component, while in other domains, such as natural language processing, feature sets can consist of word embeddings. A common anomaly taxonomy that is described in several works [1,4,20] categorizes anomalies into *Point Anomalies* and *Contextual Anomalies*.

Point Anomalies. A point anomaly is a single data sample that can be considered anomalous compared to the rest of the data [4]. The values in its feature-set therefore significantly differ from the values in the feature-sets of normal samples $\mathcal{N}$. Figure 1 illustrates two examples of point anomalies. The first example shows two anomalies that are not located in the defined area for normal samples. Their feature-set is the position in 2-dimensional space. The second example depicts point anomalies in a time series. As the normal feature-set is defined by $y \in [1, 2]$, the anomalies are characterized by their feature-set $y \notin [1, 2]$.

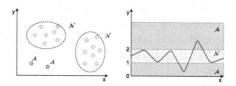

Fig. 1. Two examples for point anomalies. On the left side: point anomalies in 2D space. On the right side: Point anomalies in a time series.

Figure 2 depicts two point anomalies in written text. The first anomaly is trivially described through the feature-set of words. The anomalous sentence *Node failed to initialize* has no overlapping with the feature-sets of the remaining sentences. The second example is more fine-grained since only some words indicate an anomalous sample: All sentences share the same or a similar prefix, only the subsequent description (*ready, connected, 5 nodes, an error*) resolves the question of anomalous behavior.

Kernel started	System is ready
System is ready	System is connected
System started	System has an error
Node fail to initialize	System has 5 nodes

Fig. 2. Two examples for point anomalies in written text.

Contextual Anomalies. Samples that are anomalous in a specific context only are called contextual anomalies [4], and are also known as conditional anomalies [21]. Samples that belong to this type of anomalies can have the same feature-set (behavioral properties) as normal samples, but are still anomalous within a specific context defined by their contextual properties.

Figure 3 illustrates this situation: The behavioral properties of the anomalous points (y-values/sentences) themselves are not indicating an anomaly. However, the context of the anomalous samples defined by their contextual properties (x-values/sentence order) is different, as the normally observable strict pattern is interrupted. Furthermore, Fig. 3 illustrates two variants of a contextual anomaly in written text. In the left example, the *Send mail* statement is only an anomaly because of its context, namely because *Start mail service* and *End mail service*

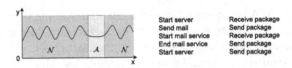

Fig. 3. One example of a contextual anomaly in time series and two examples for contextual anomalies in written text.

appear after the *Send mail* statement. Since the mail service must be started before sending any mails, this ensemble of statements exemplifies a contextual anomaly. The second example is similar to the time series example. The statements *Receive package* and *Send package* alternate constantly. The anomaly is described by the fact that this alternating pattern is interrupted by a second *Send package* statement.

3   Related Work

We next discuss related works with regards to defining and categorizing anomaly types, and subsequently debate concrete text-based anomaly detection methods.

Categorization of Anomaly Types. An important aspect of any anomaly detection technique is the prospective nature of target anomalies. In [4], the authors differentiate between *point anomalies*, where an individual data sample is anomalous in comparison to the rest, *contextual anomalies*, where a data sample is only considered anomalous in specific contexts, and *collective anomalies*, where a single data sample is only anomalous when occurring as part of a collection of related data samples, not individually. This classification is also reused in [3]. Similarly, the authors of [20] identify the classes *one-point anomaly*, *contextual anomaly*, and *sequential data anomaly*, and define them in the same way. A work on outlier/anomaly detection in time series data distinguishes between *point outliers* and *subsequence outliers* [1], which are defined as previously sketched. In addition, they introduce the notion of *outlier time series*, where entire time series can be anomalous and are only detectable in the case of multivariate time series. The so far highlighted types of anomalies are hence the basis for our taxonomy of anomalies in logs. While out of our scope, log messages are further distinguishable into event log messages and state log messages [15] and can also be written in a distributed manner, which introduces additional challenges.

Text-Based Anomaly Detection Methods. In order to exemplify the diverse log anomaly types as well as illustrate the shortcomings of commonly employed methods, we make use of multiple data mining and deep learning techniques in our evaluation. The PCA algorithm [11] is often employed for dimensionality reduction right before the actual detection procedure [10]. Invariant Miners [14] retrieve structured logs using log parsing, further group log messages according to log parameter relationships, and eventually mine program invariants from the established groups in an automated fashion which are then used to perform anomaly detection on logs. The fact that anomalies are usually few and considerably different is exploited with Isolation Forests [13], an ensemble of isolation trees, where anomalies are isolated closer to the root of a tree and thus identified. DeepLog [5] utilizes an LSTM and thus interprets a log as a sequence of sentences. It uses templates [8], performs anomaly detection per log message, and constructs system execution workflow models for diagnosis purposes. A2Log [22] utilizes a self-attention neural network to obtain anomaly scores for log messages and then performs anomaly detection via a decision boundary that was set based on data augmentation of available normal training data.

4   Classifying Anomalies in Log Data

The following chapter introduces our taxonomy for anomalies in log data. Furthermore, we present a method for classifying the anomalies in labeled datasets according to this taxonomy. With our classification method, system administrators are enabled to investigate their datasets and use the obtained insights to choose an appropriate anomaly detection algorithm.

4.1   Preliminaries

Logging is commonly employed to record the system executions by log instructions. Each instruction results in a single log message, such that the complete log is a sequence of messages $L = (l_i : i = 1, 2, \ldots)$. There is a commonly used separation in *meta-information* and *content*. The meta-information can contain various information, for example, timestamps or severity levels. The content is free text that describes the current execution and consists of a static and a variable part.

Tokenization. This splits text into its segments (e.g., words, word stems, or characters). The smallest indecomposable unit within a log content is a token. Consequently, each log content can be interpreted as a sequence of tokens: $s_i = (w_j : w_j \in V, j = 1, 2, 3, \ldots)$, where w is a token, V is a set of all known tokens commonly referred to as the *vocabulary*, and j is the positional index of a token within the token sequence s_i.

Templates. The tokenized log messages can be further processed into *log templates*, a very common technique employed in various log anomaly detection methods [5,16]. Thereby, the tokens corresponding to the static part of a log message are forming the log template t_i for the i-th log message. Each unique log template is then identifiable via a log template id x and referred to as t^x. All remaining tokens form the set of attributes a_i for the respective log message l_i. For example the log messages: `Start mail service at node wally001` and `Start printer service at node wally005` can be described trough the template `'Start * service at node *'` with attribute sets `[mail, wally001]` and `[printer, wally005]` respectively. Thus, each log message can be described through a log template and the attributes.

4.2   Anomaly Taxonomy for Log Data

In this chapter, we present our taxonomy for log data anomalies. This taxonomy relies on the categorization into *Point Anomalies* and *Contextual Anomalies*. Furthermore, we distinguish the *Point Anomalies* between two types of point anomalies, namely *Template Anomalies* and *Attribute Anomalies*.

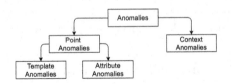

Fig. 4. Taxonomy for anomalies in log data.

Figure 4 depicts our taxonomy. In the context of log data, a *Point Anomaly* is an anomalous log message that is described through the log message itself. That is, the log message could be classified as anomalous by only investigating the respective log message and without observing its context. The anomalous behavior of a log message is therefore described either by the corresponding template or by a specific word or number (an attribute) in the log message. We hence define a *Template Anomaly* to be characterized by the template of the respective log message. In contrast, an *Attribute Anomaly* is described through the attributes that are extracted during the template generation process.

The second type of log data anomalies are *Contextual Anomalies*. In this case, the context, in other words the surrounding log messages, determines anomalous behavior. The content of an individual log message is hence only relevant with respect to the log messages in its surrounding. In our preliminary taxonomy, we consider only single-threaded event logging scenarios for contextual anomalies. So far, distributed logging as well as state log messages [15], along with the corresponding challenges, are not yet covered.

4.3   Anomaly Classification Method

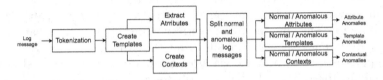

Fig. 5. Mining process of the different anomaly types.

The process of classifying types of anomalies in respect to our taxonomy is illustrated in Fig. 5. Each log message is first split into sequences of tokens in order to mine the log template. After all log templates are generated, we extract the attributes for each log message and calculate the context for each log line. The context c_i for each log message l_i relies on log template ids and is modelled as a set of template ids

$$c_i = \{t_j^x : j = i - a, \ldots, i - 1, i + 1, \ldots, i + b]\}, \tag{1}$$

where a and b are the boundaries of the context. For example, we calcualte the context of the 10th log message with boundaries $a = 2$ and $b = 1$ as $c_{10} = \{l_8, l_9, l_{11}\}$. The template of the log message whose context is created is not considered, as described in Eq. 1. After deriving templates, attributes and contexts, we divide the dataset into a set of normal $\mathcal{N}$ and anomalous log messages $\mathcal{A}$ based on labels determined by experts or automated processes [23]. Next, by utilizing these two sets and the previously calculated entities, we derive a score for each anomaly log message and each anomaly type. The scores represent how strong the respective anomaly type is pronounced. Each score is in $[0, 1]$, with 1 referring to the strongest manifestation.

Template Anomalies. The template anomaly α is calculated for each template id x. To get all templates for a specific template id x, we write $t^x(\cdot)$.

$$\alpha(t^x) = \frac{|t^x(\mathcal{A})|}{|t^x(\mathcal{A})| + |t^x(\mathcal{N})|} \tag{2}$$

Attribute Anomalies. The attribute anomaly β is calculated for each log message l_i. Since each log message can have multiple attributes, a score for each attribute is calculated and the attribute anomaly is then represented by the maximal score. Here, $a_j(\cdot)$ gets all the same tokens as a_j from the corresponding normal or anomalous set.

$$\beta(a_i) = \max\left(s : \forall a_j \in a_i . s = \frac{|a_j(\mathcal{A})|}{|a_j(\mathcal{A})| + |a_j(\mathcal{N})|}\right) \tag{3}$$

Contextual Anomalies. The contextual anomaly γ is calculated for each log message l_i. To get all the same contexts, for a specific context c_i, for each log message, we write $c_i(\cdot)$.

$$\gamma(c_i) = \frac{|c_i(\mathcal{A})|}{|c_i(\mathcal{A})| + |c_i(\mathcal{N})|} \tag{4}$$

Thus all scores can be calculated by dividing the occurrences of an event in the anomalous set by the occurrences of this event across both sets. As a result, α, β, and γ do not make an exact assignment to anomaly types but create a score that indicates how strongly it behaves to a particular anomaly type. Hence, a log line can also have several anomaly types.

Table 1. Dataset statistics. Templates were generated using Drain3 [8].

Dataset	Log messages		Templates		
	Normal	Anomalous	Normal	Anomalous	Intersection
Thunderbird	4 773 713	226 287	969	17	3
Spirit	4 235 109	764 891	1121	23	5
BGL	4 399 503	348 460	802	58	10

5   Evaluation

To provide an understanding on the distribution of different types of anomalies in common benchmarks according to our taxonomy, we apply our method to the Thunderbird, Spirit, and BGL datasets. We furthermore trained five state-of-the-art unsupervised log anomaly detection methods on these datasets to evaluate their performance on predicting different types of anomalies.

The evaluation datasets were recorded at different large-scale computer systems, labeled manually by experts, and presented in [17]. Table 1 contains the number of normal and anomalous log messages in each dataset, the amount of templates in these classes, and the number of intersecting templates.

- The *Thunderbird* dataset is collected from a supercomputer at Sandia National Labs (SNL) and contains more than 211 million log messages.
- The *Spirit* dataset is collected from a Spirit supercomputer at SNL and contains more than 272 million log messages.
- The *BGL* dataset is collected from a BlueGene/L supercomputer at Lawrence Livermore National Labs (LLNL) and contains 4 747 963 log messages.

From *Thunderbird* and *Spirit* we selected the first 5 million messages.

5.1   Analysis of Benchmark Datasets

We applied our approach for classifying types of anomalies to the datasets using threshold values of 0.6, 0.7, 0.8, 0.9, and 1.0. To create the contexts we choose the following boundaries: $a = 10$ and $b = 0$. Figure 6 displays the results. We can observe that, even at threshold 1.0, more than 99% of all anomalies in the Thunderbird and Spirit datasets are being classified as template anomalies. This can be explained by the fact that the intersection of normal and abnormal templates is very small. Additionally, 226 071 of all anomalies in Thunderbird have the same template - that is 99.9%. Similarly, 99.4% of all log anomalies in Spirit belong to one of the templates shown in Listing 1. The case is very similar for BGL, although log templates are more heterogeneous in this dataset. Only at threshold 1 the amount of log messages classified as template anomalies drops to around 80%.

Listing 1: Most anomalies in Spirit belong to these two templates (380 271 each).

```
kernel: hda: drive not ready for command
kernel: hda: status error: status=<:HEX:> { }
```

Until threshold 0.7 almost 100% of all anomalies in Thunderbird are classifieds as attribute anomalies, meaning all template anomalies are also attribute anomalies in this case. As the "anomalous" attributes are also contained in some normal log messages, the amount of attribute anomalies drops to zero for higher thresholds. For Spirit, we can observe that only one of the two most important log templates shown in Listing 1 contains an attribute, which explains why the number of attribute anomalies is around 50%. At higher thresholds the number of attribute anomalies drops to zero. For the BGL dataset no attribute anomalies were identified, not even at low thresholds. However, more than 91% of all anomalies in BGL are classified as contextual anomalies for thresholds between 0.6 and 0.9. This is significantly more than in the Thunderbird ($13-2\%$) and Spirit ($63-14\%$) datasets.

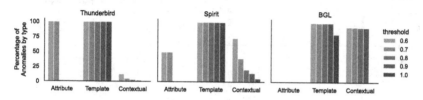

Fig. 6. Percentage of anomalies by type at different thresholds. Anomalies can be classified into multiple categories at the same time.

Concluding, for all three benchmark datasets algorithms that focus on detecting template anomalies are expected to perform very well. Additionally identifying attribute anomalies may be helpful, but since most attribute anomalies are also template anomalies, the expected benefit is limited. Approaches that identify anomalies by observing the context seem promising on datasets like BGL, but are not expected to perform well on Thunderbird and only to a certain degree on Spirit.

Anomalies Outside the Taxonomy. Our method does not guarantee that a given anomaly can be attributed to at least one of the classes in the proposed taxonomy - especially at high thresholds. The severity of this problem was evident to varying degrees in the datasets. For Thunderbird, only 2 of 226 287 messages could not be classified. For Spirit, it was 28 out of 764 891 for thresholds up to 0.9. For a threshold of 1.0, we could not classify 1113 log messages, which is still only 0.15%. For the BGL dataset, for thresholds of 0.6, 0.7, 0.8, and 0.9, we could not classify 524, 831, 2646, and 2646 of 348460 messages, respectively. However, 68 372 protocol messages, 19.6% of all anomalous protocol messages, remained unlabeled at a threshold of 1.0. Future work should either improve our classification method or describe additional types of anomalies that our proposed taxonomy does not yet cover.

5.2 Evaluation of Unsupervised Learning Methods

We trained five unsupervised anomaly detection algorithms to predict the different kinds of anomalies at a threshold of 0.7 in all three datasets. The goal is to identify which kinds of anomalies are easy or hard to predict and also which methods perform well on which anomalies. In particular, we chose two deep learning approaches Deeplog [5] and A2Log [22], and three data mining approaches PCA [10], Invarant Miners [14], and Isolation Forest [13]. We evaluated all methods on four different train/test splits of 0.2/0.8, 0.4/0.6, 0.6/0.4, and 0.2/0.8 to test the robustness of the different methods.

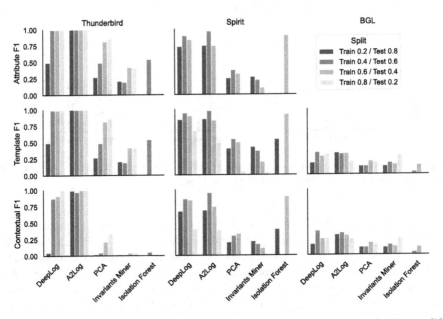

Fig. 7. F1 scores for predicting attribute, template, and contextual anomalies at different train/test splits at threshold 0.7. BGL contains no attribute anomalies.

All results are depicted in Fig. 7. We can observe that the two deep learning approaches outperform the data mining approaches across all experiments. Isolation Forrest seems to be extremely sensitive to certain "sweet spots" in train/test splits, but does not prove to be a robust method and is generally the worst-performing method. It is, hence, excluded from any further analysis in the following paragraphs. For the Thunderbird dataset, DeepLog and A2Log manage to correctly classify almost all kinds of anomalies. A2Log performs generally better, even when only 20% of the data is available as training data. This might be attributed to the fact that DeepLog bases its predictions on templates only, while A2Log also takes attribute information into account. The non-deep learning methods achieve worse F1 scores on attribute and template anomalies and fail to predict most contextual anomalies. On the Spirit dataset, the performance of all methods is generally worse. However, the deep learning methods still achieve F1 scores of around 0.75, 0.85, and 0.7 for attribute, template, and contextual anomalies, respectively. On BGL, all methods obtain low F1 scores for template and contextual anomalies: DeepLog around 0.27, A2Log around 0.3, PCA around 0.14, and Invariants Miner around 0.15. BGL does not contain any attribute anomalies.

It can be concluded, that for unsupervised methods template anomalies are the easiest to predict. It can be suspected that attribute anomalies that are *no* template anomalies are amongst the hardest to predict. However, this is hard to show, as attribute anomalies and aemplate anomalies are highly correlated in all datasets. Contextual anomalies were only predicted reliably by the two deep learning methods, but are generally harder to detect than template anomalies.

6   Conclusion

In this paper, we present a taxonomy for different kinds of log data anomalies and introduce a method for applying this taxonomy on labelled datasets. Using this method, we analyze the three common benchmark datasets Thunderbird, Spirit, and BGL. While the vast majority of anomalies are template anomalies, BGL also contains a large number of contextual anomalies. Attribute anomalies are highly correlated with template anomalies in all datasets. We furthermore evaluated the ability to detect different kinds of anomalies of five state-of-the-art unsupervised anomaly detection methods: DeepLog, A2Log, PCA, Invariants Miner, and Isolation Forrest. Our results show, that template anomalies are the easiest to predict, which explains the good performance of approaches like DeepLog. In general, deep learning-based approaches outperform data mining-based approaches, especially when it comes to detecting contextual anomalies.

We hope that our taxonomy will enable researchers and IT Operators to better understand their datasets and help them to pick suitable anomaly detection algorithms. Future work should investigate the log messages our approach fails to classify, potentially hinting towards further classes that are currently not present in the taxonomy.

References

1. Blázquez-García, A., Conde, A., Mori, U., Lozano, J.A.: A review on outlier/anomaly detection in time series data. arXiv preprint arXiv:2002.04236 (2020)
2. Breier, J., Branišová, J.: Anomaly detection from log files using data mining techniques. In: Information Science and Applications. Springer, Singapore (2015). https://doi.org/10.1007/978-981-33-6385-4
3. Chalapathy, R., Chawla, S.: Deep learning for anomaly detection: a survey. arXiv preprint arXiv:1901.03407 (2019

4. Chandola, V., Banerjee, A., Kumar, V.: Anomaly detection: a survey. ACM Comput. Surv. **41**(3) (2009)
5. Du, M., Li, F., Zheng, G., Srikumar, V.: DeepLog: anomaly detection and diagnosis from system logs through deep learning. In: SIGSAC (2017)
6. Gulenko, A., Acker, A., Kao, O., Liu, F.: AI-governance and levels of automation for AIOps-supported system administration. In: ICCCN. IEEE (2020)
7. Guo, H., Yuan, S., Wu, X.: LogBERT: Log anomaly detection via BERT. In: International Joint Conference on Neural Networks, IJCNN. IEEE (2021)
8. He, P., Zhu, J., Zheng, Z., Lyu, M.R.: Drain: an online log parsing approach with fixed depth tree. In: ICWS. IEEE (2017)
9. He, S., He, P., Chen, Z., Yang, T., Su, Y., Lyu, M.R.: A survey on automated log analysis for reliability engineering. ACM Comput. Surv. **54**(6) (2021)
10. He, S., Zhu, J., He, P., Lyu, M.R.: Experience report: system log analysis for anomaly detection. In: ISSRE. IEEE (2016)
11. Jolliffe I.: Principal component analysis. In: Lovric, M. (ed.) International Encyclopedia of Statistical Science. Springer, Heidelberg. https://doi.org/10.1007/978-3-642-04898-2_455
12. Li, X., Chen, P., Jing, L., He, Z., Yu, G.: SwissLog: robust and unified deep learning based log anomaly detection for diverse faults. In: ISSRE. IEEE (2020)
13. Liu, F.T., Ting, K.M., Zhou, Z.H.: Isolation forest. In: 2008 eighth ieee international conference on data mining. IEEE (2008)
14. Lou, J.G., Fu, Q., Yang, S., Xu, Y., Li, J.: Mining invariants from console logs for system problem detection. In: USENIX Annual Technical Conference (2010)
15. Nagaraj, K., Killian, C.E., Neville, J.: Structured comparative analysis of systems logs to diagnose performance problems. In: NSDI. USENIX Association (2012)
16. Nedelkoski, S., Bogatinovski, J., Acker, A., Cardoso, J., Kao, O.: Self-supervised Log Parsing. In: Dong, Y., Mladenić, D., Saunders, C. (eds.) ECML PKDD 2020. LNCS (LNAI), vol. 12460, pp. 122–138. Springer, Cham (2021). https://doi.org/10.1007/978-3-030-67667-4_8
17. Oliner, A., Stearley, J.: What supercomputers say: A study of five system logs. In: DSN (2007)
18. Pang, G., Shen, C., Cao, L., Hengel, A.V.D.: Deep learning for anomaly detection: a review. ACM Comput. Surv. **54**(2) (2021)
19. Santos, G.L., et al.: Analyzing the it subsystem failure impact on availability of cloud services. In: ISCC. IEEE (2017)
20. Sebestyen, G., Hangan, A., Czako, Z., Kovacs, G.: A taxonomy and platform for anomaly detection. In: AQTR. IEEE (2018)
21. Song, X., Wu, M., Jermaine, C.M., Ranka, S.: Conditional anomaly detection. IEEE Trans. Knowl. Data Eng. **19**(5) (2007)
22. Wittkopp, T., et al.: A2Log: attentive augmented log anomaly detection. In: HICSS (2022)
23. Wittkopp, T., Wiesner, P., Scheinert, D., Acker, A.: LogLAB: attention-based labeling of log data anomalies via weak supervision. In: Hacid, H., Kao, O., Mecella, M., Moha, N., Paik, H. (eds.) ICSOC 2021. LNCS, vol. 13121, pp. 700–707. Springer, Cham (2021). https://doi.org/10.1007/978-3-030-91431-8_46
24. Xu, W., Huang, L., Fox, A., Patterson, D., Jordan, M.I.: Detecting large-scale system problems by mining console logs. In: SIGOPS. ACM (2009)
25. Yang, L., et al.: Semi-supervised log-based anomaly detection via probabilistic label estimation. In: ICSE. IEEE (2021)
26. Zhang, X., et al.: Robust log-based anomaly detection on unstable log data. In: ESEC/FSE (2019)

Little Help Makes a Big Difference: Leveraging Active Learning to Improve Unsupervised Time Series Anomaly Detection

Hamza Bodor[1,2], Thai V. Hoang[1(✉)], and Zonghua Zhang[1]

[1] Paris Research Center, Huawei Technologies France, Boulogne-Billancourt, France
{thai.v.hoang,zonghua.zhang}@huawei.com
[2] École des Ponts ParisTech, Marne-la-Valle, France
hamza.bodor@ponts.org

Abstract. Key Performance Indicators (KPI), which are essentially time series data, have been widely used to indicate the performance of telecom networks. Based on the given KPIs, a large set of anomaly detection algorithms have been deployed for detecting the unexpected network incidents. Generally, unsupervised anomaly detection algorithms gain more popularity than the supervised ones, due to the fact that labeling KPIs is extremely time- and resource-consuming, and error-prone. However, those unsupervised anomaly detection algorithms often suffer from excessive false alarms, especially in the presence of concept drifts resulting from network re-configurations or maintenance. To tackle this challenge and improve the overall performance of unsupervised anomaly detection algorithms, we propose to use active learning to introduce and benefit from the feedback of operators, who can verify the alarms (both false and true ones) and label the corresponding KPIs with reasonable effort. Specifically, we develop three query strategies to select the most informative and representative samples to label. We also develop an efficient method to update the weights of Isolation Forest and optimally adjust the decision threshold, so as to eventually improve the performance of detection model. The experiments with one public dataset and one proprietary dataset demonstrate that our active learning empowered anomaly detection pipeline could achieve performance gain, in terms of F1-score, more than 50% over the baseline algorithm. It also outperforms the existing active learning based methods by approximately 6%–10%, with significantly reduced budget (the ratio of samples to be labeled).

Keywords: Active learning · Anomaly detection · Time series data

1 Introduction

Anomaly detection has always been one of the grand challenges, yet an essential capability, in building resilient computer and communication networks. Being able to detect anomalies will not only guarantee a timely warning of potential

© Springer Nature Switzerland AG 2022
H. Hacid et al. (Eds.): ICSOC 2021 Workshops, LNCS 13236, pp. 165–176, 2022.
https://doi.org/10.1007/978-3-031-14135-5_13

failures in the systems, but also ensure a quick remediation and error correction, which may save a lot of unnecessary expenses. As a matter of fact, Key Performances Indicators (KPIs), which are essentially time series data collected over time, have been widely used to assess the health status of networks and services. Any network failures or unexpected incidents can lead to the significant deviation of KPIs from their normal patterns. Therefore, KPI based anomaly detection algorithms aim at detecting those deviations with respect to the time series characteristics (e.g., seasonality, trend) or statistical features (e.g., min, max, mean). For example, the commonly seen anomalies include, but not limited to, spike, dip, continuous bursts, sudden or gradual trend change.

To date, many anomaly detection algorithms have been proposed, including both supervised and unsupervised ones. It is commonly recognized that the supervised anomaly detection algorithms perform better than the unsupervised ones if the labels are sufficiently provided. However, this assumption does not always hold true considering the fact that labeling tons of KPIs is extremely time- and effort-consuming and error-prone. Unsupervised ones are therefore preferred over the supervised ones in practice. One question naturally arising here is that, can we balance the trade-off between labeling effort and detection performance? In other words, human operators only pay a reasonable amount of effort to label the KPIs of interest (e.g., the ones lead to false positives), and then guide the algorithm towards a better detection behavior. This is particularly interesting considering the fact that unsupervised anomaly detection algorithms often suffer from excessive false alarms, especially in the presence of concept drifts potentially resulting from network routine updates or legitimate changes.

In fact, the aforementioned question has found some answers in the community [4,18], which share the relevant theoretical foundation with active learning. With the same question in mind, in this paper, we intend to present a new active learning based solution to improve the performance of an unsupervised anomaly detection algorithm.

Specifically, our contributions are four-fold: (1) we develop three query strategies to obtain the most informative and representative positive samples for labeling; (2) we propose a light-weight model update strategy to efficiently derive more accurate anomaly scores and optimal decision threshold, solving the parameterization issue of unsupervised learning methods, and eventually contributing to the improved detection performance; (3) the proposed methods are systematically integrated into an unsupervised anomaly detection algorithm (Isolation Forest), clearly illustrating a feasible approach to introducing human operator's feedback into the closed-loop pipeline for improving its adaptability and performance; (4) a set of experiments with two different datasets are carried out for comparative studies with the state-of-the-art approaches, demonstrating the strong generalization capability of detecting various anomalies in different time series dataset.

The remainder of this paper is organized as follows. We firstly review the related work in Sect. 2 and introduce the unsupervised solution to time series anomaly detection problem in Sect. 3. Section 4 describes in detail our active learning solution with experimental results in Sect. 5. Section 6 finally concludes the paper.

2   Related Work

Supervised Learning. Statistical models such as ARIMA [21] have been traditionally used to model the normal behaviors of time series by training on "clean" data. The deviation from the model forecast is then used as a measure of abnormality. However, directly thresholding this deviation is usually insufficient in real-world applications. Liu et al. [9] used statistical models for feature extraction and then a classifier such as Random Forest (RF) [2] to detect the anomalies. More recently, deep learning algorithms, such as LSTMs [11,12], have been introduced to work as feature extractors in supervised anomaly detection.

Unsupervised Learning. There has been a growing interest in unsupervised methods for time series anomaly detection in order to overcome the lack of labeled data in real world scenarios. For example, Luminol [8], developed by LinkedIn, segments time-series into chunks and uses the frequency of similar chunks to calculate the anomaly scores. SPOT and DSPOT [16] use extreme value theory to model distribution tail in order to detect outliers in time series. Microsoft [13] uses spectral residual (SR) concept from signal processing to develop their SR-based anomaly detector. More recently, deep learning-based methods [12] have been also employed to detect anomalies in unsupervised settings. For example, DONUT [20] uses variational auto-encoder to detect anomalies from seasonal KPIs.

Active Learning. There are two major active learning approaches for anomaly detection. The first approach, such as the one proposed in [5], usually solves a semi-supervised learning problem (SSAD) that uses both labeled and unlabeled points in its underlying formulation. When no labels are available, the models are first trained on unlabeled data in unsupervised settings. They are subsequently updated by incorporating labeled points from feedback into the learning problem. In [17], the authors used variants of SSAD model for benchmarking and found that there is no one-fit-all strategy for one-class active learning. Recently, Amazon developed NCAD based on deep semi-supervised learning [14] for time series anomaly detection [3].

The second approach to active anomaly detection is based on ensemble learning. The base learners are usually tree-based, such as Isolation Forest (iForest) [10], RS-Forest [19], or Robust Random Cut Forest (RRCF) [6]. They are firstly trained on unlabeled data and then be updated using labeled points from the feedback. The update can be in the form of adjusting the weights of trees [18], weights of trees' nodes [4] or trees' edges [15] in order to improve the performance of the base models on these labeled points.

For sample selection, the existing methods explored three strategies. *Top anomalies* implies the selection of points that have the maximum anomaly scores. *Top diverse* strategy, which is similar to the previous one, but requires the maximization of certain "diversity" measure in the group of selected points. *Random* strategy selects points randomly and is usually used as a baseline.

3   Unsupervised Anomaly Detection

Let $\mathbf{X} = (x_1, x_2, .., x_n)$ be a time series, which is a sequence indexed in time with $x_i \in \mathbb{R}^d$. In this work, we focus on univariate time series data where $d = 1$. A time series anomaly detector usually takes $\mathbf{X}$ as input and outputs a sequence $\mathbf{Y} = (y_1, y_2, ..., y_n)$ of the same length, where $y_i = 1$ if x_i is anomalous or $y_i = 0$ otherwise. An anomaly detection pipeline is usually composed of two main modules: feature extraction and anomaly detection model.

Features Extraction. A feature extractor projects the input sequence $\mathbf{X} = (x_1, x_2, .., x_n)$ into a feature space of dimension d_f so that it becomes easier to distinguish anomalous points from normal ones. Each point $x_i \in \mathbf{X}$ is then represented by a vector x_i' of size d_f. In this work, we extract features in online mode and uses sliding windows of size $w = 5$. More specifically, for a given timestamp t, we calculate some measures from the window $[x_{t-w}, ..., x_{t-1}]$ and subtract them from the current value x_t. Among $d_f = 6$ features used in this work and listed in Table 1, five of them are statistical features widely used in time series anomaly detection. The last one is the saliency map calculated from one-day-length subsequence [13].

Table 1. The 6 features used to represent each point in the feature space.

Feature	Description
max	Difference with maximum value on window of size w
min	Difference with minimum value on window of size w
mean	Difference with mean of window of size w
naive	Difference from the previous value
linear_residual	Fit a linear model on a window of size w and compute the residual at the current point
saliency_map	Spectral saliency at the current point

Anomaly Detection Model. Similar to [4], we use iForest as the anomaly detection model because it is an unsupervised model being composed of trees and is much faster than RRCF. The initial iForest model is trained on the pool of unlabeled points. An ensemble of trees also facilitates the model update in active learning.

Ensemble learning methods, especially the tree-based ones such as iForest, are well suited for active learning anomaly detection. This is because anomalies usually exhibit abnormal behaviors and their representations are mostly scattered in the feature space, which is in contrast to normal points whose representations form high density clusters. The separation boundaries between the representation of normal and abnormal points in the feature space are thus *non-homogeneous*, which can be well-represented by a properly trained iForest model. In addition, since iForest is widely used in outlier and anomaly detection, having a mechanism to enhance it using active learning would benefit the whole community.

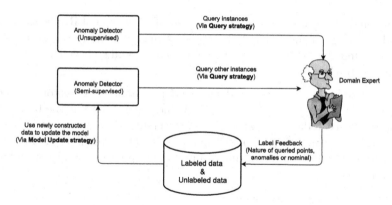

Fig. 1. Design workflow: use active learning to improve unsupervised anomaly detection

4 Active Anomaly Detection

4.1 Design Assumptions

While there are different variants of active learning scenarios in the literature, our focus in this paper is on the *pool-based active learning* [7] only. Specifically, we assume that a (typically large) number of unlabeled data points, referred to as the *unlabeled pool*, is available and accessible to the learning process. An analyst or a domain expert is also available to provide a ground-truth label for any point in this pool upon request. The requests to domain expert are in "batch", consisting of all "interesting" points within the given budget. Figure 1 illustrates the active learning process that has three main components, (1) start with a fully unsupervised model; (2) select points according to the given budget and query strategy; (3) update the model based on domain expert's feedback. This process can be iterative, either upon a request, or is automatically triggered when the model performance gets worse than a certain threshold.

Our active learning pipeline is essentially inspired from [4] and [18]. They both use tree-based ensemble unsupervised models as the anomaly detectors. Wang et al. [18] handles the ensemble at the tree level, and the model update focuses on adjusting the weights of trees in the ensemble. Das et al. [4], on the other hand, works at the nodes of constituting trees and updates the model by adjusting node weights. These two approaches, however, have high complexity.

- Das et al. [4] uses iForest as the base model, which is relatively fast. However, its model updating process is computationally expensive since node-level features and scores for each point need to be computed, either in training or inference. In addition, it solves an underlying optimization problem in order to find the best node weights (**NW**), which can be too expensive for a model with a high number of nodes.
- Wang et al. [18] employs a very fast and straightforward model update strategy which consists of adjusting tree weights according to the scores of anoma-

lous points. It, however, uses a slightly modified version of RRCF as the base model. It's well known that RRCF is slow since it is designed for streaming context and the model auto-adjusts for each incoming stream value in order to adapt to the new data distribution.

In order to build a very fast and easy-to-deploy anomaly detector with active learning, we propose an active learning pipeline using iForest as the base model. iForest model is updated by adjusting the weights of its constituting and by seeking the best value for its "offset" parameter. The remaining of this section will describe in details different query and model update strategies we adopt in this work.

4.2   Query Strategy

There exists a number of query methods in the literature, and they usually follow a common formulation. Given $\mathcal{U}$ the set of unlabeled points, b the given budget, and an interest function $f : \mathcal{U} \to \mathbb{R}$ used as a measure of "informativeness" or "utility" of requesting the label for each point $x \in \mathcal{U}$, a query strategy aims at selecting $x_{i=1,...,b} \in \mathcal{U}$ in order to maximize $\sum_{i=1}^{b} f(x_i)$. The choice of a query strategy thus usually reduces to the choice of an interest function f. In this work, we use three query strategies for our active learning pipeline. Algorithm 1 presents how points are selected according to $\mathcal{U}$, b, and f.

Algorithm 1. select_points($\mathcal{U}$, b, f)

Input: $\mathcal{U}$ (unlabeled dataset), b (budget), f (interest function)
Set $\mathbf{Q} \leftarrow \emptyset$
while $|\mathbf{Q}| < b$ **do**
  Let $\mathbf{x} \leftarrow \underset{x \in \mathcal{U} \backslash \mathbf{Q}}{\operatorname{argmax}} f(x)$
  Set $\mathbf{Q} \leftarrow \mathbf{Q} \cup \{\mathbf{x}\}$
end while
return Q

- Top anomalous selection **(TA)** (Fig. 2(a)): Also called "greedy strategy" in the context of anomaly detection, it selects points that have the highest anomaly scores. Let $\mathcal{U}\mathcal{AD}(x)$ be the anomaly score of x, the interest function corresponding is defined as $f(x) = \mathcal{U}\mathcal{AD}(x)$.
- Close to decision boundary selection **(CTDB)** (Fig. 2(b)): Points that are the closest to the decision boundary of the anomaly detector are selected. The region near the decision boundary is expected to contain the most difficult points to classify. Mathematically, if we denote δ the threshold used to classify points, the interest function is defined as $f(x) = -(\mathcal{U}\mathcal{AD}(x) - \delta)^2$.
- **TA + CTDB** (Fig. 2 (c)): This is a combination of the two above strategies by using half of the budget for **TA** and the other half for **CTDB**. This combination gives more diversity in the selected samples and is expected to better update the unsupervised model.

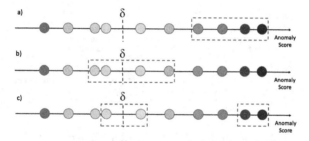

Fig. 2. Illustration of query strategies using a dataset of 10 points sorted according to their anomaly scores (higher scores → right). With a budget $b = 4$, points inside the rectangle are selected according to a) TA, b) CTDB and c) TA + CTDB strategies.

4.3   Model Update Strategy

For model update, we rely on the anomaly scores obtained by the iForest model. After querying a set of points for their labels, the unsupervised iForest model can be updated using one of the following strategies:

- Tree weights update (**TW**): iForest is a tree-based model and, in the original formulation of its scoring function, its trees contribute equally to the calculation of anomaly score for each input point. We adopt the strategy proposed in [18] to adjust the contributions or weights of iForest trees so that if a tree turns out to be more accurate in its anomaly scoring of anomalous queried points, it contributes more to the calculation of anomaly score.
- Offset update (**O**): iForest has a critical hyperparameter called *contamination ratio*. It indicates the proportion of outliers in the dataset and is used to determine the threshold (or offset in iForest terminology) on the scores of the samples. In practice, this ratio is usually guessed based on application context, and it usually turns out to be very difficult to set it properly. We propose to "learn" this offset value based on the feedback for queried points. Finding the best offset from feedback can be done in various ways, such as a simple rule-based thresholding or training a classifier and then using its decision boundary as the learned offset. We have tested and found that the rule-based thresholding has performed better than a linear SVM. Algorithm 2 presents the procedure to calculate offset value from feedback.

Algorithm 2. calculate_offset($\mathcal{L}$, **S**)

Input: $\mathcal{L}$ (dataset of n labeled points) and **S** (anomaly scores of points in $\mathcal{L}$ by iForest model)
Set $\mathcal{L}_a \leftarrow \{x \in \mathcal{L} : x$ is labelled anomalous$\}$
Set $\mathcal{L}_n \leftarrow \{x \in \mathcal{L} : x$ is labelled normal$\}$
offset $= (\min(\{\mathbf{S}(x) : x \in \mathcal{L}_a\}) + \max(\{\mathbf{S}(x) : x \in \mathcal{L}_n\}))/2$
return offset

- (**TW+O**): This is a combination of the two aforementioned strategies by applying **TW** and **O** in sequence.

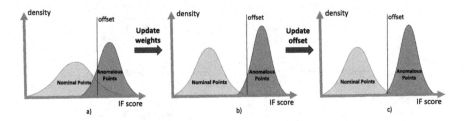

Fig. 3. Illustration of the evolution of score distribution and offset of an iForest model under tree weights and offset update strategies.

Intuition Behind the Two-Step Model Update Strategy: Figure 3 illustrates the impact of the above-stated model update strategies on the score distribution and offset of an iForest model. At the beginning, the score distributions of nominal and anomalous points produced by the unsupervised iForest model have a "large" overlap and the offset value is not properly set to "well" separate these two distributions (Fig. 3(a)). By updating the weights of iForest trees, these two distributions are pushed further away from each other, causing anomalous and nominal samples to have higher and lower scores, respectively (Fig. 3(b)). It should be noted that updating the weights of iForest trees has no impact on the offset of the iForest model. This offset value is further adjusted using Algorithm 2 so that it can better separate the two score distributions and, consequently, the adjusted iForest model can better distinguish anomalous points from nominal ones (Fig. 3(c)). Experimental evidence for the impact of each model updating strategy is given in Sect. 5.3.

5   Performance Evaluation

5.1   Experimental Settings

Datasets: Our evaluation experiments are conducted on two datasets. The first one (AIOPS) is public and is commonly used for performance evaluation of univariate time series anomaly detection algorithms. It is released by the AIOps2018 competition[1] and consists of 29 KPIs collected from some internet companies in China. In our experiments, we keep the same train/test splitting from the *Final* subset of this dataset to facilitate the comparison with existing works. The second dataset (HUAWEI) is private and collected from Huawei production environment. It is composed of 8 univariate telecom core network KPIs from different network elements. Each KPI is split into two halves, the first for training and the second for testing. These two datasets contain KPIs with a wide range of time series characteristics and anomalous patterns. Anomalous points and segments are labeled by domain experts and annotated as positive points, whereas nominal ones are designated as negative points. Some KPIs from these two datasets are shown in Fig. 4. Table 2 summarizes some statistics of the two datasets.

---

[1] https://github.com/NetManAIOps/KPI-Anomaly-Detection.

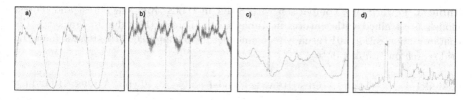

Fig. 4. Some KPIs from AIOps ((a) and (b)) and Huawei ((c) and (d)) datasets. Anomalous points are marked using red color.

Table 2. Statistics of benchmark datasets

Dataset	#KPIs	#Points	#Anomalous points	Sampling interval
AIOps	29	5922913	134114 (2.26%)	1 or 5 min
Huawei	8	119744	1188 (0.99%)	5 min

Metrics: For performance evaluation and comparison purpose, we adopt the evaluation protocol suggested by [22] and commonly used by the community. In this protocol, the F1-score is not calculated directly based on point-wise matching between the labels and detection results. A delay parameter k is introduced to adjust the detection results before F1-score calculation. According to human experts and for a contiguous anomaly segment, it is acceptable if the algorithm can detect and trigger an alert within a delay of k points. More precisely, we mark a segment of continuous anomalies as correctly detected if a point from this segment that is within k points from the segment's beginning is detected. In our experiments, the delay for both AIOps and Huawei datasets is $k = 7$, as recommended by the AIOps competition and used in other works.

5.2 Supervised and Unsupervised Anomaly Detection

To demonstrate the effectiveness of our anomaly detection pipeline, we compare it with several SoTA anomaly detection methods in Table 3. In addition to Isolation Forest (iForest), we also use Random Forest (RF) and run our pipeline presented in Sect. 3 in a supervised setting in order to establish the performance upper bound when all points are labeled in our active anomaly detection (i.e., query budget = 100%). RF is selected because it is also a tree-based ensemble model.

It can be seen that our best RF and iForest models achieve the best results among all supervised (sup.) and unsupervised (un.) methods, respectively. Thus, with a relatively simple and right set of features, combined with the right parameterization of popular supervised and unsupervised models, we can achieve the SoTA performance on benchmark datasets. This observation is valuable in a practical viewpoint, in which simple and explainable features and models are usually preferable over more complex ones. Also, the performance of iForest drops significantly when it uses, for example, an inappropriate value for its *contamination* parameter, among others. We assume that this type of performance degradation due to parameterization also applies to all other methods. This issue, however, can be handled by active learning, as will be shown in the next section.

Table 3. Performance comparison with SoTA methods. Random Forest and iForest models are trained with features in Table 1. For iForest, the value in bracket is the contamination ratio: 0.01 (usually recommended), 0.03, and 0.007 (best value found by grid search on training data).

Model	Type	AIOPS	HUAWEI	Model	Type	AIOPS	HUAWEI
SPOT [16]	un.	21.7	—	NCAD (un.) [3]	un.	76.6	—
DSPOT [16]	un.	52.1	—	NCAD (sup.) [3]	sup.	79.2	—
DONUT [20]	un.	72.0	—	RF (best)	sup.	**81.2**	**72.6**
SR [13]	un.	62.2	40.5	iForest (0.01)	un.	73.3	51.8
SR-CNN [13]	un.	77.1	—	iForest (0.03)	un.	51.3	40.8
SR-DNN [13]	sup.	81.1	—	iForest (best)	un.	**78.4**	**65.4**

It should be noted that NCAD [3] does not strictly follow the adopted evaluation protocol. It uses a more relaxed one without a delay restriction, which is thus equivalent to the adopted protocol with $k = \infty$. SR-CNN is a semi-supervised model that requires 65 million anomaly-free simulated points to train its CNN model for saliency map thresholding [13]. Finally, the gap in performance between RF and a "perfect" model can be explained partially by the lack of coherence in dataset labeling and by the limited representative and expressive power of the feature set and model.

5.3   Active Anomaly Detection

We evaluate our active anomaly detection pipeline on the two benchmark datasets and provide the results in Table 4. We are able to compare our approach with [4] using its open source implementation[2]. Since the implementation of [18] is not open, we've implemented it using an open source implementation of RRCF [1]. Our implementation turns out to be too slow due to the high complexity of RRCF. We couldn't obtain experimental results in a reasonable amount of time, and thus decided to not compare our method with [18] here.

In all experiments, iForest (0.03) presented in Sect. 5.2 is used as the baseline unsupervised model. Each value in Table 4 represents the F1-score obtained by using a unique combination of query strategy, model update strategy, and query budget. It can be seen that, even starting with a weak unsupervised model and using only 1% query budget, our active learning pipeline improves the performance by 56.51%, reaching 80.29% F1-score for AIOPS dataset. Similarly, for HUAWEI dataset, the performance is improved by 75.14% and reaches 71.37% F1-score. This performance is better than the performance achieved by [4] and iForest (best) and is very close to the performance of RF (best) shown in Table 3. This clearly demonstrates the effectiveness of our active learning pipeline.

Among query strategies, greedy selection (**TA**) outperforms random and **CTDB**. The random strategy does not really improve the performance. This is

---

[2] https://github.com/shubhomoydas/ad_examples.

Table 4. Experimental results on active anomaly detection. B/L indicate the iForest (0.03) model. 1%, 5%, 25% and 50% are the allowed query budgets. Each time value indicates the training and inference time to generate all results of the same row (B/L, 1%, 5%, 25% and 50%) using 10 Intel Xeon CPUs (E5-2690 v3 @ 2.60GHz).

| | | | AIOps | | | | | HUAWEI | | | | |
	Query strategy	Model update Strategy	B/L	1%	5%	25%	50%	Time (s)	B/L	1%	5%	25%	50%	Time (s)
[4]	TA	NW	51.30	48.87	53.93	53.40	**71.82**	6964	40.75	60.21	42.55	69.22	**70.74**	208
	Random			27.91	54.90	50.27	62.30	8133		3.08	51.66	47.49	61.36	252
Ours	TA	TW		51.77	51.55	51.42	51.44	427		60.73	60.41	60.17	60.17	29.3
		O		**80.29**	75.02	60.37	53.82	533		70.86	71.34	57.55	53.10	32.3
		TW+O		80.24	74.88	60.18	53.67	520		**71.37**	71.29	55.83	52.47	33.8
	CTDB	TW		50.93	50.97	51.22	51.34	453		59.31	60.17	60.17	60.17	29.2
		O		52.10	52.78	53.68	52.01	516		59.49	60.10	57.55	53.10	32.5
		TW+O		52.69	53.20	53.49	51.75	500		60.47	61.18	55.83	52.47	33.8
	TA + CTDB	TW		51.76	51.55	51.34	51.35	509		60.17	60.41	60.10	60.17	33.2
		O		77.03	75.47	67.16	60.32	610		47.53	70.90	63.32	57.55	37.4
		TW+O		52.19	52.24	53.04	53.64	610		60.79	60.91	61.91	55.83	38.3

similar to the observations reported in previous works [4, 18] and demonstrates the necessity of a good query strategy in an active learning pipeline. Among model update strategies, **TW** has almost no impact, regardless the amount of query budget. **O** and **TW+O** lead to the best improvement for AIOps and HUAWEI datasets, respectively. This demonstrates the importance of our proposed offset update (**O**) strategy. Finally, using our active learning approach, the required query budget to reach the best improvement is small compared to [4]. Our approach needs about 1% whereas [4] needs 25–50% of the training data.

In terms of computational complexity, our approach is about 14× and 6× faster than [4] on AIOps and HUAWEI datasets. This confirms the utility of our simple model update strategies, compared to the more complex ones used in [4].

6   Concluding Remarks

This paper proposed an efficient active learning based approach to systematically integrating the feedback and expert knowledge of network operators into unsupervised anomaly detection pipeline. In particular, we proposed three effective query strategies to assist operator in labeling those KPI samples that lead to alarms, including both true and false ones. A lightweight model update algorithm, which consists of updating the weights of trees and the adjustment of decision threshold in Isolation Forest model, has been also developed to improve the performance and efficiency. The experiments with two datasets have validated the performance advantages over baseline Isolation Forest and existing active learning based method in terms of detection performance and computational efficiency. Despite the claimed advantages, we believe sample query strategy and model update algorithm, as well as their integration with other unsupervised anomaly detection algorithms (e.g., RRCF), still have room to be further improved.

References

1. Bartos, M.D., Mullapudi, A., Troutman, S.C.: RRCF: implementation of the robust random cut forest algorithm for anomaly detection on streams. J. Open Sour. Softw. **4**(35), 1336 (2019)
2. Breiman, L.: Random forests. Mach. Learn. **45**(1), 5–32 (2001)
3. Carmona, C.U., Aubet, F.X., Flunkert, V., Gasthaus, J.: Neural contextual anomaly detection for time series. arXiv:2107.07702 (2021)
4. Das, S., Islam, M.R., Jayakodi, N.K., Doppa, J.R.: Active anomaly detection via ensembles: insights, algorithms, and interpretability. arXiv:1901.08930 (2019)
5. Görnitz, N., Kloft, M., Rieck, K., Brefeld, U.: Toward supervised anomaly detection. J. Artif Intell. Res. **46**, 235–262 (2013)
6. Guha, S., Mishra, N., Roy, G., Schrijvers, O.: Robust random cut forest based anomaly detection on streams. In: ICM (2016)
7. Hanneke, S., et al.: Theory of disagreement-based active learning. Found. Trends Mach. Learn. **7**(2–3), 131–309 (2014)
8. Linkedin: Luminol: anomaly detection and correlation library. https://github.com/linkedin/luminol
9. Liu, D., et al.: Opprentice: towards practical and automatic anomaly detection through machine learning. In: Proceedings of the 2015 Internet Measurement Conference (2015)
10. Liu, F.T., Ting, K.M., Zhou, Z.H.: Isolation forest. In: ICDM (2008)
11. Malhotra, P., Vig, L., Shroff, G., Agarwal, P.: Long short term memory networks for anomaly detection in time series. In: ESANN (2015)
12. Pang, G., Shen, C., Cao, L., Hengel, A.V.D.: Deep learning for anomaly detection: A review. ACM Comput. Surv. **54**(2), 1–38 (2021)
13. Ren, H., et al.: Time-series anomaly detection service at Microsoft. In: KDD (2019)
14. Ruff, L., et al.: Deep semi-supervised anomaly detection. In: ICLR (2020)
15. Siddiqui, M.A., et al.: Feedback-guided anomaly discovery via online optimization. In: KDD (2018)
16. Siffer, A., Fouque, P.A., Termier, A., Largouet, C.: Anomaly detection in streams with extreme value theory. In: KDD (2017)
17. Trittenbach, H., Englhardt, A., Böhm, K.: An overview and a benchmark of active learning for outlier detection with one-class classifiers. Exp. Syst. Appl. **168**, 114372 (2020)
18. Wang, Y., et al.: Practical and white-box anomaly detection through unsupervised and active learning. In: ICCCN (2020)
19. Wu, K., Zhang, K., Fan, W., Edwards, A., Philip, S.Y.: RS-forest: a rapid density estimator for streaming anomaly detection. In: ICDM (2014)
20. Xu, H., et al.: Unsupervised anomaly detection via variational auto-encoder for seasonal KPIs in web applications. In: WWW (2018)
21. Yu, Q., Jibin, L., Jiang, L.: An improved ARIMA-based traffic anomaly detection algorithm for wireless sensor networks. Int. J. Distrib. Sens. Netw. **12**(1) (2016)
22. Zhao, N., Zhu, J., Liu, R., Liu, D., Zhang, M., Pei, D.: Label-less: a semi-automatic labelling tool for KPI anomalies. In: INFOCOM (2019)

MMRCA: MultiModal Root Cause Analysis

Gary White[(✉)], Jaroslaw Diuwe, Erika Fonseca, and Owen O'Brien

Huawei Ireland Research Centre, Townsend Street, Dublin 2 D02 R156, Ireland
{gary.white,jaroslaw.diuwe,erika.fonseca,owen.obrien}@huawei.com

Abstract. Cloud systems are becoming increasingly complex and more difficult for human operators to manage due to the scale and interconnectedness of microservices. Increased observability and anomaly detection are able to alert when something has gone wrong, however a fault can propagate throughout the cloud leading to a large number of alerts. It is difficult for operators to manage these large number of alerts and to differentiate between the symptoms that have propagated due to the fault and the actual root cause. In this paper we present a MultiModal Root Cause Analysis algorithm called MMRCA. This algorithm leverages data from traces, topology, configurations and metrics to accurately predict the root cause of the fault. Our approach consists of a three step pipeline of topology reduction, metric causality and metric reduction. The experimental results show that MMRCA can accurately detect the root cause in a number of different data sets, while maintaining an efficient use of resources and scaling to a large deployment.

Keywords: Root cause analysis · Cloud computing · Causality analysis · Unsupervised learning · MultiModal data

1 Introduction

Artificial Intelligence for IT Operations (AIOps) is an increasingly popular field, utilizing research in the areas of machine learning and big data for the management of IT operations [5]. There are a number of topics of interest in this area e.g., anomaly detection, self-healing, self-adaptation, root cause analysis, log analysis, predictive maintenance and many more [14]. This has lead to dedicated workshops such as AIOps@ICSOC, to bring together researchers focused on these specific problems [3].

With the increasing popularity of AIOps and the scale of cloud systems, there has been a large increase in the amount of instrumentation data being collected. Accurate anomaly detection algorithms can be used to identify sudden changes in metrics that indicate that a fault is about to happen [17]. However, when a fault does happen a large amount of alerts can be generated, which makes it difficult to identify what was the root cause of the fault. An automated root cause analysis approach can take this list of anomalies and identify what was the

© Springer Nature Switzerland AG 2022
H. Hacid et al. (Eds.): ICSOC 2021 Workshops, LNCS 13236, pp. 177–189, 2022.
https://doi.org/10.1007/978-3-031-14135-5_14

actual root cause that led to the failure. We now summarize the challenges posed in the construction of the desired industry-grade root cause analysis algorithm:

- *Lack of labels:* In production-level business scenarios, the systems often process millions of metrics, traces and configurations. There is no easy way to label data on this scale manually. Moreover, if the multimodal data is in a dynamic environment where the distribution is constantly changing, then the model will need to be retrained frequently on the new data. Labelling this data can introduce a significant delay and cost with a need to continuously update the models. This makes supervised models insufficient for the industrial scenario.
- *Efficiency and accuracy:* In production scenarios, a monitoring system must process millions of data points in near real-time. This can lead to a large number of alerts being generated and also a large amount of data to process to identify the root cause. Furthermore, if the root cause analysis algorithm is deployed on a production node, then it may not use a lot of computing resources to respect the quota assigned for customers. Therefore, even though models with large time complexity may achieve good accuracy, they are often of little use in a production scenario, due to their overhead.

In this paper, we propose MMRCA, a multimodal root cause analysis algorithm for the cloud. We also show how the algorithm can be deployed as part of an event driven pipeline, in combination with anomaly detection and fault prediction. MMRCA is a data-driven root cause analysis approach that does not assume the shape of the distribution, requires very little parameter tuning and is able to accurately detect the root cause. It leverages different modes of data, such as key-value configurations, graph topologies and time series metric data. The algorithm has three main steps. In the first step, the topology reduction algorithm uses topology, trace and configuration data to identify the services and nodes that were most likely to have been the root cause of the problem. The second step of the algorithm uses this reduced topology to identify any causal relationships between the trigger of the root cause analysis and the metrics that have been observed. The third component of the algorithm is used to remove any static or random metrics and to cluster the remaining metrics. These processed metrics are passed to the second step, which greatly reduces amount of metrics that need to be analysed for causality, shortening the time to return the root cause, while maintaining accuracy.

The rest of the paper is organised as follows: Sect. 2 outlines the related work. Section 3 presents our proposed MMRCA algorithm and the data streaming pipeline used. Section 4 describes the experimental approach used to evaluate our algorithm and Sect. 5 presents the results of those experiments. Finally, Sect. 6 concludes the paper and outlines some future work.

2   Related Work

Existing root cause analysis approaches can be categorised based on the data sources that they use. A number of the existing approaches have focused on the use of a single or of a couple of data sources. In this section, we present a survey of the related literature that has focused on the root cause analysis problem.

2.1 Topology

Knowledge of up-to-date physical topology can be useful to help identify how faults are propagating through the network and where the root cause may have been introduced. Given the dynamic nature of modern networks this topology information needs to be updated regularly in an automated way [1]. Topology information allows for a data-driven approach for root cause analysis [12]. A service dependency model can be built up using the topology information to identify the root cause of the problem in an automated way [20].

2.2 Traces

Topology-based approaches assume that the adjacent services with abnormal invocation are more likely to be the root cause. However, due to the complex dependencies and fault propagation among microservices, anomaly invocations between adjacent mircosevices are not sufficient to reflect the location of root causes [11]. Trace-based root cause analysis approaches overcome this limitation by correlating all the microservices involved in a trace instead of just the adjacent ones [23].

2.3 Metrics/Logs

The temporal nature of metrics allows for directional causality between the metrics to be established [15]. A number of machine learning methods have been developed, such as self-organising maps, local outlier factor and k-means, which can easily be applied to establish the root cause using metric data [8]. Recent approaches have focused on the use of deep learning, such as stacked bi-directional self-attention LSTM networks to identify the root cause in metric data [22].

2.4 Configuration

Configuration issues cause the largest percentage (31%) of high-severity support requests, based on an empirical study in commercial and open source systems [21]. Therefore, it is very important to take into account configuration changes in root cause analysis as they can often lead to failures. Cloud applications are highly configurable so there are a lot of different options that can be specified. Recent approaches have focused on automating analytics for these systems [18].

2.5 Multimodal

In various disciplines, information about the same phenomenon can be acquired from different types of detectors [9]. There are a number of issues with combining multimodal data, such as different resolutions, alignment and incompatible size [9], but it also provides a great opportunity to leverage another view on how the system is behaving [7]. Some RCA approaches have started to leverage multiple sources of information, such as topology, traces and metrics in the root cause analysis system for Alibaba datacenters [4]. Other approaches have focused on the fusion of metrics and topology data for root cause analysis [19].

A number of industry cloud providers, such as Google and Microsoft as well as cloud instrumentation specialists, such as Dynatrace and AppDynamics have been researching this space to deal with the growing complexity of cloud systems. Table 1 shows a summary of the main features of the approaches and how they compare to our MMRCA approach. We also add some academic approaches to the table, such as Grano [19] and MLCloud [8], though MLCloud only focuses on using metric data to identify the root cause. We can see that a lot of the industry approaches take into account multiple data sources, such as topology, traces and metrics. However, our MMRCA approach is the only one to take into account topology, traces, metrics, configurations as well as being able to return the root cause in under three minutes.

Table 1. State of the art root cause analysis approaches

Approach	Topology	Traces	Metrics/Logs	Configuration	<3 min
Google [2]	✓	✓	✓	✗	✗
Microsoft [10]	✓	✓	✓	✗	✗
Alibaba [4]	✓	✓	✓	✗	✗
Dynatrace [6]	✓	✓	✓	✓	✗
AppDynamics [13]	✓	✓	✓	✓	✗
MLCLoud [8]	✗	✗	✓	✗	✗
Grano [19]	✓	✗	✓	✗	✗
MMRCA	✓	✓	✓	✓	✓

3   Multimodal Root Cause Analysis (MMRCA)

3.1   Terminology and Problem Statement

This section presents the terminology used in this paper for the root cause, the problem statement and the algorithmic choices made when creating this algorithm. As cloud data centers continue to grow in scale and complexity, errors can propagate when a fault happens leading to a large cascade. It is difficult for cloud operators to know which of these are symptoms of the root cause and which are the actual root cause. The problem that we tackle in this paper is given a fault trigger to identify what was the actual root cause of this fault by leveraging topology, trace, metric and configuration data and to return the result in under 3 min.

3.2   MMRCA Algorithm

Figure 1 shows an overview of the individual components of the MMRCA algorithm and how they fit into the data streaming pipeline. The main steps of the algorithm, indicated by the blue boxes are Topology Reduction, Metric Causality and Metric Reduction. Topology reduction queries the Topology API, Traces and Configuration store to build up the dynamic, directed and weighted graph used to identify the services that have deviated from their past behaviour. Metric reduction queries the metrics DB and removes the static and random metrics, before clustering the remaining metrics which are then passed to metric causality. Metric causality then temporally aligns the metrics and identifies the root cause using causality analysis between the metrics and additional weight for recent configuration changes.

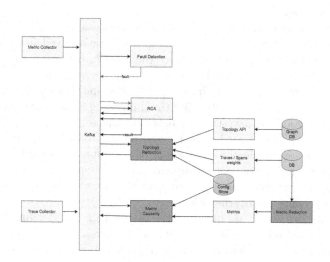

Fig. 1. Overview of the MMRCA data streaming pipeline

We now explain in more detail each step of MMRCA:

1. Topology Reduction
   - Figure 2 shows the main stages of the topology reduction algorithm. We first parse the trace data then conduct data analysis and feature extraction to identify the most important features.
   - We then query the topology graph to identify how the services are linked and the nodes they are deployed on. This topology information is then combined with the trace weights to build up a dynamic, directed and weighted graph.
   - Once the graph has been created, deviation in response time, response code and throughput are calculated over the previous time window.
   - A weight is then added to services that have had a recent change in configuration. This is combined with the weight for the services that have deviated from their previous behaviour.
   - The top-k services with the largest weight are then selected and passed to the metric causality algorithm, with the node and instance metrics. This greatly reduces the amount of metrics that need to be analysed.

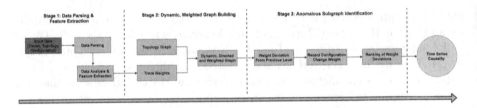

Fig. 2. Stages of topology reduction algorithm

2. Metric Causality

- There are four main stages in the metric causality algorithm as shown in Fig. 3. The first stage is to parse the data received from the metric reduction algorithm and to extract the most important features.
- The data is then analysed and temporally aligned, this is an important step as different nodes can have different starting times and different sampling rates. We temporally align the data using some rounding combined with the common lowest factor between data sets.
- The temporally aligned data is then passed to the temporal causation stage. We experimented with a number of approaches for temporal causation including transfer entropy, cosine distance and granger causality. Granger causality was found to give the most accurate causal results, while keeping the overhead low.
- The metrics are then ranked by P-value, with a low P-value indicating a large casual relationship. A weight is also given for recent configuration changes to the services that are being tracked by the metrics. The top 5 most likely root causes with their confidence are then given as output.

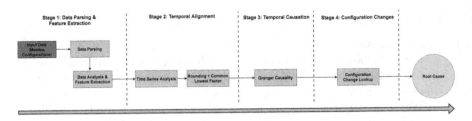

Fig. 3. Stages of metric causality algorithm

3. Metric Reduction

- In production systems there are a large number of metrics that can remain static. These metrics will not be useful in identifying the root cause as a static metric cannot have a causal relationship. The metrics are first parsed and analysed, as shown in Fig. 4. The static and random metrics are then removed as they cannot have a causal relationship with the fault. The algorithm is run before a fault is triggered to build up the list static and random metrics. We use the standard deviation of the metric value to evaluate whether a metric is static. To evaluate whether a metric is random we use the KV test and auto-correlation.

- Once we have removed the static and random metrics we then cluster the remaining metrics. These clusters allow us to group similar metrics. We evaluated a number of algorithms for clustering and found the best results using the gap statistic to identify the correct number of clusters and k-means clustering to cluster the data.
- Once the data has been clustered, it can be sampled using a percentage sampling of clusters. The time series causality algorithm uses the percentage sampling of clusters to identify which clusters are the most likely to contain the root cause based on their P-values. This improves accuracy as the most likely root cause metrics are analysed first.

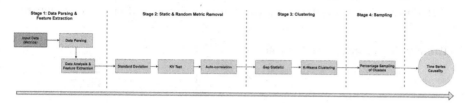

Fig. 4. Stages of metric reduction algorithm

4 Experimental Setup

4.1 Data Set Characteristics

We provision an environment using Docker containers and Ansible playbooks to allow for a repeatable and realistic experimentation environment. We use Apache JMeter to load the services in the deployment and inject faults into the system. We allow the fault to propagate through the system and use one of the propagated faults as the input to root cause analysis. This allows us to evaluate whether we can trace back the fault to a known ground truth root cause.

4.2 Evaluation Metrics

We use a combination of metrics to evaluate that our approach is accurate and efficient. To evaluate the accuracy, we use the precision, recall, and F1 score of the actual and identified root cause. We also calculate the mean average precision @ k (mAP@k), with k = 5, using Formula 1:

$$mAP@k = \frac{1}{N} \sum_{i=1}^{N} AP@k_i \qquad (1)$$

To evaluate the resource cost we measure the memory usage and the scalability of the algorithm with an increasing number of metrics and services.

5   Results

This section presents the results that we have used to verify the accuracy and scalability of our approach. We break down the individual components of the algorithm including Topology Reduction in Sect. 5.1, Metric Causality in Sect. 5.2, Metric Reduction in Sect. 5.3 and finally the end-to-end results in Sect. 5.4.

5.1   Topology Reduction

We evaluated how to construct the data in a graph structure to allow us to identify the nodes that have deviated from past behaviour. We can use the trace data and the inductive causation (IC) or PC algorithm to infer the causality between the connections in the service graph. Another approach that we evaluate is to combine the trace, topology and config data to identify the services in the call chain that have deviated from their usual behaviour after a config change.

Table 2. Topology reduction accuracy

Algorithm	Precision	Recall	F1	MAP@5
IC algorithm	0.52	0.81	0.62	0.60
PC algorithm	0.66	0.82	0.73	0.68
Stan dev.	0.98	0.98	0.98	1
Config + Stan dev.	1	1	1	1

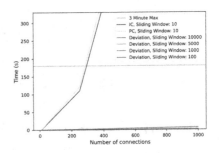

Fig. 5. Topology reduction scale

Table 2 shows that combining the topology and traces data and using the standard deviation has led to improved precision, recall, F1 score and MAP@5 score compared to the IC and PC algorithm. We can also see from Fig. 5 that the standard deviation algorithm is able to scale much better than using an inductive causation-based approach. The standard deviation algorithm is able to scale well, even with a large sliding window and a large number of connections. This is important due to the large number of services and connections that can be deployed in a cloud environment.

5.2   Metric Causality

One of the major issues with metric causality is to be able to deal with metrics that have different sampling rates and starting times, which means that the metrics are not aligned. Figure 6 shows the results of metric causality for three different data alignments. We experimented with Dynamic Time warping (DTW) [16], common lowest factor (CLF) and adding rounding to common lowest factor. We can see from Fig. 6a that in the ideal case granger causality on its own will only work when the sampling rate is the same as it requires the same number of observations in both data sets. We can see that CLF + Granger, Config + Rounding + CLF + Granger and DTW + Granger all have similar performance as the difference in sampling rate varies. Figure 6b shows how CLF + Granger performs poorly when there is no direct overlap between the metrics because they have a different starting time and the same sampling rate, so there is never a common lowest factor. Adding some rounding to the metrics helps to solve this issue as shown in Fig. 6b and Fig. 6c, where Rounding + CLF + Granger is able to achieve the same accuracy as DTW + Granger, with configuration information further improving performance.

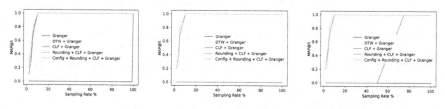

(a) Same Starting Time and Sampling Rate

(b) Different Starting Time and Same Sampling Rate

(c) Different Starting Time and Sampling Rate

Fig. 6. Impact of sampling rate and starting time on metric causality

Figure 7 shows the scalability of the different approaches. One of the problems with the DTW approach is that it does not scale well, which means that we can only pass through a limited number of metrics in the 3 min window. If we do not have enough time to pass through all of the metrics before having to return the root cause then this will impact the accuracy of the algorithm. We can see how the CLF approaches to temporal alignment are able to scale much better even when adding in configuration and rounding parameters.

5.3   Metric Reduction

Table 3 shows the impact of metric reduction on the number of metrics that we have to process at runtime when running the actual root cause analysis algorithm. Without metric reduction we have to process all of the 11,970 metrics, with metric reduction we can remove 93.4% of the metrics and only have to process 790 at runtime. This leads to a big saving in the amount of time that we have to spend on metric causality to identify the casual relationship between the fault and the metrics. We can also see from the results that there is no case where the root cause is removed by using metric reduction.

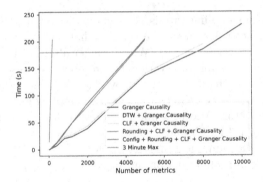

Fig. 7. Scale of metric causality algorithms

Table 3. Impact of metric reduction

	Without metric reduction	With metric reduction
Number of metrics	11970	790
% Reduction in metrics	0	93.4
% Root cause removed	0	0

5.4 End-to-End

Table 4 shows the end-to-end accuracy of the root cause analysis approach and the impact of the different components of the algorithm. Metric causality on its own is able to achieve an MAP@5 of 0.65. The reason for the lower MAP score is that because of the lack of topology reduction and metric reduction so much more metrics have to be processed to return the root cause within 3 min. The next step, where we combine topology reduction with metric causality, improves the accuracy and MAP@5 scores to 0.93. This reduces the amount of metrics that have to be processed to only the services that have deviated from past behaviour and the node and instance metrics. Finally, adding the Metric Reduction step helps to further improve the accuracy by removing any of the random and static metrics from being processed and clustering similar metrics together. Metric causality can then sample these clusters and start with the metrics that have the lowest P-value. This ensures that the most likely root cause candidates are processed first helping to improve the MAP@5 accuracy within the 3 min window to 0.96. We can see how this shows improvement against the current state of the art Grano approach which achieved an MAP@5 accuracy of 0.9.

Table 5 shows the scalability of the algorithm as the deployment size increases. We can see how the individual component decisions have allowed the algorithm to scale very well and even as the number of services doubles in size there is only a slight increase in the time taken and the topology space used. The metric space remains constant as we select the same number of services that have deviated from their previous behaviour using topology reduction.

Table 4. End-to-end accuracy

Algorithm	Precision	Recall	F1 score	MAP@5
Metric causality	0.6	0.7	0.65	0.65
Grano [19]	0.88	0.9	0.89	0.9
Topology reduction + MC	0.9	0.9	0.9	0.93
TR + MC + Metric reduction	0.95	1	0.97	0.96

Table 5. End-to-end scale

Services	Time (s)	Metric space (MB)	Topology space (MB)
12	10	4.4	81
16	12	4.4	81
32	14	4.4	83

6 Conclusion and Future Work

Our paper proposes MMRCA, a Multimodal Root Cause Analysis algorithm that leverages data from traces, topology, metrics and configurations. Our approach is designed from the ground up to be fast, accurate and resource efficient. We locate the general area of the root cause of the problem using trace, topology and configuration information. We preprocess the metrics to remove static and random metrics and cluster the remaining metrics. We then temporally align the metrics and evaluate the causal relationship between the trigger of the fault and the metrics collected before the fault. The end-to-end results show the improvement in MAP@5 as the mutimodal data is used and the improvement against the state of the art Grano algorithm, while maintaining time and space complexity.

For future work, we aim to evaluate other additional signals and instrumentation that can be used as a signal for root cause analysis. We are also investigating using the output of our root cause analysis algorithm as input to an automated remediation system that would allow for remediation actions after identifying the root cause.

References

1. Bejerano, Y., Breitbart, Y., Garofalakis, M., Rastogi, R.: Physical topology discovery for large multisubNet networks. In: IEEE INFOCOM 2003. Twenty-second Annual Joint Conference of the IEEE Computer and Communications Societies. vol. 1, pp. 342–352 (2003)
2. Beyer, B., Jones, C., Petoff, J., Murphy, N.R.: Site Reliability Engineering: How Google Runs Production Systems. O'Reilly Media, Inc., Sebastopol (2016)
3. Bogatinovski, J., et al.: Artificial intelligence for it operations (AIOPS) workshop white paper. arXiv preprint arXiv:2101.06054 (2021)

4. Cai, Z., Li, W., Zhu, W., Liu, L., Yang, B.: A real-time trace-level root-cause diagnosis system in Alibaba datacenters. IEEE Access **7**, 142692–142702 (2019)
5. Dang, Y., Lin, Q., Huang, P.: AIOPS: real-world challenges and research innovations. In: 2019 IEEE/ACM 41st International Conference on Software Engineering: Companion Proceedings (ICSE-Companion). pp. 4–5. IEEE (2019)
6. Dynatrace: root cause analysis. https://www.dynatrace.com/support/help/how-to-use-dynatrace/problem-detection-and-analysis/problem-analysis/root-cause-analysis/. Accessed 4 Apr 2022
7. Gao, J., Li, P., Chen, Z., Zhang, J.: A survey on deep learning for multimodal data fusion. Neural Comput. **32**(5), 829–864 (2020)
8. Josefsson, T.: Root-cause analysis through machine learning in the cloud (2017)
9. Lahat, D., Adali, T., Jutten, C.: Multimodal data fusion: An overview of methods, challenges, and prospects. Proc. IEEE **103**(9), 1449–1477 (2015)
10. Levy, S., et al.: Predictive and adaptive failure mitigation to avert production cloud VM interruptions. In: 14th USENIX Symposium on Operating Systems Design and Implementation (OSDI 20). pp. 1155–1170, November 2020
11. Li, Z., et al.: Practical root cause localization for microservice systems via trace analysis. In: 2021 IEEE/ACM 29th International Symposium on Quality of Service (IWQOS), pp. 1–10 (2021)
12. Lindner, B.S., Auret, L.: Data-driven fault detection with process topology for fault identification. IFAC Proc. **47**(3), 8903–8908 (2014)
13. Maerz, C.: Root cause analysis. https://www.appdynamics.com/blog/product/how-to-monitor-root-cause-analysis/. Accessed 4 Oct 2021
14. Masood, A., Hashmi, A.: AIOps: predictive analytics & machine learning in operations. In: Cognitive Computing Recipes, pp. 359–382. Apress, Berkeley, CA (2019). https://doi.org/10.1007/978-1-4842-4106-6_7
15. Muñoz, P., de la Bandera, I., Khatib, E.J., Gómez-Andrades, A., Serrano, I., Barco, R.: Root cause analysis based on temporal analysis of metrics toward self-organizing 5g networks. IEEE Trans. Veh. Technol. **66**(3), 2811–2824 (2017). https://doi.org/10.1109/TVT.2016.2586143
16. Senin, P.: Dynamic time warping algorithm review. Inf. Comput. Sci. **855**(1–23), 40 (2008)
17. Shahid, A., White, G., Diuwe, J., Agapitos, A., O'Brien, O.: SLMAD: statistical learning-based metric anomaly detection. In: Hacid, H., et al. (eds.) ICSOC 2020. LNCS, vol. 12632, pp. 252–263. Springer, Cham (2021). https://doi.org/10.1007/978-3-030-76352-7_26
18. Tuncer, O., et al.: ConfEx: towards automating software configuration analytics in the cloud. In: 2018 48th Annual IEEE/IFIP International Conference on Dependable Systems and Networks Workshops (DSN-W), pp. 30–33. IEEE (2018)
19. Wang, H., Nguyen, P., Li, J., Kopru, S., Zhang, G., Katariya, S., Ben-Romdhane, S.: GRANO: interactive graph-based root cause analysis for cloud-native distributed data platform. Proc. VLDB Endow. **12**(12), 1942–1945 (2019)
20. Yan, H., Breslau, L., Ge, Z., Massey, D., Pei, D., Yates, J.: G-RCA: a generic root cause analysis platform for service quality management in large IP networks. IEEE/ACM Trans. Netw. **20**(6), 1734–1747 (2012)
21. Yin, Z., Ma, X., Zheng, J., Zhou, Y., Bairavasundaram, L.N., Pasupathy, S.: An empirical study on configuration errors in commercial and open source systems. In: Proceedings of the Twenty-Third ACM Symposium on Operating Systems Principles, pp. 159–172 (2011)

22. You, C., Wang, Q., Sun, C.: sBiLSAN: stacked bidirectional self-attention LSTM network for anomaly detection and diagnosis from system logs. In: Arai, K. (ed.) Intelligent Systems and Applications, pp. 777–793 (2022)
23. Yuan, C., et al.: Automated known problem diagnosis with event traces. In: Proceedings of the 1st ACM SIGOPS/EuroSys European Conference on Computer Systems, vol. 40, pp. 375–388, April 2006

IAD: Indirect Anomalous VMMs Detection in the Cloud-Based Environment

Anshul Jindal[1](✉)[ID], Ilya Shakhat[2], Jorge Cardoso[2,3][ID], Michael Gerndt[1][ID], and Vladimir Podolskiy[1][ID]

[1] Chair of Computer Architecture and Parallel Systems,
Technical University of Munich, Garching, Germany
{anshul.jindal,v.podolskiy}@tum.de, gerndt@in.tum.de
[2] Huawei Munich Research Center, Huawei Technologies Munich, Munich, Germany
{ilya.shakhat1,jorge.cardoso}@huawei.com
[3] University of Coimbra, CISUC, DEI, Coimbra, Portugal

Abstract. Server virtualization in the form of virtual machines (VMs) with the use of a hypervisor or a Virtual Machine Monitor (VMM) is an essential part of cloud computing technology to provide infrastructure-as-a-service (IaaS). A fault or an anomaly in the VMM can propagate to the VMs hosted on it and ultimately affect the availability and reliability of the applications running on those VMs. Therefore, identifying and eventually resolving it quickly is highly important. However, anomalous VMM detection is a challenge in the cloud environment since the user does not have access to the VMM.

This paper addresses this *challenge of anomalous VMM detection in the cloud-based environment without having any knowledge or data from VMM* by introducing a novel machine learning-based algorithm called **IAD**: **I**ndirect **A**nomalous VMMs **D**etection. This algorithm solely uses the VM's resources utilization data hosted on those VMMs for the anomalous VMMs detection. The developed algorithm's accuracy was tested on four datasets comprising the synthetic and real and compared against four other popular algorithms, which can also be used to the described problem. It was found that the proposed *IAD* algorithm has an average F1-score of 83.7% averaged across four datasets, and also outperforms other algorithms by an average F1-score of 11%.

Keywords: Anomaly detection · Cloud computing · VMM · Hypervisor

1 Introduction

Cloud computing enables industries to develop and deploy highly available and scalable applications to provide affordable and on-demand access to compute and storage resources. Server virtualization in the form of virtual machines (VMs) is an essential part of cloud computing technology to provide infrastructure-as-a-service (IaaS) with the use of a hypervisor or Virtual Machine Monitor

© Springer Nature Switzerland AG 2022
H. Hacid et al. (Eds.): ICSOC 2021 Workshops, LNCS 13236, pp. 190–201, 2022.
https://doi.org/10.1007/978-3-031-14135-5_15

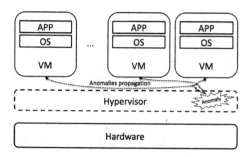

Fig. 1. An example showcasing the propagation of anomalies in a Type-1 hypervisor or VMM to the virtual machines (VMs) hosted on it.

(VMM) [12]. Users can then deploy their applications on these VMs with only the required resources. This allows the efficient usage of the physical hardware and reduces the overall cost. The virtualization layer, especially the hypervisors, is prone to temporary hardware errors caused by manufacturing defects, a sudden increase in CPU utilization caused by some task or disconnection of externally mounted storage devices, etc. The VMs running on these VMMs are then susceptible to errors from the underneath stack, as a result, can impact the performance of the applications running on these VMs [7,8]. Figure 1 shows an example propagation of anomalies in a virtualization stack using a type-1 hypervisor to the VM hosted on it. These anomalies may lead to the failure of all VMs and, ultimately, the applications hosted on them.

In the development environment, these anomalous VMMs are relatively easily detectable by analyzing the logs from the hypervisor dumps. But in the production environment running on the cloud, anomalous VMMs detection is a challenge since a cloud user does not have access to the VMMs logs. Additionally, many anomalous VMM detection techniques have been proposed [11,13,15]. However, these works either require the monitoring data of the hypervisor or inject custom probes into the hypervisor. Therefore, the usage of such solutions becomes infeasible. Furthermore, due to the low downtime requirements for the applications running on the cloud, detecting such anomalous VMMs and their resolutions is to be done as quickly as possible.

Therefore, this challenge is addressed in this paper for detecting anomalous VMMs *by solely using the VM's resources utilization data hosted on those VMMs* by creating a novel algorithm called **IAD**: **I**ndirect **A**nomalous VMMs **D**etection. We call the algorithm indirect since the detection must be done without any internal knowledge or data from the VMM; it should be solely based on the virtual machine's data hosted on it. The key contributions are:

- We present an online novel machine learning-based algorithm **IAD** for accurate and efficient detection of anomalous VMMs by solely using the resource's utilization data of the VM's hosted on them as the main metric (Sect. 3).
- We evaluate the performance of the *IAD* on two different aspects: 1) Anomalous VMMs finding accuracy (Sect. 5.1), and 2) Anomalous VMMs finding efficiency

Table 1. Symbols and definitions.

Symbol	Interpretation
n	Number of time ticks in data
d	Number of virtual machines hosted on a VMM
X_t	The percentage utilization of a resource (for example, CPU or disk usage) by a VM at a time t
X_t^j	The percentage utilization of a resource at a time t for j^{th} VM
$\{c_t^1, c_t^2, ..., c_t^m\}$	a set of m $\leq d$ VMs with change point at time tick t
w	Window size
minPercentVMsFault	Minimum % of total number of VMs on a VMM which must have a change point for classifying the VMM anomalous

and scalability (Sect. 5.2), and compare it against five other popular algorithms which can also be applied to some extent on the described problem.

– We evaluate the *IAD* algorithm and other five popular algorithms on synthetic and two real datasets.

Paper Organization: Section 2 describes the overall problem statement addressed in this paper along with an illustrative example. The design and details of the proposed *IAD* algorithm are presented in Sect. 3. Section 4 provides experimental configuration details along with the algorithms and the datasets used in this work for evaluation. In Sect. 5, the evaluation results are presented. Finally, Sect. 6 concludes the paper and presents an outlook.

2   Problem Definition

This section presents the overall problem definition of indirectly detecting anomalous VMMs in a cloud-based environment. Table 1 shows the symbols used in this paper.

We are given $X = n \times d$ dataset, with n representing the number of time ticks and d the number of virtual machines hosted on a VMM. X_t^j denotes the percentage utilization of a resource (for example, CPU or disk usage) at a time t for j^{th} VM. Our goal is to detect whether the VMM on which the d virtual machines are hosted is anomalous or not. Formally:

Problem 1. (Indirect Anomalous VMM Detection)

– **Given** *a multivariate dataset of n time ticks, with d virtual machines (X_t^j for $j = \{1, \cdots, d\}$ and $t = \{1, \cdots, n\}$) representing the CPU utilization observations of VMs hosted on a VMM.*

– **Output** *a subset of time ticks or a time tick where the behavior of the VMM is anomalous.*

One of the significant challenges in this problem is the online detection, in which we receive the data incrementally, one time tick for each VM at a time, i.e., $X_1^j, X_2^j, \cdots$, for the jth VM. As we receive the data, the algorithm should

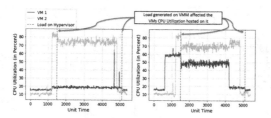

Fig. 2. Examples showing CPU utilization of two virtual machines hosted on a VMM. The left sub-figure shows an application running only on VM 2, while the right sub-figure shows the application running on both VMs. We can see a significant decrement in the CPU utilization of the two VMs when an anomaly (high-CPU load) is generated on the VMM (shown by dotted red lines). (Color figure online)

output the time ticks where the behavior of the VMM is observed as anomalous. However, without looking at the future few time ticks after time t, it would be impractical to determine whether at time point t, the VMM is anomalous or not since the time ticks $t + 1, t + 2, \cdots$, are essential in deciding whether an apparent detection at time t was an actual or simply noise. Hence, we introduce a window parameter w, upon receiving a time tick $t + w$, the algorithm outputs whether at time t the VMM showcased anomalous behavior or not. Additionally, as the change points for VMs hosted on VMM could be spread over a specific duration due to the effect of the actual fault being propagating to the VMs and the granularity of the collected monitoring data, therefore, using an appropriate window size can provide a way for getting those change points.

2.1 Illustrative Example

Here we illustrate the problem with two examples in Fig. 2 showcasing the CPU utilization of two virtual machines hosted on a VMM. In the left sub-figure, an application is running only on VM 2, while in the right, an application is running on both VMs. During the application run time, an anomaly, i.e., high CPU load, was generated on the hypervisor for some time (shown by dotted red lines). During this time, we can observe a significant drop in the CPU utilization by the application (affecting the performance of the application) of the two VMs (especially when an application is running on the VM). The load on a VMM affects all or most of the VMs hosted on it, which ultimately can significantly affect the performance of the applications running on the two VMs; therefore, we call such a VMM anomalous when the load was generated on it.

3 Indirect Anomaly Detection (IAD) Algorithm

This section presents our proposed Indirect Anomaly Detection (IAD) algorithm along with the implemented system for evaluating it. The overall system workflow diagram is shown in Fig. 4 and mainly consists of two parts: the main *IAD Algorithm*, and the *Test Module* for evaluating the algorithm.

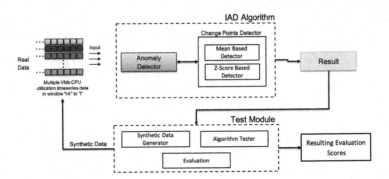

Fig. 3. High-level system workflow of the implemented system for evaluating IAD algorithm and the interaction between its components in a general use case.

3.1 IAD Algorithm

Our principal intuition behind the algorithm is that if a time tick t represents a change point for some resource utilization (such as CPU utilization) in most VMs hosted on a VMM; then the VMM is also anomalous at that time tick. This is based on the fact that a fault in VMM will affect most of the VMs hosted on it, and therefore those VMs would observe a change point at a similar point of time (in the chosen window w (Table 1)) in their resource's utilization. IAD algorithm consists of two main parts, described below:

Change Points Detector: We first explain how the change point, i.e., time tick where the time series changes significantly, is calculated. Recall from Sect. 2 that, we have introduced a window parameter w, upon receiving the time tick $t + w$, the *Change Points Detector* outputs whether the time tick t is a change point or not. Given a dataset X^j of size w for j^{th} VM, this component is responsible for finding the change points in that VM. This can be calculated in two ways: Mean-based detector and Z-score-based detector.

- **Mean-based Detector:** In this detector, a *windowed_mean*, i.e., the mean of all the values in the window, and the *global_mean*, i.e., the mean of all the values until the current time tick is calculated. Since the IAD algorithm is designed for running it in an online way, therefore not all the values can be stored. Thus *global_mean* is calculated using Knuth's algorithm [5,9]. We then calculate the absolute percentage difference between the two means: *windowed_mean* and *global_mean*. If the percentage difference is more significant than the specified threshold (by default is 5%), then the time tick t for j^{th} VM is regarded as the change point.
- **Z-score-based Detector:** This detector is based on the calculation of the Z-scores [4,6]. Similar to the Mean-based detector, here also a *windowed_mean*, i.e., the mean of all the values in the window, and the *global_mean*, i.e., the mean of all the values until the current time tick is calculated. We additionally calculate the *global_stand_deviation*, i.e., the standard deviation of all

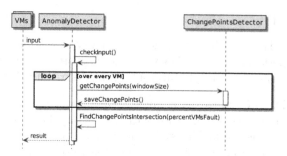

Fig. 4. Indirect Anomaly Detection (IAD) algorithm workflow sequence diagram

the values until the current time tick. Since the IAD algorithm is designed for running it in an online way, *global_stand_deviation* is calculated using Welford's method [9]. These statistics are then used for the calculation of the z-scores for all the data points in the window using Eq. 1.

$$z\_scores = \frac{(windowed\_mean - global\_mean)}{\frac{global\_stand\_deviation}{\sqrt{w}}} \tag{1}$$

If the Z-scores of all windowed observations are greater than the defined threshold ($3 \times global\_stand\_deviation$) then the time tick t for j^{th} VM is regarded as the change point.

In the main algorithm, only *Z-Ssore-based Detector* is used as it provides higher accuracy and has fewer false positives.

Anomaly Detector. This component receives the input resource utilization data X of size $n \times d$ where d is the number VMs hosted on a VMM along with the `minPercentVMsFault` (Table 1)) as the input parameter. We first check the input timeseries of w length for 1) zero-length timeseries and 2) if the input timeseries of all VMs are of the same length or not. If any of the two initial checks are true, then we quit and don't proceed ahead. We assume that all the VM's resources utilization data is of the same length only. After doing the initial checks, each of the VM's windowed timeseries belonging to the VMM is sent to the *Change Points Detector* for the detection of whether the time tick t is a change point or not. If the percentage number of VMs ($\{c_t^1, c_t^2, ..., c_t^m\}$ out of d) having the change point at time tick t is greater than the `minPercentVMsFault` input parameter, then the VMM is reported as anomalous at time tick t. The above procedure is repeated for all time ticks. Figure 4 shows the workflow sequence diagram of the IAD algorithm. Furthermore, the developed approach can be applied for multiple VMMs as well.

3.2   Test Module

This component is responsible for generating the synthetic data and evaluating the algorithm performance by calculating the F1-score on the results from the algorithm. It consists of multiple sub-component described below:

- **Synthetic Data Generator:** It takes the number of VMMs, number of VMs per VMM, percentage of the VMs with a fault; as the input for generating synthetic timeseries data. This synthetic data follows a Gaussian distribution based on the input parameters. This component also automatically divides the generated data into true positive and true negative labels based on the percentage of the VMs with a fault parameter.
- **Algorithm Tester:** It is responsible for invoking the algorithm with various parameters on the synthetic data and tune the algorithm's hyperparameters.
- **Evaluation:** The results from the algorithm are passed as the input to this sub-component, where the results are compared with the actual labels, and the overall algorithm score in terms of F1-score is reported.

4   Experimental Settings

We design our experiments to answer the following questions:

Q1. Indirect Anomaly Detection Accuracy: how accurate is IAD in the detection of anomalous VMM when compared to other popular algorithms?
Q2. Anomalous VMMs Finding Efficiency and Scalability: How does the algorithm scale with the increase in the data points and number of VMs?

4.1   Datasets

For evaluating the IAD algorithm, we considered four types of datasets listed in Table 2 along with their information and are described below:

Synthetic: This is the artificially generated dataset using the *Test Module* component described in Sect. 3.

Experimental-Synthetic Merged: This is a dataset with a combination of experimental data and synthetic data. We created two nested virtual machines on a VM in the Google Cloud Platform to collect the experimental dataset. The underneath VM instance type is n1-standard-4 with four vCPUs and 15 GB of memory, and Ubuntu 18.04 OS was installed on it. This VM instance acts as a host for the above VMs. *libvirt* toolkit is used to manage and create nested virtualization on top of the host machine. Kernel-based Virtual Machine (KVM) is used as a VMM. The configuration of the two nested VMs are i) 2vCPU and 2 GB memory, ii) 1vCPU and 1 GB memory. Cloud-native web applications were run on these two VMs. Monitoring data from the two VMs and underneath host is exported using the Prometheus agent deployed on each of them to an external virtual machine. *stress-ng* is used for generating the load on the VMM. Based on this infrastructure, we collected a dataset for various scenarios and combined it with the synthetic data.

Table 2. Datasets used in this work for evaluating the algorithms.

Dataset Name	Anomalous VMMs	Non-anomalous VMMs	VMs Per VMM	TimeTicks per VM
Synthetic	5	5	10	1000
Exp-Synthetic merged	42	17	2 (experimental) 8 (synthetic)	5400
Azure[1]	16	10	10	5400
Alibaba[14]	10	10	10	5400

[†]These are modified for our usecase.

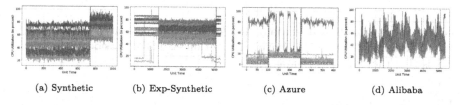

(a) Synthetic   (b) Exp-Synthetic   (c) Azure   (d) Alibaba

Fig. 5. An example profile of an anomalous VMM having 10 VMs in all the datasets used in this work for evaluation.

Azure Dataset: This dataset is based on the publicly available cloud traces data from Azure [1]. We used the VMs data from it and created random groups of VMs, with each group representing the VMs hosted on a VMM. Afterward, we feed these timeseries groups in our synthetic data generator for randomly increasing or decreasing the CPU utilization of the VMs within a VMM based on the input parameters to create anomalous and non-anomalous VMMs.

Alibaba Dataset: This dataset is based on the publicly available cloud traces and metrics data from Alibaba cloud [14]. A similar method as the *Azure Dataset* was also applied to form this dataset.

Figure 5 shows an example profile of an anomalous VMM for all the datasets.

4.2   Evaluated Algorithms

We compare IAD to the five other algorithms listed in Table 3 along with their input dimension and parameters. ECP is a non-parametric-based change detection algorithm that uses the E-statistic, a non-parametric goodness-of-fit statistic, with hierarchical division and dynamic programming for finding them [3]. BnB (Branch and Border) and its online version (BnBO) are also non-parametric change detection methods that can detect multiple changes in multivariate data by separating points before and after the change using an ensemble of random partitions [2]. Lastly, we use the popular anomaly detection algorithm: isolation forest for detecting anomalous VMM [10]. The primary isolation forest (IF) works on the input data directly, while we also created a modified version of it called the isolation forest features (IFF), which first calculates several features

Table 3. The details of the algorithms used in this work for evaluation, along with their input dimension and parameters.

Algorithm	Input dimension	Parameters
IAD	n × d	w, minPercentVMsFault
ECP [3]	n × d	Change points, Min. points b/w change points
BNB [2]	n × d	w, number of trees, threshold for change points
BNBOnline [2]	n × d	w, number of trees, threshold for change points
IF [10]	n × d	Contamination factor (requires training)
IFF [10]	n × features	Contamination factor (requires training)

such as mean, standard deviation, etc., for all values within a window on the input dataset and then apply isolation forest on it. The downside of the IF and IFF is that they require training.

4.3   Other Settings

We have used F1-Score (denoted as F1) to evaluate the performance of the algorithm. Evaluation tests have been executed on 2.6 GHz 6-Core Intel Core i7 MacBook Pro, 32 GB RAM running macOS BigSur version 11. We implement our method in Python. For our experiments, hyper-parameters are set as follows. The window size w is set as 1 min (60 samples, with sampling done per second), threshold k as 5%, and percentVMsFault f as 90%. However, we also show experiments on parameter sensitivity in this section.

5   Results

Our Initial experiments showed that 1) CPU metric is the most affected and visualized parameters in the VMs when some load is generated on the VMM; 2) All or most VMs are affected when a load is introduced on the VMM.

5.1   Q1. Indirect Anomaly Detection Accuracy

Table 4 shows the best F1-score corresponding to each algorithm evaluated in this work (Sect. 4.2) and on all the datasets (Sect. 4.1). We can observe that *IAD* algorithm outperforms the others on two datasets, except for the Experiment-Synthetic dataset (BNB performed best with F1-Score of 0.90) and Alibaba dataset (IFF performed best with F1-Score of 0.66. However, if one wants to find an algorithm that is performing well on all the datasets (Average F1-score column in Table 4), in that case, *IAD* algorithm outperforms all the others with an average F1-score of 0.837 across all datasets.

Furthermore, we present the detailed results of the algorithms on all four datasets varying with the number of VMs and are shown in Fig. 6. One can observe that *IAD* performs best across all the datasets, and its accuracy increases

Table 4. F1-score corresponding to each algorithm evaluated in this work (Sect. 4.2) and on all the datasets (Sect. 4.1)

Algorithm	Synthetic	Exp-Synthetic	Azure	Alibaba	Average F1-score
IAD	**0.96**	0.86	**0.96**	0.57	**0.837**
ECP	0.67	–	0.76	0.51	0.64
BNB	0.62	**0.90**	0.8	0.33	0.662
BNBOnline	0.87	0.81	0.86	0.4	0.735
IF	0.76	0.83	0.76	0.2	0.637
IF Features (IFF)	0.76	0.83	0.76	**0.66**	0.75

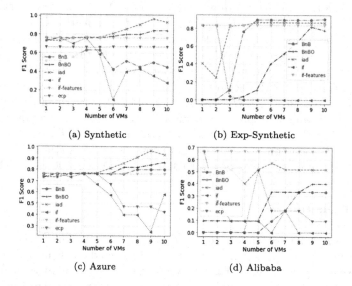

(a) Synthetic   (b) Exp-Synthetic

(c) Azure   (d) Alibaba

Fig. 6. F1-score variation with the number of VMs corresponding to each algorithm evaluated in this work (Sect. 4.2) and on all the datasets (Sect. 4.1)

with the increase in the number of VMs. Additionally, after a certain number of VMs, the F1-score of *IAD* becomes stable. This shows that if, for example, we have the synthetic dataset, then the best performance is possible with VMs ≥ 9. Similarly, in the case of the Azure dataset, while for the Exp-Synthetic dataset, one needs at least five VMs, and for the Alibaba dataset, seven VMs for the algorithm to perform well.

5.2   Q2. Anomalous VMMs Finding Efficiency and Scalability

Next, we verify that our algorithm's detection method scale linearly and compare it against other algorithms. This experiment is performed with the synthetic dataset, since we can increase the number of VMs per VMM in it. We linearly increased the number of VMs from 1 to 100 and repeatedly duplicated

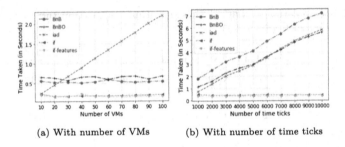

(a) With number of VMs   (b) With number of time ticks

Fig. 7. Algorithm's detection method scalability with respect to different parameters.

our dataset in time ticks by adding Gaussian noise. Figure 7 shows various algorithm's detection method scalability for different parameters. One can observe that *IAD's* detection method scale linearly in terms of both the parameters. However, when the number of VMs are scaled to 100, *IAD* takes a longer time as compared to others, but it provides results under 2.5 s which if we see is not that much considering the accuracy we get with that algorithm. However, on the time ticks parameter, *BNB*, *BNBOnline* and *IAD* performed similar to each other, while *IF* and *IFF* provides results under 1 s, but its accuracy is worse as compared to the others on all the datasets, and it has the extra overhead of training. *ECP* algorithm's results are not shown, since it requires more than an hour for performing the detection with 100 VMs and 100,000 time ticks.

6   Conclusion

We propose *IAD* algorithm for indirect detection of anomalous VMMs by solely using the resource's utilization data of the VM's hosted on them as the primary metric. We compared it against the popular change detection algorithms, which could also be applied to the problem. We showcased that *IAD* algorithm outperforms all the others on an average across four datasets by 11% with an average accuracy score of 83.7%. We further showcased that *IAD* algorithm scale's linear with the number of VMs hosted on a VMM and number of time ticks. It takes less than 2.5 s for *IAD* algorithm to analyze 100 VMs hosted on a VMM for detecting if that VMM is anomalous or not. This allows it to be easily usable in the cloud environment where the fault-detection time requirement is low and can quickly help DevOps to know the problem is of the hypervisor or not.

The future direction includes using other metrics like network and storage utilization to enhance the algorithm's accuracy further.

References

1. Cortez, E., Bonde, A., Muzio, A., Russinovich, M., Fontoura, M., Bianchini, R.: Resource central: understanding and predicting workloads for improved resource management in large cloud platforms. In: Proceedings of the 26th Symposium on Operating Systems Principles, SOSP 2017, pp. 153–167. Association for Computing Machinery, New York, NY, USA (2017). https://doi.org/10.1145/3132747.3132772

2. Hooi, B., Faloutsos, C.: Branch and border: Partition-based change detection in multivariate time series. In: SDM (2019)
3. James, N.A., Matteson, D.S.: ecp: an r package for nonparametric multiple change point analysis of multivariate data (2013)
4. Jindal, A., Gerndt, M., Bauch, M., Haddouti, H.: Scalable infrastructure and workflow for anomaly detection in an automotive industry. In: 2020 International Conference on Innovative Trends in Information Technology (ICITIIT), pp. 1–6 (2020). https://doi.org/10.1109/ICITIIT49094.2020.9071555
5. Knuth, D.E.: The Art of Computer Programming, vol. 2 3rd edn. Seminumerical Algorithms. Addison-Wesley Longman Publishing Co., Inc., New York (1997)
6. Kochendörffer, R.: Kreyszig, E.: Advanced Engineering Mathematics. Wiley, New York, London 1962. ix + 856s. 402,abstract, p. 79.-. Biometrische Zeitschrift **7**(2), 129–130 (1965). https://doi.org/10.1002/bimj.19650070232, https://onlinelibrary.wiley.com/doi/abs/10.1002/bimj.19650070232
7. Le, M., Tamir, Y.: ReHype: enabling VM survival across hypervisor failures. In: Proceedings of the 7th ACM SIGPLAN/SIGOPS International Conference on Virtual Execution Environments, VEE 2011, pp. 63–74. Association for Computing Machinery, New York, NY, USA (2011). https://doi.org/10.1145/1952682.1952692
8. Li, M.L., Ramachandran, P., Sahoo, S.K., Adve, S.V., Adve, V.S., Zhou, Y.: Understanding the propagation of hard errors to software and implications for resilient system design. In: ASPLOS 2008 (2008)
9. Ling, R.F.: Comparison of several algorithms for computing sample means and variances. J. Am. Stat. Assoc. **69**(348), 859–866 (1974). https://doi.org/10.1080/01621459.1974.10480219, https://www.tandfonline.com/doi/abs/10.1080/01621459.1974.10480219
10. Liu, F.T., Ting, K.M., Zhou, Z.H.: Isolation forest. In: 2008 Eighth IEEE International Conference on Data Mining, pp. 413–422 (2008). https://doi.org/10.1109/ICDM.2008.17
11. Nikolai, J., Wang, Y.: Hypervisor-based cloud intrusion detection system. In: 2014 International Conference on Computing, Networking and Communications (ICNC), pp. 989–993 (2014)
12. Parashar, M., AbdelBaky, M., Rodero, I., Devarakonda, A.: Cloud paradigms and practices for computational and data-enabled science and engineering. Comput. Sci. Eng. **15**(4), 10–18 (2013). https://doi.org/10.1109/MCSE.2013.49
13. Reinhardt, S.K., Mukherjee, S.S.: Transient fault detection via simultaneous multithreading. In: Proceedings of the 27th Annual International Symposium on Computer Architecture, ISCA 2000, pp. 25–36. Association for Computing Machinery, New York, NY, USA (2000). https://doi.org/10.1145/339647.339652
14. Shan, Y., Huang, Y., Chen, Y., Zhang, Y.: LegoOS: a disseminated, distributed OS for hardware resource disaggregation. In: Proceedings of the 13th USENIX Conference on Operating Systems Design and Implementation, pp. 69–87. OSDI'18, USENIX Association, USA (2018)
15. Xu, X., Chiang, R.C., Huang, H.H.: Xentry: hypervisor-level soft error detection. In: 2014 43rd International Conference on Parallel Processing, pp. 341–350 (2014). https://doi.org/10.1109/ICPP.2014.43

STRAPS

Comparing Graph Data Science Libraries for Querying and Analysing Datasets: Towards Data Science Queries on Graphs

Genoveva Vargas-Solar[1]([⊠]), Pierre Marrec[2], and Mirian Halfeld Ferrari Alves[3]

[1] CNRS, Univ Lyon, INSA Lyon, UCBL, LIRIS, UMR5205, 69622 Paris, France
genoveva.vargas-solar@cnrs.fr
[2] Ecole Natoinal Supérieur de Lyon, Lyon, France
pierre.marrec@ens-lyon.fr
[3] University of Orléans, LIFO, Orléans, France
mirian@univ-orleans.fr

Abstract. This paper presents an experimental study to compare analysis tools with management systems for querying and analysing graphs. Our experiment compares classic graph navigational operations queries where analytics tools and management systems adopt different execution strategies. Then, our experiment addresses data science pipelines with clustering and prediction models applied to graphs. In this kind of experiment, we underline the interest of combining both approaches and the interest of relying on a parallel execution platform for executing queries.

1 Introduction

Vast collections of heterogeneous data containing observations of phenomena have become the backbone of scientific, analytic and forecasting processes for addressing problems in domains like Connected Enterprise, Digital Mesh, and Internet-connected things and Knowledge networks. Observations can be structured as networks that have interconnection rules determined by the variables (i.e., attributes) characterising each observation.

The graph concept is a powerful mathematical concept with associated operations that can be implemented through efficient data structures and exploited by applying different algorithms. Note that relations among observations and interconnection rules are often not explicit, and it is the role of the analytics process to deduce, discover and eventually predict them. Take, for example, a graph-based representation of the plot of the famous Saga Game of Thrones. In this graph, it is possible to (i) Ask simple queries like the number of characters of the Saga?; (ii) build communities to know which are the build communities the houses in which some characters are organised and how influential they are?; (iii) compute the popularity of characters and observe its evolution; (iv) build maps

This work was funded by the Quasimodo action in U. Orléans and the DOING action of the GDR MADICS.

H. Hacid et al. (Eds.): ICSOC 2021 Workshops, LNCS 13236, pp. 205–216, 2022.
https://doi.org/10.1007/978-3-031-14135-5_16

to describe the geographical distribution of the countries; (v) be more ambitious and predict who can be the next final King or Queen?

If we group the querying techniques, we can do it across two families. The first one concerns querying as we know it in comics and information research. Here the principle is that pipelines explore and analyse the data to profile it quantitatively and with the objective of either modelling, prediction or recommendation. In the first case, the results have a notion of completeness and probabilistic approximation. While in the other family, the results have an associated degree of error, and they may be data and queries or data samples. In this paper, we tackled exploratory queries that tackle data collections that are expanding or where the structure provides little knowledge about the data. These queries run step-by-step like pipelines, and the tasks often apply statistical, probabilistic, or data mining and artificial intelligence processing functions. Methodologies are still to come to integrate data management with the execution of algorithms that are often greedy. Of course, we are not the only ones interested in this type of query in https://www.overleaf.com/project/6136432e28f08785fdc175bcits design and execution.

Existing technology, including graph stores with different models and properties and querying facilities and analytics libraries with built-in graph analytics algorithms, provide tools for exploring graphs. The question is, which conditions the different solutions are better adapted to address different analytics queries. This paper describes an experimental approach for profiling and identifying existing tools' characteristics and how they are different and complementary.

This paper presents an experimental study to compare the graph analysis approaches with the graph management and query approach. We show that the purely analytical approach achieves better execution's performance than the data management system approach for relatively small datasets. Besides, we created a model that predicts the future interactions of the characters from The Song of Ice and Fire with learning tools.

Networkx allows graph processing rather than graph database management. The graphs have a dictionary architecture, in particular, to store strings and put attributes on the edges. Oriented and non-oriented graphs are two different objects in this package. We used notebook environments to code in Python with Networkx. Initially, the notebook was hosted in Kaggle. This comparison aims to see the difference in performance on data analysis functions between a package used to analyse graphical databases (Networkx) and software that allows the management of these databases (Neo4j).

The remainder of the paper is organized as follows. Section 2 discusses related work. Section 3 describes our study strategy including the experimental settings, datasets and discusses the obtained results. Section 4 concludes the paper and discusses future work.

2   Related Work

The most classic solutions are graph stores and systems that provide built-in graph operations organised in two families as shown in Fig. 1. Those systems implement well-known graph operations like community detection (shown in number one in Fig. 1) centrality where we find, for example, page rank and betweenness (number 2 in the Fig. 1), similarity (number 3 in the Fig. 1) and pathfinding and search like standard networks.

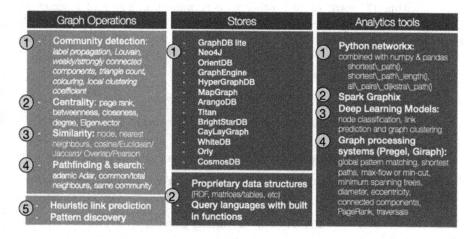

Fig. 1. Graph management and processing systems

The second family is based on data mining and machine learning techniques (shown in number 5 in Fig. 1) like heuristic link prediction and pattern discovery. Stores provide persistence support. There are many prominent commercial systems shown in number 1 in Fig. 1. They provide proprietary data structures and more or less declarative query languages with built-in functions like the ones presented before. Finally, analytics tools provide also solutions for processing graphs. For example, Python Networkx showed in number 1 in Fig. 1, Spark Graphix (number 2 in Fig. 1), deep learning models for node classification, link prediction and graph clustering (number 3 in Fig. 1) and graph processing systems like Pregel and Giraph.

3   Experimentally Comparing the Execution of Data Science Operations on Graphs on DBMS or All-in-One Programs

Our work aims to study the difference in performance on data analysis functions between a package that is used to analyse graphical databases (Networkx) and software that allows the management of these databases (Neo4j) when they are used for defining data science pipelines.

3.1   Graph and Experimental Setting

For the dataset, we have chosen the data of the five books of the saga *The song of ice and fire* which has been extensively studied and represented in graph form. It is an epic novel, and as such, the characters are organised in houses represented by kings and queens who are lords or ladies of the regions. Knights engage in battles in different places, and of course, there are deaths in these battles.

Characters of The Wise The song of Ice and Fire created by Andrew Beveridge [1]. To create this dataset, he looked in the books to see which characters appear within 15 words of each other to determine the degree of interaction. By adapting it to our needs, we got a graph with about 800 nodes and 3000 relations. This adaptation allows for a graph that is not too big to keep the calculation times decent (see Fig. 2).

(:Person)-[:INTERACTS]→(:Person)

Graph enriched with data on houses, battles, commanders, kings, knights, regions, locations, and deaths

:Person nodes, representing the characters (800 nodes),
:INTERACTS relationships, representing the characters' interactions (3000 edges per character)

An interaction occurs each time two characters' names (or nicknames) appear within 15 words of one another in the book text [2]

Fig. 2. Game of Thrones graph

We selected five functions that are implemented on Neo4j and Networkx:

- *Page rank* measures the importance of each node within a graph, based on the number of incoming relationships and the importance of the corresponding source nodes. The underlying assumption is that a page is only as important as the pages that link to it. We assumed Neo4j and Networkx used the same implementation of these algorithms as in [2]. We configured the damping = 0.85 and max iterations = 20 in both environments.
- *Betweenness centrality* is a way of detecting the amount of influence of a node over the flow of information in a graph. It is often used to find nodes that serve as a bridge from a subgraph to another. The algorithm calculates unweighted shortest paths between all pairs of nodes in a graph. Each node receives a score based on the number of shortest paths that pass through the node. Nodes that more frequently lie on shortest paths between other nodes will have higher betweenness centrality scores.
- *Label propagation* finds communities in a graph using the graph structure alone as its guide and does not require a pre-defined objective function or prior information about the communities. The intuition behind the algorithm is that a single label can quickly become dominant in a densely connected group of nodes but will have trouble crossing a sparsely connected region. Labels will get trapped inside a densely connected group of nodes, and those

nodes that end up with the same label when the algorithms finish can be considered part of the same community. Both environments implement the same version of the algorithm[1].

– *Breadth-First Search* is a graph traversal algorithm that, given a start node, visits nodes in order of increasing distance. Multiple termination conditions are supported for the traversal, based on either reaching one of several target nodes, reaching a maximum depth, exhausting a given budget of traversed relationship cost, or just traversing the whole graph. The output of the procedure contains information about which nodes were visited and in what order.

– *Minimum Spanning Tree* is a kind of pathfinding algorithm. It starts from a given node and finds all its reachable nodes and the relationships that connect them with the minimum possible weight. Prim's algorithm [3] is one of the simplest and best-known minimum spanning tree algorithms.

The study focused on the execution times by minimizing the delays that do not depend on the algorithm.

Since we were looking at the way the environments execute the pipelines that analyse the graphs, it is important to compare both approaches from an architectural point of view that determines execution conditions (see Fig. 3). In the case of Python, when we use Jupyter notebooks. A client machine with a browser has access to a file system that holds the data, and it has access to a Jupyter server that has access to Python interpreters. At runtime, the data is all loaded into RAM; this poses some resource management problems.

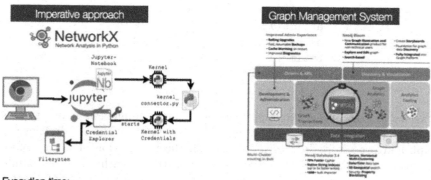

Execution time:
- The first iteration of the algorithm
- The average time of the following iterations where the Neo4j's cache memory reduces the computation time

Fig. 3. Graph analytics execution environments

For the Networkx Python environment we built the following graph. The specification of the graphh in Neo4J is given within the expression of the query.

---

[1] https://neo4j.com/docs/graph-data-science/current/algorithms/label-propagation/.

```
import networkx as nx
G=nx.Graph(name="Game of Networks")
n=len(table['Source'])
for i in range(n):
    G.add_edge(table['Source'][i],table['Target'][i],
               weight=table['weight'][i])
```

The following expressions compare the code used in Python programs and Cypher to express data science queries, exploring the graph and answering these queries.

- *Q-1: Which are the most influencial characters of the novel?*
  - Cypher expression:
    ```
    CALL gds.alpha.betweenness.stream({
        nodeQuery: 'MATCH (p) RETURN id(p) AS id',
        relationshipQuery: 'MATCH (p1)-[]-(p2)
        RETURN id(p1) AS source, id(p2) AS target'})
    YIELD nodeId,centrality
    return gds.util.asNode(nodeId).name
        as user, centrality
        order by centrality DESC limit 1
    ```

  - Python program using Networkx method nx.betweenness_centrality(G).
    ```
    list=[]
    for i in range(100):
        a=time()
        nx.betweenness\_centrality(G)
        b=time()
        liste.append(b-a)
    ```

Of course, Python promotes imperative query programming, assuming that the underlying infrastructure provides enough main memory space to retrieve the graph and process it. In the case of Neo4J, the preparation of main memory allocation, the tuning of specific parameters of the algorithm like the number of iterations, the precision objective to define a termination condition must be executed before the code shown above. Neo4J also works with graphs in main memory when applying data science functions. The graphs are views of persistent graphs defined using Cypher. The view can provide a subset of nodes respecting some restriction, and relations can be directed/labelled or not. It is up to the programmer to store the view and results upon the termination of the process.

Results. In the case of Neo4J, the graphs are stored and can be queried declaratively, but when applying graph analysis functions, the system requires the user to manage the memory and the routing of the graph pieces to the execution space. So we compared their behaviour concerning execution time: particularly the time cost of the first iteration of the algorithms, and then calculating the average execution time for the following iterations to look at the advantage of having a cache in the case of Neo4J no-cache in the case of Python.

Algorithm	Time Neo4j (s)	Time Networkx(s)
Betweenness centrality	0,32 (0,08 after)	1,6
Page rank (unweighted)	0,36 (0,1 after)	0,09
Page rank(weighted)	0,36 (0,1 after)	0,12
Label Propagation	0,17 (0,05 after)	1e-6 (1e-5 for the first one)
Breadth First Search	38	3e-7 (2e-6 for the 2 first ones)
Minimum spanning tree	0,52 (0,06 after)	7e-3

Fig. 4. Comparison of execution results

3.2   Graph Data Science Pipeline

We implemented data science queries as pipelines that combine graph matching queries and aggregations (see Fig. 5). The first group of tasks includes resource estimation (main memory) and data preparation. The second group of tasks include exploratory, modelling and prediction operations and results assessment. However, data preparation has not been considered in our performance comparison since we are interested in comparing data science operations execution. Using this pipeline, we could solve a set of analytics questions by implementing notebooks in Python and Cypher queries.

- Q_1 - *Which are the houses that challenge the thrones and how influential are they?* This question was answered applying centrality algorithms namely page rank and betweeness centrality.
- Q_2 - *Which are the most popular characters?* This question was answered applying centrality algorithms namely page rank and betweeness centrality.
- Q_3 - *Which are the houses that challenge the thrones?* We used the community detection family to answer this question.
- Q_4 - *Who are the leading characters in Game of Thrones?* The notoriety of characters was analysed with the breadth first search and minimum spanning tree.

For Networkx, we performed many tests, so the uncertainty is low. There are significant order-of-magnitude differences for specific algorithms, such as the width path. Indeed, if we did not limit the maximum depth to 5, the algorithm did not finish (or its execution was very long). Afterwards, it is expected that there is a difference because Neo4j also manages the graph in real-time, whereas for Networkx, we had to recreate the graph each time we launched the notebook.

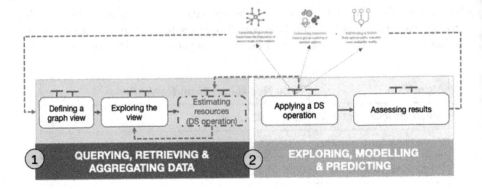

Fig. 5. Graph analytics general pipeline

3.3   Link Prediction

For the link prediction part integration of two platforms with a parallel programming model with Spark[2]. So, for the prediction, we compared different strategies by including properties of characters represented by the node to discover links that would be hidden. Secret relations to beat a king or conquer a house. So our pipeline developed different complementary branches with richer analysis to discover as many new relations as possible. Figure 6 shows the general pipeline implemented for discovering links among the novel characters according to different sets of attributes.

Data Preparation. The dataset is divided into 2 parts. The first contained about 60% of the data and served as a learning set. The second contains 40% of the data, and it is used to test the model's performance.

Each of the two sets comprises a certain number of pairs of nodes connected by an edge and the same number of nodes not connected by an edge. Provided that, in general, there are far fewer existing edges than possible edges, the number of unconnected node pairs in each set had to be reduced beforehand.

Specifying Characteristics. We created characteristics for the nodes in the graph to correctly classify the edges so that our forest of trees could correctly. The choice of these characteristics is the tricky part of link prediction because the suitable characteristics depend a lot on the graph's topology. These characteristics are often values calculated by a graph analysis algorithm such as Page Rank but can also be more specific functions such as the number of neighbours in common. A good feature is a value that allows the model to classify pairs of vertices correctly.

Model 1: Predicting Links Using the Attribute Number of Neighbours. We started by predicting links with the number of neighbours in common as the only characteristic as a criterion. As seen in Fig. 7 characters with few neighbours in

---

[2] https://github.com/gevargas/doing-graph-datascience-queries.

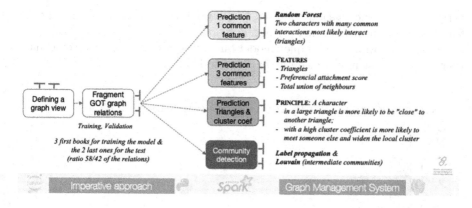

Fig. 6. Link prediction pipeline

common are pretty unlikely to interact with each other, and characters with many neighbours in common interact with each other. This observation gives a good clue about the usefulness of this characteristic in differentiating between interactions that will and will not exist. After training the model with just this characteristic, the model is already much better than a random classifier.

Model 2: Predicting Links According to Several Characters Characteristics. We added 8 characteristics looking for a model with better prediction performance:

1. The number of neighbours in common.
2. The number of neighbours.
3. The preferential link score which is a coefficient calculated by multiplying for each pair of nodes the number of neighbours each one has.
4. The number of triangles in which the nodes are. More precisely, the maximum number of triangles for a node and the minimum number of triangles.
5. The clustering coefficient. Here we also have the maximal and minimum coefficient.
6. The community detection by Leuven and Label Propagation (same Louvain and same Partition). This is simply a Boolean value that indicates whether two nodes are in the same community calculated by Louvain or by Label Propagation.

The Scikit learn package in Python provides a function to display the importance of the different features in the model. Accuracy and memorization have been greatly improved, and the accuracy is still relatively high. The area under the curve is now very close to 1.

Finally, we have a model that makes predictions with an acceptable success rate. It does not perform as well as one would want to use it on a large scale. (In any case, it is a model that predicts interactions in a series of books, so the usefulness is quite limited). Nevertheless, it has the merit of having an acceptable performance for such a small data set. It also provides a method for finding good features and improving a link prediction model.

Here, if we wanted a simpler model but still quite efficient, we could have kept only the first model (see Fig. 7) characteristics.

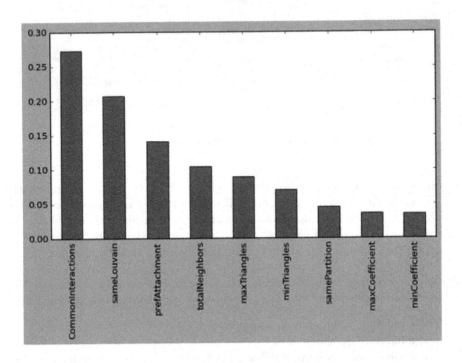

Fig. 7. First model characteristics

Results. During training, we tested different characteristics, which seemed logical considering that our graph is a graph of the relationships among people. For example, two people who are not related but with a large number of neighbours. The characters in common seem to be more likely to interact in the future.

Figure 8 summarises the different assessment results with the different experimental settings tested for link prediction. These tasks correspond to the assessment part of our data science pipeline. The link prediction query could be defined in a general manner as a template since it was designed as an abstract pipeline. Then, different pipeline instances adopting different strategies for defining the graph view led to our experimental panel.

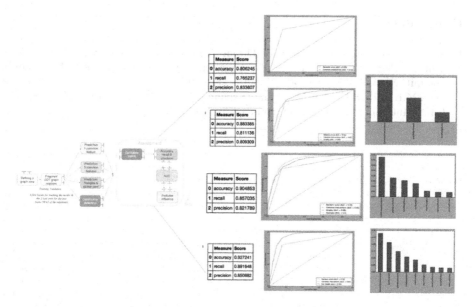

Fig. 8. Link prediction models assessment

4 Conclusions and Future Work

This paper described and reported on an experimental comparative study to compare the imperative and declarative paradigms for programming data science pipelines on graphs. Imperative approaches rely on libraries and execution environments with no built-in options for managing graph views, resources allocation and graph persistence. In contrast, declarative approaches relying on underlying graph management systems profit from the manager's strategies for managing the graphs on disk and main memory. Our link prediction experiment showed that using the graph management system for creating views can be very elegant and sound; Then, given the cost of the algorithm, relying upon a parallel execution framework as Spark provides a more natural way of dealing with main memory allocation.

Based on these observations about graphs and other related work, our current work addresses the efficient execution of pipelines applied to the analysis of graphs. We are deploying data science pipelines on target architectures such as the cloud and GPUs provide large-scale computing, memory and storage resources to further develop our experiments. There is room for querying and exploiting data through data science queries managed by the environment as first-class citizens for future work.

References

1. Beveridge, A., Shan, J.: Network of Thrones. Math Horiz. **23**(4), 18–22 (2016)
2. Brin, S., Page, L.: The anatomy of a large-scale hypertextual web search engine. Comput. Netw. ISDN Syst. **30**(1–7), 107–117 (1998)
3. Choi, M.B., Lee, S.U.: A prim minimum spanning tree algorithm for directed graph. J. Inst. Internet Broadcast. Commun. **12**(3), 51–61 (2012)

Gamification for Healthier Lifestyle – User Retention

Shabih Fatima[✉], Juan Carlos Augusto, Ralph Moseley, and Povilas Urbonas

Research Group on Development of Intelligent Environments, Department of Computer Science, Middlesex University, London, UK
{SS3283,PU035}@live.mdx.ac.uk, {J.Augusto,R.Moseley}@mdx.ac.uk

Abstract. Gamification is gaining in popularity and is increasingly showing potential to benefit humans. Numerous applications have been developed but more research is required on the key motivational factors responsible for user adherence. The purpose of this research is to investigate strategies to encourage people in adopting a healthy lifestyle using gamification and identifying the key factors responsible for retaining users. The research will not be limited to certain activities such as gym or sports but motivates people to emphasize on small activities, for example, doing household work, walking and many others. This research endeavors to develop an application using Unity and C# which is a real-time platform supported by iOS and Android. The research focuses on Behavior Change Techniques and Self-Determination Theory to tackle the issue of user adherence.

Keywords: Gamification · User retention · Behavior change techniques · Game application · Healthy lifestyle

1 Introduction

Nowadays, a sedentary lifestyle has become a major issue and is a worldwide problem. According to World Health Organisation [1], 39 million children under age of 5 are obese in 2020, more than 340 million of age group 5–19 are overweight and 39% of adults are overweight and 13% are obese i.e. 650 million adopt unhealthy lifestyle in 2016. The main reason for this is owing to minimal mobility and exercise, causing an adverse effect on human wellbeing and leading to incurable ailments such as diabetes, obesity and cardiac issues. It has become a challenge to adopt a healthy lifestyle and stay motivated in the longer term.

The contemporary lifestyle health risks are increasing due to the health behaviour of the people's individual behaviour. The change in behaviour can significantly improve well-being. However, the behaviour such as physical activity, exercise and diet require motivation and behaviour change. In the area of health and well-being, gamification has been increasing in importance [2]. Gamification is defined as "use of game design elements within non-game contexts" [3]. The advantage of using gamification is to enhance user engagement and user experience. Gamification is used in various industries such as the health sector [4], social media [3], education [5] and many more. Game-based technology, also known as gamification, is used to promote intrinsic motivation.

H. Hacid et al. (Eds.): ICSOC 2021 Workshops, LNCS 13236, pp. 217–227, 2022.
https://doi.org/10.1007/978-3-031-14135-5_17

It increases the level of involvement and motivation. There are two types of motivation, namely, the intrinsic motivation rises from doing things 'for own sake' while extrinsic motivation rises from external factors [2]. Extrinsic motivation which includes feedback such as money or vocals while intrinsic motivation includes group quests [6].

According to Kasurinena and Knutas [7], Crowdsourcing and Game of Health is used in the context of gamification. Crowdsourcing is an online task to socially interact where a group of people participate in a task of different levels to gain a mutual advantage. An Example of crowdsourcing is image labelling. On the other hand, Game for Health is to enhance the fitness level of the user by motivating users for health-based activity. It is designed to train the user to be physically active while playing games. It promotes exercise gaming to improve the user's way of living.

However, user adherence is a key factor in the domain of gamification. More insight is required for the driving factors responsible for user retention [8]. The goal of this study is to focus on user retention and encourage them to live a physically active life. The next section provides a state of the art summary of relevant work on user adherence, we explain which retention features we have considered and how we implemented them in the current system. We finalize explaining the various mechanisms we used for continuous assessment of the product and the embedded retention strategies.

2   Literature Review

This section focuses on the research closely related to user retention. The research conducted by Rose et al. [9], designed a gamified mobile health application known as 'MYSUGR' to examine behavior of a group of diabetic patients over 12 weeks. The outcome of this research was positive. Improvement in the blood result was observed. However, the challenge faced by them was to keep users engaged. The retention rate of the app users was 88% but dropped after 28 weeks to 70%. Hence, further investigation is required on factors contributing to user retention.

Findings reported by Stinson et al. [8] suggest that the 'mhealth' pain assessment tool has positive outcome on cancer patients. The quality of life and user participation was improved in teenagers aged between 9–18 over 2-week observation. The tool helped teenagers in pain management. Although the outcome was positive, it still has some drawbacks. The limitation of this study was that more analysis was required for the fundamental factors behind user's engagement, for instance, rewarding systems.

Nathália et al. [10] research developed two types of an m-Health application for hypertension monitoring, one with the game mechanics and other without it, 14 patients with hypertension were categorized in the evaluation stage into four groups such as no gamification with assistance, no gamification and no assistance, gamification with assistance and gamification with no assistance to verify user engagement in health care.

The outcome of the result shows that gamification favoured the engagement and promoted intrinsic motivation in the users. The group with gamification elements managed to control their health in comparison to the group without gamification.

2.1 Motivational Factors/Gamification Elements

Some of user adherence features used by other researchers are as follows:

Competition- Competition caused by leader-boards can create social pressure to increase the player's level of engagement and sense of not being alone [3].

Setting Goals- According to Munson et al. [11] research, setting primary and secondary goals has a positive impact on the application users. If one goal becomes unattainable, the other may still be a realistic goal. If app users are having a good week, they may push themselves to achieve the goal.

Online trophies- Munson et al. [11] research came to the conclusion that Online trophies and ribbons failed to engage most participants, which raises questions about how such rewards can be designed to encourage users to stay active for a long term.

Liking- According to Ozanne et al. [12], Liking behavior can be used for various reasons. One of the reasons is bonding, it is used for congratulating or showing support to others by liking.

2.2 Behavior Change Techniques

BCTs are techniques in which individuals change their behavior to adapt to a healthy lifestyle. It's a key factor for motivation. Michie et al. [13] listed 93 BCTs which are categorized in 16 groups such as Reward, Feedback and Monitoring, Repetition and substitution, Goals and Planning. Self-monitoring was the most effective group among other groups. However, the combination of BCTs results in an increase in physical activity. BCTs are useful and can be used for a long time [14].

2.3 Self-determination Theory

According to Sailer et al. [3] and Shi et al. [15], Self-Determination Theory (SDT) is a theory in which individuals motivate themselves through intrinsic and extrinsic motivation. It is self-engagement where a person feels motivated. In SDT, there are three factors for motivation: 1) autonomy is freedom of taking decisions of your own choice without any pressure, 2) competence is a feeling of success and achievement, 3) Relatedness is feeling connected to others. However, Bovermann et al. [16] discuss four factors of SDT: a) Autonomy, b) Competency, c) Relatedness and d) Purpose which means desire to make something meaningful. 'Purpose' is intrinsic motivation which is to do activities willingly and retain them for a longer term. The SDT factors are necessary to understand which give insight on how the gamification engagement and motivation works. Therefore, this motivation is a key factor for user retention. The study is conducted using SDT and BCT strategies to keep users engaged in physical activity using the *OnTheMove!* App.

2.4 Problem Definition

Many applications such as MyFitnessPal, FitBit, Pokemon Go and Runkeeper [17] have been developed to overcome this problem but still user retention is a major challenge and needs more research. Pokemon Go failed to keep its users engaged for the long term.

57% of users left the app due to boredom and 29% due to not being able to reach a high goal [18].

Therefore, to overcome the problem of obesity and to keep user engagement for a longer term, the application is designed for this research to have a user adherence feature using techniques such as BCT and SDT.

3   Implementing Motivational Factors in OnTheMove!

The aim of this research is to overcome the problem of user retention and implement the factors that keep users motivated and engaged for the long term.

3.1   Overview of OnTheMove! Application

The *OnTheMove!* application has been designed using BCTs and self -determination theory. Some of the BCTs and SDT used in this app are Goal Settings, Habit Formation, Self-Monitoring Behavior, Rewards and Competitions. The application mainly focuses on user retention and encourages users to stay active using small physical activities. It gets user's steps and then converts it into virtual coins. (Fig. 1). The coins can be used to buy avatars or real rewards. Some of the bonus rewards are daily, weekly, monthly and

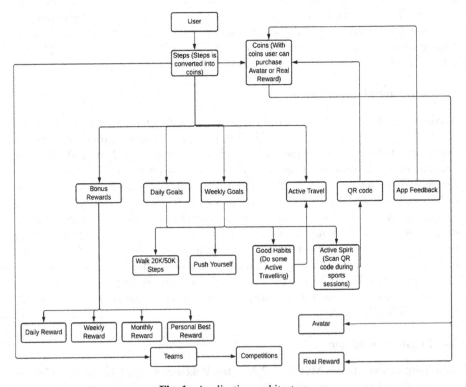

Fig. 1. Application architecture

"personal best" reward to keep users motivated. The users can also transfer their steps to different goals or team competitions. There are different types of incentives, some are increased rewards for more activity, but we also have the age factor to bring into account where old age users get benefits for their efforts. One of the main tools used in the app is QR code which is complementary to step counting and used to associate rewards to user behaviors. It allow users to be rewarded for more than steps counting, for instance, activities which cannot measure well with steps such as swimming or perhaps lifting weights.

Some of the features of the application:

Goal Setting- Goal Setting is to self-regulate and monitor the behaviour. It also satisfies the user's need for autonomy and purpose. Users can set and change their daily and weekly goals. (Fig. 2). The experiment conducted by Munson et al. [11] states that the outcome of having primary and secondary goals was positive. If a user fails to achieve one goal, they can still work on the second goal.

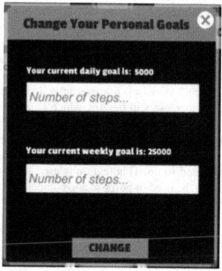

Fig. 2. Daily goal and change personal goal

Daily Streak- Daily Streak is used in the app to form new habits. This habit formation which becomes a long term habit after repetition of behaviour. Daily Streak is when the user is active and opens the app seven days regularly. They get some rewards. If the user will miss one day of completing a daily goal or challenge, the streak will be broken, and the user will start over again. The 7-day streak should keep the user motivated to complete any goal or challenge daily in order to collect bigger rewards once completing the 7-day streak.

Weekly Reward- The weekly reward is implemented to help users to develop a new habit of staying active. This is for user retention where after a certain period of time users get stuck to the routine due to behavior change. When the user is active throughout the week and has more than 10000 steps in a week, the weekly reward window appears on Monday and the user can claim their reward. This weekly reward will keep the user motivated throughout the week. The reward given to users on the basis of fidelity and effort.

Reward Effort- The main aim of this feature is to reward the effort of the users. This is to reward fidelity or loyalty of the users. If a person is physically inactive and more engaged in activities like watching TV or sitting and playing video games for a long time, they are determined to change their lifestyle. This feature will encourage them by rewarding the user for the effort and courage to change the lifestyle from sedentary to active. However, the aim is to make people physically active not necessarily by running, swimming or jogging. The user will get rewarded for changing their lifestyle. This is done by using the concept *"Personal Best"*. The term Personal Best is used in athletics. The algorithms are used to know the best time users have ever done the specific task such as running, jogging etc. *For example.* for a beginner it could be 35' for 5K, but after running for a couple of years it could be 20' (depending on age). If one day the person does S steps and next time s/he increases her/his latest PB by I%, with a fixed I%, this I coefficient will have to start big and diminish with time. For example, if a person's first walk is 100ms it should not be that difficult to walk another 100 ms (which is a 100% increase!) however after a year or so if that person manages to walk 10K, a 100% increase will be 20K which is too big of a jump. To start with some minimum distance D (easy but meaningful, can be 500ms) and first reward 100% improvement of that initial PB, then reduce the expectation in a factor of a constant C, i.e., (100-C)%, then (100-2C)%, then (100-3C)%, etc. Example if $C = 10$ we go 100%, 90%, 81%, 73%, 64%, and so on. This feature is used in this app to praise and encourage users which satisfy the need of purpose and competence.

Real Rewards- Real rewards are provided by the stakeholders which includes hot meals, Sandwich, Gym and swimming sessions. (Fig. 3). This reward raises extrinsic motivation. It gives a sense of achievement and appreciation.

Team Competition- In the *OnTheMove!* App users can create, join and view teams. Also, they can create and join competition. The competitions can be between two or more organizations. (Fig. 4). The Team Competition is used to give users the feeling of competence and relatedness with other users.

Age Band- Users can add their age. Steps are incremented according to users age e.g. if users age is 51 and steps walked is 112 then $112 * 5.1 = 571.2$ steps (Fig. 5). If the users are above the age of 40, the steps are adjusted accordingly. This is for user adherence and to make it achievable for older users. In addition, to reward them for their efforts.

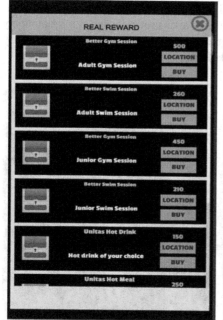

Fig. 3. Real reward   **Fig. 4.** Team competition and like

```
If (int.Parse(userData[11]) >= 40)
{
    todayStepsDisplays.text = (int.Parse(userData[5]) * int.Parse(userData[11]) / 10).ToString();
    dailyStepCount = (int.Parse(userData[5]) * int.Parse(userData[11]) / 10);
}
else
{
    todayStepsDisplays.text = userData[5];
    dailyStepCount = int.Parse(userData[5]);
}
```

Fig. 5. Screenshot of code from Age Band feature

Liking- To keep users motivated, a Like feature has been implemented. The users can like the steps of other users to congratulate or show their support. (Fig. 4). This feature has been implemented to give users a sense of social support and relatedness.

QR Code- The QR code feature is implemented to reward efforts of users, for instance, PE teachers reward students with positive attitudes even if they have perhaps not been the fastest in a race or reward people with more efforts to change and stick to a healthy lifestyle (Fig. 7). Therefore, this feature is to reward people who have shown positive behaviors such as trying the hardest, or being more cooperative with the rest of the team, or the one that has improved the most. The QR code helps in habit formation. This feature not only helps in developing new habits but also rewards competence which helps in user retention.

Avatar Customization- The user can customize the avatar skin color, dress, glasses, hair including other features. (Fig. 6). This feature gives a sense of personalization, autonomy and relatedness. If the user has not walked, the avatar gets upset and to make it happy the user has to walk 10% of the average of daily steps.

Fig. 6. Avatar customization   **Fig. 7.** Main scene of the application

4 Research Design and Methodology

The methodology used to achieve the objectives of this research follows the User-centered Intelligent Environments Development Process [19]. A state of the art analysis through the technical literature has been used to get initial insights of the research area and to update with new developments. Our project discusses the challenges faced by other researchers to gain involvement of citizens and to keep users engaged for a longer term. Intertwined with knowledge acquisition we performed several iterations of system development with continuous improvement. The feedback has been gathered from colleagues, users and stakeholders using questionnaires, surveys, validation exercises, workshops and competitions to improve the application.

Workshop with Colleagues- At the start of this project the workshop was conducted in Middlesex University with Colleagues from the Psychology Department, Management Leadership and Organizations Department and London Sports Institute. The aim of this workshop was to get understanding and feedback from colleagues by sharing knowledge in their area of expertise. The term 'user adherence' was introduced in this workshop which is used interchangeably in this paper and is a part of the internal project terminology, it does not affect game users.

Workshop with Stakeholders- A workshop was conducted with the stakeholders in November 2019. This workshop took place in London with 8 participants including GLL Manager. The workshop started with the presentation of the app by Bene [20] who was working mainly for a teenage group. Then it was followed by the discussion of how to improve the app and make it available for all the users. A list of requirements was set to work on. The aim of this workshop was to gather feedback and key requirements. The feedback from stakeholders is taken continuously after implementing the features. The requirement list is updated after each iteration which is every month. Currently, we are tackling 77 requirements. Some of them are achieved and some are still under development.

Pilot with Users- This work evolved from Bene [20], an MRes student who worked on a thesis which mainly focuses on teenagers. However, this is now for all users, regardless of their age. The data is collected from the users of different age groups such as children, teenagers and adults. The Age Band feature is implemented for different age groups which has been discussed in the previous section. The first pilot was conducted in Unitas, London with 25 users participating in two weeks 'Step Challenge'. The users with more steps wins the competition. The first, second and third winners got a prize. The purpose of this challenge was to gather feedback from users and to enhance the functionality of the application. Most users enjoyed the real reward and avatar feature of the application.

5   Conclusion

In conclusion, this study conducts a comprehensive analysis on the fundamental elements that contribute to encouraging and retaining the users to become more physically active and focus on their health. The strategy is to keep people engaged by using gamified app and inspiring them to stay fit. The main factor in gamification is user adherence. Hence, in this research we developed an application to focus on user retention. Based on BCT and SDT, the key extrinsic and intrinsic motivational elements such as autonomy, purpose, competence and relatedness have been implemented to boost motivation in users to adapt to an active lifestyle for the long term.

The user retention features implemented in '*OnTheMove!*' are Bonus Rewards, Real Rewards, Like, Personal Best, Competitions between different teams, Goals and Age Band. This app is for all age groups. An age band coefficient is used for people above the age of 40. It gives a feeling of relatedness with younger users. Moreover, it aims to ensure that the activities are achievable for them. Moreover, QR code is used for the activities that cannot be counted. It is to ensure that the users are rewarded for behavior change. It also emphasizes on the social interaction of application users to get a better result. These system features have been considered after various activities to gather feedback from various project stakeholders. Some of these activities are ongoing, for example, a pilot conducted in the London Borough of Barnet.

Acknowledgements. This project has benefited from the input of many colleagues: M.Sc. Ondrej Benes, Dr Nicola Payne (Psychology Department), Dr Anne Elliot (London Sports Institute), Dr Simon Best (Management Leadership and Organisations Department),). We would like to thank Laurence Oliver (Greenwich Leisure Ltd.), Gillan Kelly (Greenwich Leisure Ltd.), Andrew Gilbert (Greenwich Leisure Ltd.), Jalpa Assani (Greenwich Leisure Ltd.), Alesia Carrington (Barnet Council) for their contribution.

References

1. World Health Organization. https://www.who.int/news-room/fact-sheets/detail/obesity-and-overweight. Accessed 09 June 2021
2. Johnson, D., Deterding, S., Kuhn, K.A., Staneva, A., Stoyanov, S., Hides, L.: Gamification for health and wellbeing: a systematic review of the literature. Internet Interv. **6**, 89–106 (2016). https://doi.org/10.1016/j.invent.2016.10.002
3. Sailer, M., Hense, J.U., Mayr, S.K., Mandl, H.: How gamification motivates: an experimental study of the effects of specific game design elements on psychological need satisfaction. Comput. Hum. Behav. **69**, 371–380 (2017). https://doi.org/10.1016/j.chb.2016.12.033
4. Jones, B.A., Madden, G.J., Wengreen, H.J.: The FIT game: preliminary evaluation of a gamification approach to increasing fruit and vegetable consumption in school. Prev. Med. **68**, 76e79 (2014). https://doi.org/10.1016/j.ypmed
5. Landers, R.N., Landers, A.K.: An empirical test of the theory of gamified learning: the effect of Leaderboards on time-on-task and academic performance. Simul. Gaming **45**(6), 769e785 (2014). https://doi.org/10.1177/1046878114563662
6. Tóth, Á., Tóvölgyi, S.: The introduction of gamification: a review paper about the applied gamification in the smartphone applications 000213–000218 (2016). https://doi.org/10.1109/CogInfoCom.2016.7804551
7. Kasurinena, J., Knutas, A.: Publication trends in gamification: a systematic mapping study. In: Computer Science Review, vol. 27, pp. 33–44, February 2018
8. Stinson, N.S., et al.: development and testing of a multidimensional iphone pain assessment application for adolescents with cancer. J. Med. Internet Res. **15**. e51 (2013). https://doi.org/10.2196/jmir.2350
9. Rose, K.J., Koenig, M., Wiesbauer, F.: Evaluating success for behavioral change in diabetes via mHealth and gamification: MySugr's keys to retention and patient engagement. Diab. Technol. Ther. **15**, A114 (2013). https://doi.org/10.1089/dia.2012.1221
10. Cechetti, N.P., et al.: Developing and implementing a gamification method to improve user engagement: a case study with an m-Health application for hypertension monitoring. Telematics Inform. **41**, 126–138 (2019)
11. Munson, S.A., Consolvo, S.: Exploring goal setting, rewards, self-monitoring, and sharing to motivate physical activity. In: 2012 6th International Conference on Pervasive Computing Technologies for Healthcare and Workshops, Pervasive Health 2012, pp. 25–32 (2012). https://doi.org/10.4108/icst.pervasivehealth.2012.248691
12. Ozanne, M., Navas, A.C., Mattila, A.S., Van Hoof, H.B.: An investigation into Facebook "liking" behavior an exploratory study. Soc. Media Soc. **3**(2) (2017). https://doi.org/10.1177/2056305117706785
13. Michie, S., et al.: The Behavior Change Technique Taxonomy (v1) of 93 hierarchically clustered techniques: building an international consensus for the reporting of behavior change interventions. Ann. Behav. Med. **46** (2013). https://doi.org/10.1007/s12160-013-9486-6. A Publication of the Society of Behavioral Medicine

14. Pickering, K., et al.: Gamification for physical activity behaviour change. Perspect. Public Health **138**(6), 309–310 (2018)
15. Shi, L., Cristea, A.I., Hadzidedic, S., Dervishalidovic, N.: Contextual gamification of social interaction – towards increasing motivation in social E-learning. In: Popescu, E., Lau, R.W.H., Pata, K., Leung, H., Laanpere, M. (eds.) ICWL 2014. LNCS, vol. 8613, pp. 116–122. Springer, Cham (2014). https://doi.org/10.1007/978-3-319-09635-3_12
16. Bovermann, K., Bastiaens, T.J.: Towards a motivational design? Connecting gamification user types and online learning activities. Res. Pract. Technol. Enhanc. Learn. **15**(1), 1–18 (2020). https://doi.org/10.1186/s41039-019-0121-4
17. Souza-Júnior, M., Queiroz, L., Correia-Neto, J., Vilar, G.: Evaluating the use of gamification in m-health lifestyle-related applications. In: Rocha, Á., Correia, A., Adeli, H., Reis, L., Mendonça Teixeira, M. (eds.) New Advances in Information Systems and Technologies. AISC, vol. 445, pp. 63–72. Springer, Cham (2016). https://doi.org/10.1007/978-3-319-31307-8_7
18. Rasche, P., Schlomann, A., Mertens, A.: Who is still playing Pokémon Go? A Web-Based Survey. JMIR Serious Games. **5**(2), e7 (2017). https://doi.org/10.2196/games.7197.In: Eysenbach, G., (ed.)
19. Augusto, J., Kramer, D., Alegre, U., Covaci, A., Santokhee, A.: The user-centred intelligent environments development process as a guide to co-create smart technology for people with special needs. Univ. Access Inf. Soc. **17**(1), 115–130 (2017). https://doi.org/10.1007/s10209-016-0514-8
20. Bene, O.: Gamification to Encourage Increase on Healthier Physical Activity in Younger Users, M.Sc Thesis, Middlesex University, London (2019)

Streaming and Visualising Neuronal Signals for Understanding Pain

Javier Alfonso Espinosa-Oviedo[(✉)]

University Lumiere Lyon 2, ERIC, Lyon, France
javier.espinosa-oviedo@univ-lyon2.fr

Abstract. This paper presents our stream processing, and visualization system adapted to the requirements of the neuroscience domain. We propose to build a visual stream processing system for supporting the analysis and exploration of data streams in real-time by exploiting human's natural ability for discovering patterns. Our work combines stream processing and data storage techniques with data visualization theory. We study strategies for visualizing different types of data considering constraints related to real-time and data volume.

1 Introduction

As with other experimental sciences, neuroscience supports refutes or validates hypotheses by conducting experiments on living organisms. For instance, by connecting electrical sensors to a cat's spinal cord and monitoring its neurons activities, neuroscientists can determine whether capsaicin (chilli pepper active component) has the same effect as anaesthesia in the presence of pain [8]. In a typical neuroscience experiment, a neuroscientist is responsible for:

- preparing the subject;
- connecting and calibrating sensors;
- collecting and storing the experiment data (e.g. file, database);
- applying algorithms and statistics for discovering meaningful patterns

Because of the resulting data volume and the complexity of the algorithms used for finding patterns, the data analysis is usually done post-mortem. The complexity of setting a neuroscience experiment is given by

- particular and expensive equipment, juridical protocols concerning experiments using animals, gathering together field experts,
- its duration (e.g., 8 h), and uniqueness (e.g., every subject has its characteristics),

Therefore, neuroscientists require novel tools for processing and exploring data in real-time to better control the progress of an experiment.

This work was done in the Barcelona Supercomputing Centre in collaboration with the CINVESTAV in Mexico.

H. Hacid et al. (Eds.): ICSOC 2021 Workshops, LNCS 13236, pp. 228–235, 2022.
https://doi.org/10.1007/978-3-031-14135-5_18

Although there has been a lot of progress in automatic knowledge discovery (e.g., deep learning), we believe humans play a central role in the data analysis task. Therefore, this paper proposes building a visual stream processing system to support the analysis and exploration of data streams in real-time by exploiting humans' natural ability to discover patterns. Our work combines stream processing and data storage techniques [9,16] with data visualization theory. We study strategies for visualizing different types of data considering constraints related to real-time and data volume. This paper presents our stream processing, and visualization system adapted to the requirements of the neuroscience domain.

The remainder of the paper is organized as follows. Section 2 discusses related work regarding stream processing systems and visualisation approaches. Section 3 describes the visual neuronal Stream Processing System proposed to implement a neuroscience data centred experiment. Section 4 concludes the paper and discusses future work.

2   Related Work

Existing work related to our work concerns two domains: (i) stream processing approaches and systems; and (ii) visualisation techniques related to streams. The following gives a synthetic discussion on current trends and open issues associated with these areas.

2.1   Stream Processing

Stream processing refers to data processing in motion or computing on data directly as it is produced or received. In the early 2000s, academic and commercial approaches proposed stream operators for defining continuous queries (windows, joins, aggregation) that dealt with streams [6,10]. These operators were integrated as extensions of database management systems. Streams were stored in a database, a file system, or other forms of mass storage. Applications would query the data or compute over the data as needed. These solutions evolved towards stream processors that receive and send the data streams and execute the application or analytics logic. A stream processor ensures that data flows efficiently and the computation scales and is fault-tolerant. Many stream processors adopt stateful stream processing [1,4,5,11] that maintains contextual state used to store information derived from the previously-seen events.

Apache Storm[1] is a distributed stream processing computation framework that is distributed, fault-tolerant and guarantees data processing. A Storm application is designed as a "topology" in the shape of a directed acyclic graph (DAG) with spouts and bolts acting as the graph vertices. Edges on the graph represent named streams flows and direct data from one node to another. Together, the topology acts as a data transformation pipeline. Apache Flink is an opensource stateful stream processing framework. Stateful stream processing integrates the database and the event-driven/reactive application or analytics logic

---

[1] https://storm.apache.org.

into one tightly integrated entity. With Flink, streams from many sources can be ingested, processed, and distributed across various nodes. Flink can handle graph processing, machine learning, and other complex event processing. Apache Kafka is an open-source publish and subscribe messaging solution. Services publishing (writing) events to Kafka topics are asynchronously connected to other services consuming (reading) events from Kafka - all in real-time. Kafka Streams lacks point-to-point queues and falls short in terms of analytics. Spring Cloud Data Flow[2] is a microservice-based streaming and batch processing platform. It provides tools to create data pipelines for target use cases. Spring Cloud Data Flows has an intuitive graphic editor that makes building data pipelines interactive for developers. Amazon Kinesis Streams[3] is a service to collect, process, and analyse streaming data in real-time, designed to get important information needed to make decisions on time. Cloud Dataflow[4] is a serverless processing platform designed to execute data processing pipelines. It uses the Apache Beam SDK for MapReduce operations and accuracy control for batch and streaming data. Apache Pulsar is a cloud-native, distributed messaging and streaming platform. Apache Pulsar[5] is a high-performance cloud-native, distributed messaging and streaming platform that provides server-to-server messaging and geo-replication of messages across clusters. IBM Streams[6] proposes a Streams Processing Language (SPL). It powers a Stream Analytics service that allows to ingest and analyse millions of events per second. Queries can be expressed to retrieve specific data and create filters to refine the data on your dashboard to dive deeper.Source[7]. Event stream query engines like Elasticsearch, Amazon Athena, Amazon Redshift, Cassandra define queries to analyze and sequence data for storage or use by other processors. They rely on "classic" ETL (extraction, transformation and loading) processes and use query engines to execute online search and aggregation, for example, in social media contexts (e.g., Elasticsearch) and SQL like queries on streams (e.g. Amazon Athena, Redshift and Cassandra).

2.2   Data Visualisation

Current solutions for visualizing data depend on traditional DBMSs for storing and retrieving raw data and the use of custom visualization tools to process and render it [7]. For instance, ScalaR is a 3-layer based visualization system (GUI, web server, database) that dynamically performs resolution reduction when the expected result of a DBMS query is too significant to be effectively rendered on a screen [3]. Instead of running the original query, ScalaR inserts aggregation, sampling or filtering operations to reduce the size of the result before plotting it.

---

[2] https://spring.io/projects/spring-cloud-dataflow.
[3] http://aws.amazon.com/kinesis/data-streams/.
[4] https://cloud.google.com/dataflow.
[5] https://pulsar.apache.org/.
[6] https://www.ibm.com/cloud/streaming-analytics.
[7] https://deepsource.io.

A similar example is ForeCache [2], a general-purpose tool for exploratory browsing of large datasets based on a lightweight browser interface, and a DBMS running on a back-end server. For improving response times, ForeCache introduces the use of a cache system for pre-fetching data as the user explores a dataset. Finally, Tableau is an interactive interface to general OLAP queries.

As stated in [12,14], the decouple database-visualization tool has three main drawbacks: (i) the database is unaware of related queries and may recompute the same results (e.g., slightly panning a map will issue a query to recompute the entire map, though most results are unchanged); (ii) visualization tools duplicate basic database operations, such as filtering and aggregation; (iii) visualization tools assume that all raw data and metadata fit entirely in memory, which is not the case for large datasets.

2.3   Discussion

Stream processing systems have emerged to process (i.e., query) streams from continuous data providers (e.g. sensors, things). These systems are designed to address scalability, including (i) streams produced at a high pace and from millions of providers; (ii) computationally costly processing tasks (analytics operations); (iii) online consumption requirements.

Online analysis techniques must process streams on the fly and combine them with historical data to provide past and current analytics of observed environments. Despite solid stream processing platforms and query engines, solutions do not let programmers design their analytics pipelines without considering the conditions in which streams are collected and eventually stored.

There are some initial results for building interactive real-time visualizations over data streams [15]. Yet, these works focus mainly on visualizing time series. We study the full spectrum of data types (temporal, tabular, geo-spatial) and data visualization techniques and propose a general-purpose visualization processing system adapted to the requirements of the neuroscience domain.

3   Visual Neuronal Stream Processing System

Like other experimental sciences, neuroscience supports refutes or validates hypotheses by conducting experiments on living organisms. For instance, by connecting electrical sensors to a cat's spinal cord and monitoring its neurons activities, neuroscientists can determine whether capsaicin (chilli pepper active component) has the same effect as anaesthesia in the presence of pain [8].

We have conducted an experimental validation (see Fig. 1) using data from studies regarding pain performed in the neuroscience group at the Mexican research centre CINVESTAV[8].

Therefore, we provided a solution for harvesting data produced during an experiment and observing specific states of the vertebrae. Figure 2 gives an

---

[8] Special thanks to Diogenes Chavez from CINVESTAV Department of Physiology, Biophysics and Neuroscience for providing the datasets used in this work.

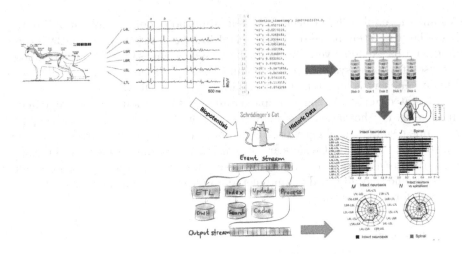

Fig. 1. Neuroscience scenario.

overview of our approach. In the figure, data are collected during a neuroscience experiment and continuously transmitted to our system. Data are processed in real-time. Then, depending on the type of analysis that a neuroscientist wants to conduct, she (i) defines queries using a set of operators and (ii) chooses the kind of visual representation required. For instance, in our approach, a neuroscientist can group the data into temporal windows of 1 h. Then, (for each window), she can choose different visualizations (e.g., point chart, histogram, start plot) to analyze the collected data's correlation.

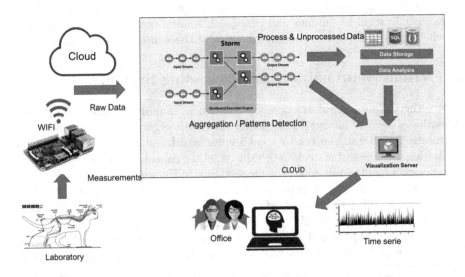

Fig. 2. Visual stream processing system for analyzing and visualizing data streams.

In this experiment, there is one microservice for collecting the data from an IoT environment (e.g. an a neuroscience experiment) and one for plotting the data. The idea is that a consumer defines the sequence of microservices for processing and then plotting the data she desires to observe (as shown in Fig. 3). For example, *Give me the evolution of pain intensity in L4ci* or *Give me the evolution of the minimum, average, maximum and intensity of pain in L4ci every 3 s*. Figure 3 shows how these queries are implemented in terms of microservices, including a window, aggregation, and plotting ones. With this experiment, it was possible to observe online the execution of the neuroscience experiment. Since streams were stored, it was possible to observe the data from other experiments, compare them to make decisions and adjust the phases of the experiment.

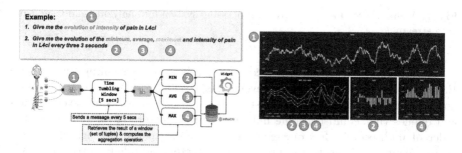

Fig. 3. Observing continuously the execution of the neuroscience experiment.

Our system is based on the notion of stream operators (e.g., fetch, sliding window, average, etc.) [13]. Figure 4 shows the general architecture of a stream operator. As shown in the figure, an operator communicates asynchronously with other operators using a message-oriented middleware. As data is produced, the operator fetches and copies the data to an internal buffer. Then, depending on the operator' logic, it applies an algorithm and sends the data to the next operator. There is one operator for collecting the data from an experiment and one for plotting the data in our approach. The idea is that a neuroscientist defines the sequence of operators for processing and then plotting the data.

For deploying our experiment, we built an IoT farm using our Azure Grant[9] and implemented a distributed version of the IoT environment to test a clustered version of Rabbit MQ. Therefore, we address the scaling-up problem regarding the number of data producers (things) for our microservices. Using Azure Virtual Machines (VM), we implemented a realistic scenario for testing scalability in terms of: (i) Initial MOM (RabbitMQ) installed in the VM_2 (ii) Producers (Things) installed in the VM_1 (iii) microservices installed in the VM_3

---

[9] The MS Azure Grant was associated with a project to perform data analytics on crowds flows in cities. It consisted of credits for using cloud resources for performing high-performance data processing.

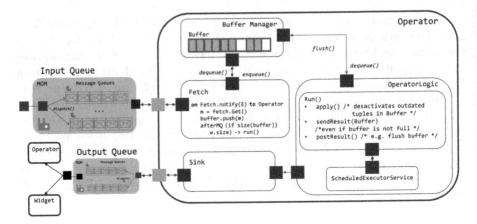

Fig. 4. Architecture of a visual stream processing system for analyzing and visualizing data streams.

In this experiment, microservices and testbeds were running on separate VMs. This experiment leads to several cases scaling up to several machines hosting until 800 things with a clustered version of Rabbit using several nodes and queues that could consume millions of messages produced at rates in the order of milliseconds. For our experiments, we varied the settings of the IoT environment according to the properties characterising different scenarios. We used fewer things and queues, and more nodes to achieve data processing in an agile way. In this scenario, we assumed that there were few connected things (just the number required for observing the sensors of the spinal cord) with a high production rate.

4   Conclusions and Future Work

We have implemented the first version of data stream operators and conducted an experimental validation using data from studies regarding pain. We highlighted the importance of having an environment both collecting and archiving signals stemming from neuronal cells and visualizing them continuously during the experiment and then post-mortem for looking for behaviour patterns. The possibility of expressing observations as queries provide agility to the experiment in vivo, making it possible to control it online.

We are currently evaluating the capacity of the system for addressing data volume (with respect to online memory consumption) and data processing performance, while visualizing time-series continuously.

References

1. Alaasam, A.B., Radchenko, G., Tchernykh, A.: Stateful stream processing for digital twins: microservice-based kafka stream DSL. In: 2019 International Multi-Conference on Engineering, Computer and Information Sciences (SIBIRCON), pp. 0804–0809. IEEE (2019)
2. Battle, L., Chang, R., Stonebraker, M.: Dynamic prefetching of data tiles for interactive visualization. In: Proceedings of the 2016 International Conference on Management of Data, pp. 1363–1375 (2016)
3. Battle, L., Stonebraker, M., Chang, R.: Dynamic reduction of query result sets for interactive visualization. In: 2013 IEEE International Conference on Big Data, pp. 1–8. IEEE (2013)
4. Carbone, P., Ewen, S., Fóra, G., Haridi, S., Richter, S., Tzoumas, K.: State management in Apache Flink®: consistent stateful distributed stream processing. Proc. VLDB Endow. 10(12), 1718–1729 (2017)
5. Cardellini, V., Nardelli, M., Luzi, D.: Elastic stateful stream processing in storm. In: 2016 International Conference on High Performance Computing & Simulation (HPCS), pp. 583–590. IEEE (2016)
6. Fragkoulis, M., Carbone, P., Kalavri, V., Katsifodimos, A.: A survey on the evolution of stream processing systems. arXiv preprint arXiv:2008.00842 (2020)
7. Idreos, S., Papaemmanouil, O., Chaudhuri, S.: Overview of data exploration techniques. In: Proceedings of the 2015 ACM SIGMOD International Conference on Management of Data, pp. 277–281 (2015)
8. Martin, M., et al.: A machine learning methodology for the selection and classification of spontaneous spinal cord dorsum potentials allows disclosure of structured (non-random) changes in neuronal connectivity induced by nociceptive stimulation. Front. Neuroinform. 9, 21 (2015)
9. Pelkonen, T., et al.: Gorilla: a fast, scalable, in-memory time series database. Proc. VLDB Endow. 8(12), 1816–1827 (2015)
10. Rao, T.R., Mitra, P., Bhatt, R., Goswami, A.: The big data system, components, tools, and technologies: a survey. Knowl. Inf. Syst. 60(3), 1165–1245 (2018). https://doi.org/10.1007/s10115-018-1248-0
11. To, Q.-C., Soto, J., Markl, V.: A survey of state management in big data processing systems. VLDB J. 27(6), 847–872 (2018). https://doi.org/10.1007/s00778-018-0514-9
12. Traub, J., Steenbergen, N., Grulich, P.M., Rabl, T., Markl, V.: I2: interactive real-time visualization for streaming data. In: EDBT, pp. 526–529 (2017)
13. Vargas-Solar, G., Espinosa-Oviedo, J.A.: H-STREAM: composing microservices for enacting stream and histories analytics pipelines. In: Hacid, H., Kao, O., Mecella, M., Moha, N., Paik, H. (eds.) ICSOC 2021. LNCS, vol. 13121, pp. 867–874. Springer, Cham (2021). https://doi.org/10.1007/978-3-030-91431-8_64
14. Wu, E., Battle, L., Madden, S.R.: The case for data visualization management systems: vision paper. Proc. VLDB Endow. 7(10), 903–906 (2014)
15. Wu, E., et al.: Combining design and performance in a data visualization management system. In: CIDR (2017)
16. Zaharia, M., Das, T., Li, H., Hunter, T., Shenker, S., Stoica, I.: Discretized streams: fault-tolerant streaming computation at scale. In: Proceedings of the Twenty-Fourth ACM Symposium on Operating Systems Principles, pp. 423–438 (2013)

Graph Analytics Workflows Enactment on Just in Time Data Centres Position Paper

Ali Akoglu[1]([✉]), José-Luis Zechinelli-Martini[2], Hamamache Kheddouci[3], and Genoveva Vargas-Solar[3]

[1] University of Arizona, Tucson, USA
akoglu@arizona.edu
[2] Fundación Universidad de las Américas Puebla, San Andrés Cholula, México
joseluis.zechinelli@udlap.mx
[3] CNRS, Univ Lyon, INSA Lyon, UCBL, LIRIS, UMR5205,
69622 Villeurbanne, France
hamamache.kheddouci@univ-lyon1.fr,
genoveva.vargas-solar@cnrs.fr

Abstract. This paper discusses our vision about multirole-capable decision-making systems across a broad range of Data Science (DS) workflows working on graphs through disaggregated data centres. Our vision is that an alternative is possible, to work on a disaggregated solution for the provision of computational services under the notion of a disaggregated data centre. We define this alternative as a virtual entity that dynamically provides resources crosscutting the layers of edge, fog and data centre according to the workloads submitted by the workflows and their Service Level Objectives.

1 Introduction

Data collections can be structured as networks that have interconnection rules determined by the variables characterising each observation. The graph is a powerful mathematical concept with associated operations that can be implemented through efficient data structures and exploited by applying different algorithms. Note that relations among observations and interconnection rules are often not explicit, and it is the role of the analytics process to deduce, discover and eventually predict them.

When the graphs become large and even too large the algorithms used to process, explore and analyse them become costly in execution time, even if several cores are used. In this case, given the characteristics of the algorithms, communication is also likely to be costly. So workflows that exploit graphs become gluttonous

---

This work is funded by the project GALILEAN, LIRIS intergroup collaboration https://galilean-project.github.io.

H. Hacid et al. (Eds.): ICSOC 2021 Workshops, LNCS 13236, pp. 236–243, 2022.
https://doi.org/10.1007/978-3-031-14135-5_19

consumers of computing resources. Our work comes into the scene in this context; we are interested in the execution conditions of graph processing workflows.

Data Science (DS) workflows pose unique challenges due to the growing complexity of processing and querying methods for big-data applications increasingly governed by analytics operations, machine learning-based workflows and models.

Considering DS workflows' complexity, heterogeneity, and dynamic behaviour, it is impossible to produce a timely computing system solution in response to many dynamically arriving streaming applications and associated queries over complex graphs with potentially millions of nodes. Meeting performance requirements of large-scale DS workflows with tasks applying greedy operations applied on graphs across the entire dynamic system execution space is a daunting task without a clear understanding of the dependencies from available data sources to information extraction algorithms, from available information to decision algorithms, from algorithms to performance requirements, and from heterogeneous computing resources to performance capabilities.

Graph processing and analysis workflows consist of tasks that include:

- deploying or retrieving graphs which are often distributed over an execution environment,
- applying algorithms of varying complexity in a distributed way, and
- retrieving the results and making them available to other processes or to the end-users.

In terms of infrastructure, the execution takes place on often heterogeneous architectures that provide computing services with different capacities to execute them. From this point of view, it is possible to access computing solutions that range from resource rich cloud based infrastructures all the way to the edge based power and resource limited resources.

In the current context, workloads are typically greedily delegated to the cloud or data centres. Still, the computing resources residing on these architectures cannot be composed in an elastic and integrated way to build ad hoc execution environments on the fly. Our work is in the context of approaches to designing alternative architectures to provide computing, storage, and memory resources that are more elastic and lightweight than greedy based approaches.

In addition to graph-based workflows, today's environments favour high-performance cloud-based platforms as a means to outsource their execution completely. These monolithically designed platforms provide various infrastructure services with heterogeneous computing resources with different capabilities to design, execute and maintain workflows.

Our vision is that it is possible to alternatively provide computational services under the notion of a disaggregated data centre. So a virtual entity that dynamically provides resources that touch the edge, fog and data centre according to the workloads submitted by the workflows and their Service Level Objectives (SLOs). Therefore, this paper discusses our vision about multirole-capable decision-making systems across a broad range of DS workflows working on graphs through an agile, autonomous, composable, and resilient "Just-in-Time Architecture" for DS Pipelines (JITA-4DS) [1].

Accordingly, the remainder of the paper is organised as follows. Section 2 discusses related work regarding existing disaggregated data centres approaches and data science workflow execution. Section 3 describes our vision and research challenges about graph data science workflows and execution on disaggregated data centres. Section 4 describes the general lines of how to build just in time virtual data centres for executing data science workflows. Section 5 concludes the paper and discusses future work.

2   Related Work

In general, querying techniques can be categorised across two families: (i) the first concerns querying as we know it in databases and information retrieval; (ii) the second, a family where workflows, namely Data Science (DS) workflows, explore and analyse the data to profile them quantitatively either with modelling, prediction, or recommendation purposes. The results of queries have an associated degree of error, and they may not only be data but also queries or data samples and models.

DS workflows need specialised architectures because of their size, dynamic behaviour, and nonlinear scaling and relatively unpredictable growth with respect to their inputs being processed. Existing IT architectures are not designed to provide an agile infrastructure to keep up with the rapidly evolving next-generation mobile, big data, and data science workflows demands. They require continuous provisioning and re-provisioning of DC resources [4,5,10] given their dynamic and unpredictable changes in the SLOs (e.g., availability response time, reliability, energy).

Existing DS environments are "one-fits-all" cloud systems that can manage and query data with different structures through built-in or user-defined operations integrated into imperative or SQL like solutions. They are provided by major vendors like Google, Amazon, IBM and Microsoft. They address the analytics and data management divide with integrated backends for efficient execution of analytics activities workflows, allocating the necessary infrastructure (CPU, FPGA, GPU, TPU) and platform (Spark, Tensorflow) services. These environments provide resources for executing DS tasks requiring storage and computing resources. DS workflows evolve from in-house executions into deployment phases on the cloud. Therefore, they need underlying elastic architectures that can provide resources at different scales. Disaggregated data centres solutions seem promising for them. Our work addresses the challenges implied when coupling disaggregated solutions with DS workflows.

3   Graph DS Workflows Execution on Disaggregated Data Centres

The research we propose aims to study the execution of DS workflows addressing graph analytics focusing on data processing, transmission and sharing across several resources. We identify research challenges to study the execution of graph analysis workflows concerning the processing, transmission and sharing of data and different resources. Our hypothesis is that it is possible to schedule its tasks on a Virtual Data Science Centre (VDS) given a workflow. We organise our study around three research questions:

R1 Is it possible to adopt a database approach and draw on query evaluation to define the execution plan(s) of workflows taking into account the data distribution/execution load?

R2 How and according to which metrics can we estimate the resource required by each task depending on the algorithm it calls and the volume of data to be processed?

R3 According to which strategies can we estimate and configure the VDS according to a given workflow execution plan?

Given the difficulty of the problem, we propose to adopt a three step data management and processing methodology as summarised below.

Disaggregated Data Centre. We start from the abstract idea of a disaggregated data centre as a possible configuration in the form of a virtual machine that provides computing, storage and RAM resources available on a Data Centre Building Block Pool. The needs of a workflow guide the configuration of VDS in terms of execution, monitoring and maintenance throughout its lifecycle.

Executing Data Science Workflows. The execution of DS workflows on graphs consists of data processing tasks to be scheduled on a disaggregated VDS. Our approach is to define execution, configuration and deployment plans that can guide the execution, to represent the strategies to allocate resources and calibrate a VDS according to the characteristics of a given data science workflow. Therefore a first challenge to address is to rewrite DS workflows into these plans. The objective is to define data dependencies among tasks and the control flow to adopt for executing them considering the distribution of the data/execution workloads.

Our study has started from pipelines using analytics graph algorithms to answer community detection problems like page rank, Louvain, more mathematical models applied to matrices representing graphs (run in the LNS, Mexico) according to previous work [2,3]. Our focus will be on the characterization of DS workflows considering (i) the type of graph processing algorithms they address; (ii) the characteristics of the graphs (data) processed and results through these algorithms.

Designing and rewriting strategies for generating execution plans implies the definition of metrics to estimate costs and SLOs in the different phases of the workflow execution cycle. The execution must be guided by dynamic and elastic provisioning of resources. The challenge to address is to estimate the resources requirements associated with each task of the execution plan according to the algorithm it calls and the data injection function estimating the volume of data to process. In this context, experiments are essential to guide and validate the proposals.

We have focused on defining the right metrics for estimating the requirements of target DS workflows as presented in our previous work [1]. We describe the SLO objectives of given DS workflows on graphs that should be fulfilled at execution time.

Estimating and Configuring Initial VDS Workflows. Our focus is in proposing a DS workflow rewriting strategy that will generate an ad hoc execution specification including (i) tasks to be executed by the workflow (classic execution plan); (ii) the corresponding specification of the underlying VDS workflow architecture (configuration plan) and (iii) the deployment plan defining the distribution of the processes from the edge to the VDS workflow. DS workflows introduce other challenges like weaving data preparation, fragmentation, and analytics operations where data dependencies and requirements across tasks must be fine-tuned and modelled.

4   Towards Just in Time Virtual Data Centres for Data Science Workflows

Our research investigates architectural support, system performance metrics, resource management algorithms, and modelling techniques to enable the design of composable (disaggregated) DCs. The goal is to design an innovative composable "Just in Time Architecture" for configuring DCs for Data Science Pipelines (JITA-4DS) and associated resource management techniques [1]. DCs utilize a set of flexible building blocks that can be dynamically and automatically assembled and re-assembled to meet the dynamic changes in workload's SLOs of current and future DC applications. DCs under our approach are composable based on vertical integration of the application, middleware/operating system, and hardware layers customized dynamically to meet application SLO (application-aware management). Thus, DCs configured using JITA-4DS provide ad-hoc environments efficiently and effectively meeting the continuous evolution of requirements of data-driven applications or workloads (e.g., data science pipelines).

A DC is based on a novel application-aware VDC Management system by dynamically invoking the appropriate actions to change the current VDC configuration to meet its objectives at runtime. To assess disaggregated DC's, we study how to model and validate their performance in large-scale settings. We rely on our novel model-driven resource management heuristics [6–8] based on metrics that measure a service's value for achieving a balance between competing goals (e.g., completion time and energy consumption). Our focus is on defining new system performance measures that combine objectives, such as execution time and energy use, that dynamically change during the day.

Initially, we propose a hierarchical modelling approach that integrates simulation tools and models. Results can be used for developing benchmarks

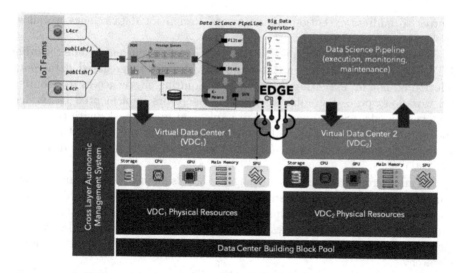

Fig. 1. Just in time architecture for data science pipelines - JITA-4DS.

that accurately characterize the requirements and SLOs of next-generation DC applications.

The Just in Time Architecture for Data Science Pipelines (JITA-4DS), illustrated in Fig. 1, is a cross-layer management system that is aware of both the application characteristics and the underlying infrastructures to break the barriers between applications, middleware/operating system, and hardware layers. Vertical integration of these layers is needed to build a customizable VDC to meet the dynamically changing data science pipelines' performance, availability, and energy consumption requirements.

JITA-4DS can build a VDC that can meet the application SLO for execution performance and energy consumption to execute data science pipelines. Then, the selected VDC is mapped to a set of heterogeneous computing nodes such as GPPs, GPUs, TPUs, special-purpose units (SPUs) such as ASICs and FPGAs, along with memory and storage units.

5 Conclusions and Future Work

This paper introduced our vision and research position regarding the design of just in time architectures for providing disaggregated resources for the execution of graph analytics workflows. The originality of our research program is promoting the provision of resources holistic system through intelligent resource management. This holistic system integrates three elements, graph processing models, associated computational resources, autonomous execution of complex and dynamic workflows.

From a more general point of view, three aspects characterise the approach. Its pioneering and promising aspect tackles the design of disaggregated data centres to address execution environments' design for data science workflows applied to graphs.

We have described the general characteristics of our current results regarding JITA-4DS. This virtualised architecture provides a disaggregated data centre solution ad hoc for executing DS workflows requiring elastic access to resources. DS workflows process graphs coordinating operators implemented by services deployed on edge. Since operators can implement greedy tasks with computing and storage requirements beyond those residing on edge, they interact with VDC services. We have set the first simulation setting to study resources delivery in JITA-4DS.

We are currently addressing challenges of VDCs management on simpler environments, on cloud resource management heuristics, (e.g., [6–9]), big data analysis, and data mining for performance prediction. To simulate, evaluate, analyze, and compare different heuristics, we will build simulators for simpler environments and combine open-source simulators for different levels of the JITA-4DS hierarchy.

Disaggregated approaches for providing data centres resources are emerging as a promising topic discussed in panels at major conferences and by leading scientists and companies. For the time being, approaches address the communication layers, but the wave is starting to touch computing and storage and platform levels. We have a first proposal for including the edge because of the characteristics of DS workflows.

To conclude, we believe that reasoning about the design and provision of alternatives to data science execution environments under a disaggregated perspective is pioneering and promising. Supporting this kind of exploratory project can encourage digital independence on the way data science experimentation is enacted and can provide solutions beyond the lab walls.

References

1. Akoglu, A., Vargas-Solar, G.: Putting data science pipelines on the edge. arXiv preprint arXiv:2103.07978 (2021)
2. Bouhenni, S., Yahiaoui, S., Nouali-Taboudjemat, N., Kheddouci, H.: A survey on distributed graph pattern matching in massive graphs. ACM Comput. Surv. (CSUR) 54(2), 1–35 (2021)
3. Brighen, A., Slimani, H., Rezgui, A., Kheddouci, H.: A distributed large graph coloring algorithm on giraph. In: 2020 5th International Conference on Cloud Computing and Artificial Intelligence: Technologies and Applications (CloudTech), pp. 1–7. IEEE (2020)
4. Chen, H., Zhang, Y., Caramanis, M.C., Coskun, A.K.: EnergyQARE: QoS-aware data center participation in smart grid regulation service reserve provision. ACM Trans. Model. Perform. Eval. Comput. Syst. 4(1) (2019). https://doi.org/10.1145/3243172
5. Kannan, R.S., Subramanian, L., Raju, A., Ahn, J., Mars, J., Tang, L.: Grand-SLAm: guaranteeing SLAs for jobs in microservices execution frameworks. In: EuroSys 2019. Association for Computing Machinery, New York (2019). https://doi.org/10.1145/3302424.3303958

6. Kumbhare, N., Akoglu, A., Marathe, A., Hariri, S., Abdulla, G.: Dynamic power management for value-oriented schedulers in power-constrained HPC system. Parallel Comput. **99**, 102686 (2020)
7. Kumbhare, N., Marathe, A., Akoglu, A., Siegel, H.J., Abdulla, G., Hariri, S.: A value-oriented job scheduling approach for power-constrained and oversubscribed HPC systems. IEEE Trans. Parallel Distrib. Syst. **31**(6), 1419–1433 (2020)
8. Kumbhare, N., Tunc, C., Machovec, D., Akoglu, A., Hariri, S., Siegel, H.J.: Value based scheduling for oversubscribed power-constrained homogeneous hpc systems. In: 2017 International Conference on Cloud and Autonomic Computing (ICCAC), pp. 120–130. IEEE (2017)
9. Machovec, D., et al.: Utility-based resource management in an oversubscribed energy-constrained heterogeneous environment executing parallel applications. In: Parallel Computing, vol. 83, pp. 48–72, April 2019
10. Xu, X., Dou, W., Zhang, X., Chen, J.: EnReal: an energy-aware resource allocation method for scientific workflow executions in cloud environment. IEEE Trans. Cloud Comput. **4**(2), 166–179 (2015)

Data Centred Intelligent Geosciences: Research Agenda and Opportunities Position Paper

Aderson Farias do Nascimento[1(✉)], Martin A. Musicante[1],
Umberto Souza da Costa[1], Bruno M. Carvalho[1], Marcus Alexandre Nunes[1],
and Genoveva Vargas-Solar[2]

[1] Universidade Rio Grande do Norte, Natal, Brazil
aderson.nascimento@ufrn.br, {mam,umberto,bruno}@dimap.ufrn.br
[2] CNRS, Univ Lyon, INSA Lyon, UCBL, LIRIS, UMR5205,
69622 Villeurbanne, France
genoveva.vargas-solar@cnrs.fr

Abstract. This paper describes and discusses our vision to develop and reason about best practices and novel ways of curating data-centric geosciences knowledge (data, experiments, models, methods, conclusions, and interpretations). This knowledge is produced from applying statistical modelling, Machine Learning, and modern data analytics methods on geo-data collections. The problems address open methodological questions in model building, models' assessment, prediction, and forecasting workflows.

1 Introduction

Massive data production is a critical aspect of experimental sciences. It has not been different for geoscience. Examples of geoscientific data include any physical observable related to the energy industry, mining, monitoring hazardous areas (e.g. effects of salt mining in populated areas), etc. Nowadays, with the relative facility and lowering the cost to collect data, the data processing to exploit their value is a challenge. It requires expertise in data maintenance and processing, data analysis, and the design of experiments of target domains for which data will provide insight and knowledge.

This paper describes and discusses our vision to develop and reason about best practices and novel ways of curating [12] data-centric geosciences knowledge (data, experiments, models, methods, conclusions, and interpretations). This knowledge is produced from applying statistical modelling, Machine Learning, and modern data analytics methods on geo-data collections. The problems address open methodological questions in model building, models' assessment, prediction, and forecasting workflows.

This work is funded by the project ADAGEO, IEA CNRS collaboration with Federal University of Rio Grande do Norte https://adageo.github.io.

© Springer Nature Switzerland AG 2022
H. Hacid et al. (Eds.): ICSOC 2021 Workshops, LNCS 13236, pp. 244–251, 2022.
https://doi.org/10.1007/978-3-031-14135-5_20

This paper is organised as follows. Section 2 discusses related work regarding existing disaggregated data centres approaches and data science workflow execution. Section 3 describes our vision and research challenges and opportunities of data centred smart geosciences. Section 4 describes examples of use cases addressed through data centred strategies using mathematical and Machine Learning or artificial intelligence algorithms. Section 5 concludes the paper and discusses future work.

2 Related Work

In France, portals like SISMER[1] and Form@Ter[2] are initiatives willing to share data about target observation in geosciences and then share analytics experiments results. Data Terra[3] is a research infrastructure dedicated to Earth System observation data. In general, the objective of these platforms and portals is to facilitate access to satellite, airborne and in-situ data collected and managed by research laboratories or federative structures, by national infrastructures, the oceanographic fleet, aircraft, balloons, and by space missions (e.g., Data Terra). They manage, archive, and share TB of data. For example, Data Terra represented 50,000 TB in 2017 and is estimated to reach 100,000 TB by 2022. Beyond multi-source data, they also share products and services through a unified portal. Data is curated with metadata, included under accepted standards like the European standard INSPIRE. The challenge is to define common bases for all data producers and make the data sets interoperable so that their resources are consistent, shareable, exploitable, and, in a multidisciplinary approach, required to study the Earth. In this sense, the ODATIS Ocean Cluster offers several services for data producers similar to data labs for referencing, hosting, dissemination and interoperability. They also provide access to computing services for running experiments (models) that require important computing resources.

At the European level, actions adopting a data science perspective, for example, the project EPOS[4] and the Alan Turing Institute extend these initiatives to European partners willing to take full advantage of the possibilities provided by analytics and data science to run experiments and contribute to solving leading problems addressed by the discipline. Indeed, with the advent of digital technologies, libraries proposing analytics models have been run on mainframes and high-performance computing centres (HPC) to produce visualisation, modelling and simulation systems to accelerate interpretation and planning.

Brazilian scientific agenda has widely installed and developed data centres like https://www.eveo.com.br/en/ and https://baxtel.com/data-center/brazil-brasil. These data centres aim to provide mainly large-scale computing resources to run experiments, for example, those regarding geosciences, particularly those key for the national economies in France and Brazil in oil and hydrocarbon exploitation, extraction of minerals, and its interaction with populated areas.

---

[1] https://data.ifremer.fr/SISMER/Missions.
[2] https://www.poleterresolide.fr.
[3] https://www.data-terra.org.
[4] https://www.epos-ip.org.

3   Towards Smart Data Centred Geosciences

Lately, geoscience researchers have been discovering the power of Machine Learning in solving problems in their field. Bergen et al. [2], for example, show that random forests were used on continuous acoustic emission in a laboratory shear experiment to model instantaneous friction and to predict time-to-failure [7,10] surveyed the applications of Machine Learning in seismology and presented five research areas in in which Machine Learning classification, regression, clustering algorithms show promise: earthquake detection and phase picking, earthquake early warning (EEW), ground-motion prediction, seismic tomography, and earthquake geodesy. In exploration geophysics, Machine Learning has been used in seismic data processing and reservoir characterization [6,9]. Clustering methods were used to identify key geophysical signatures and determine their relationship to rock types for geological mapping in the Brazilian Amazon [4]. However, many researchers in the area are still not prepared to take advantage of data-driven approaches to their analyses at scale. In this context existing projects and actions are emerging to provide specialized portals and systems that can encourage the sharing of collected data (observations), experiments, and analytics results that should even promote reproducibility.

In this context, we can see the emergence of multidisciplinary teams to collaborate in the search for computational solutions. These teams are formed by experts in geology/geosciences, computer science, statistics and physics, among others. The work of these teams usually relies on the use of mathematical/computational tools to process large amounts of data. Big data analytical techniques and Machine Learning has been used with success.

Many scientists and companies believe that they can generate fresh insight, reduce decision cycle times and steal a march on their competition by automating the search for patterns and relationships in their data. Therefore, geophysics and data science, including algorithms, mathematical models and computing, must converge for developing experiments for obtaining insight and foresight about the observations contained in data collections. Experiments represent best practices for addressing problems and questions on geophysics that must be treated as data and knowledge to be shared and reused by scientists and practitioners.

Data collections issued in situ observations shared in pivot formats are vital for developing experiments that can lead to relevant governmental, economic, and social decision making. Information about how these data have been collected, used, curated, and maintained, including the conditions in which analyses are run and associated results and their use to lead to specific policies, should be managed and shared.

Merging data-centric techniques with modelling and simulation to answer questions in geoscience and make timely, clever, and disruptive decisions can lead to a new geoscience perspective that will benefit from data curation and analytics. In our vision, three important directions can be considered described in the following lines.

Collected Data, Models and Knowledge Integration. A wide variety of geophysical data (potential fields, electromagnetic data, seismic data, weather data, etc.) has been acquired with extensive wavelength ranges from surface sensor arrays, drilled wells, satellites and many other sources. These data sets are among the most significant science data sets in use, comparable in size and complexity only to those from astronomy and particle physics. Integrating access to data collections and their curated versions under a global knowledge graph can promote its maintenance, analysis, and experimentation. It can also show the knowledge of the discipline with its vocabulary, concepts, and relations in a synthetic manner. Inspired by existing public data labs like Kaggle or CoLab of Google, it can be essential to work to extend existing portals. These portals can be revisited towards specialized data science labs on geosciences. Through these labs, scientists and practitioners can share raw data, models, and experiments' return of experience and run and reproduce other experiments with almost no requirement of interacting with specialized engineering support for accessing CPU and GPU clusters.

Curation, maintenance, exploration of data collections for bringing value to petabytes of data produced from in situ observations and also from experiments. Given that data act as a backbone in modelling phenomena for understanding their behaviour, it is critical to developing good collection and maintenance: which are available data collections? Are they complete? Which is their provenance? In which conditions were they collected? have they been processed? In which cases have they been used, and what are the associated results?

Data curation is a set of techniques to process (raw, distributed, heterogeneous) data to extract their value. It proposes methods to explore data collections using well-adapted data structures like graphs that can be explored and enriched while new data and analytics results are produced. Data curation means also keeping track of the type of experiments run on data, their results, and the conditions in which they were performed.

Maintaining a catalogue of questions and experiments related to data can help provide a new vision of data and the scientific community's knowledge. This catalogue can extend existing meta-data and associated data collections information provided by actions like ODATIS and Data Terra.

Modelling and Simulating Experiments to Answer Questions in Geoscience and Make Timely Decisions. Both data sources and models come with recognized issues that existing methodologies have difficulties coping with but which novel data science-based approaches can address. For example, features for which exact physical models are unknown (e.g., subsurface geology, earthquakes) or models that are difficult to reconcile (e.g., seismic measurements vs social media alerts). This will imply:

- Designing ad hoc experiment programming languages for enabling friendly, context-aware, and declarative construction of complex experiments in geosciences.

- Enabling the execution of experiments fusing different data collocations at different scales to maintain data, prepare experiments, and manage associated results.
- Programming experiments
  - Applying statistical methods to investigate and unveil new patterns in geophysical data, answering open problems, or leading to further research questions.
  - Building predictive models to describe better or approximate geophysical phenomena, increasing the knowledge about our planet.
  - Parallelizing algorithms for processing geophysical data, thus, allowing for the processing of very large data sets in reasonable times.

Discussion. From the Geophysics point of view, proposing best practices and ad-hoc strategies for developing data centred experiments to solve geosciences problems will impact different vital areas of the economy. For example, oil companies that ride this wave will significantly increase the current productivity of their knowledge workers, optimize business processes, and reduce operational costs in a way that is not possible through incremental change. Some companies now use algorithms to define optimal drilling locations, using automated or semi-automated systems that deliver results on much shorter cycle times than traditional methods.

From the Data Science/Data Processing perspective, this kind of multidisciplinary research can provide the ground to devise new data curation techniques, to propose a domain-specific query language, or to define new methods for processing heterogeneous data [11]. In addition, statistical knowledge is essential for extracting information from the massive amount of data we will process. New methods and models will be crucial to model data and make conclusions in a timely fashion [13].

4   Use Cases

To illustrate the type of data analytics challenges introduced by geosciences problems, we describe in this section three examples. These use cases can be solved with different techniques and can call for data science strategies for specifying solutions and deploying them in target architectures.

Estimating the Approximate Earthquake Epicentres. The understanding of earthquake occurrence in intraplate areas has been one of the most challenging tasks in Seismology [5]. Compared to border plate regions, interplate areas suffer less attenuation of seismic waves. As a consequence, a significant hazard may rise from moderate magnitude earthquakes. Understanding the earthquake generating mechanisms depends on assessing the stress field in the intraplate areas.

Seismic stations collect signals that can represent earthquakes produced in a specific area. The challenge is to determine whether signals represent earthquakes in such a case compute the epicentres. For addressing the challenge, it is possible to use mathematical, Machine Learning or artificial intelligence methods [8]. The first task to address this question is to compute the earthquake epicentre's direction using the sensor's initial movement polarity when the waves P and S are discovered. Then, compute the distance considering how the sensor moves from North-South (it should be the same as the East-West), as shown in Fig. 1.

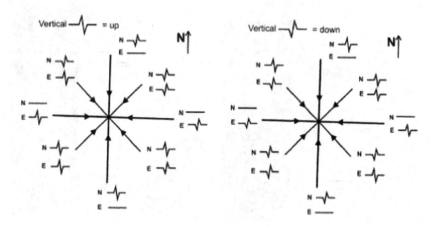

Fig. 1. Sensor movement.

Estimation of Stacking Velocity Using CDP Semblance. Semblance analysis is a technique used in the study and refinement of seismic data. Along with other methods, this technique enables the improvement of the resolution of data, even in the presence of background noise. The data yielded by semblance analysis tends to be easier to interpret when discovering the underground structure of an area (see Fig. 2).

Estimating the stacking velocities is one of the essential steps in the CMP (Common Mid Point) seismic processing. This is because the better the estimation of the stacking velocities, the better the quality of the zero-offset section obtained. Currently, the most convenient velocity analysis method consists of manually picking the stacking velocities in the velocity spectrum, using the semblance as a coherence measure. The semblance gives us a measure of multichannel coherence. It is necessary to define an analytics workflow with the following phases to perform this task: (i) transform the CDP or CMP gathered traces from the offset and time coordinates into the coherence semblances in coordinates of time and stacking velocities. (ii) Pick local maxima of these coherence semblances and assign zero offset time and corresponding stacking velocities. (iii) Correct the CDP or CMP gathers for normal moveout (NMO).

Denoising Data from Sensors. The Brazilian Seismographic Network (RSBR) operates since 2011. Station installation began in 2011 in southeast (SE) Brazil and finished in 2014 in the Amazon forest. The network integrates 84 stations (as of December 2017) operated by four institutions in different regions of Brazil.

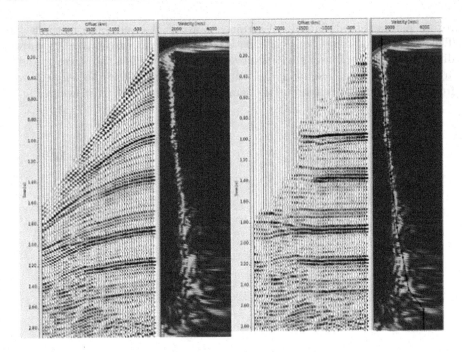

Fig. 2. NMO correction after velocity picking on the semblance. Source: [1]

Seismic stations collect signals that can represent earthquakes produced in a specific area. This data usually contains noise produced by the context where the sensor is placed and by the technology of the sensor itself. The challenge is to filter this data to make it ready to be analyzed. This consortium is responsible for the Brazilian Seismic Bulletin [3].

5   Conclusions and Future Work

This paper proposes our vision about the multidisciplinary agenda for developing data centred solutions for geosciences problems. The amount of data collected through observing the Earth and its geophysical phenomena call for agile data and knowledge curation techniques to manage both data, experiments, and results. The research agenda includes (i) integrating and describing data collected with different technology, (ii) estimating its quality, and preparing it to be used as input of different methods. Research on smart data centred geoscience also calls for curation tasks, including data tracking the way data is cleaned, the experiments that use it and the obtained results. Exploration methods and systems must be associated with curated data and knowledge to facilitate an agile understanding of this content. Finally, execution environments providing computing resources necessary for setting and deploying experiments are vital for promoting multidisciplinary global experimental sciences. The research performed within the project ADAGEO[5] is willing to address these problems through a Brazilian and French collaborative community.

References

1. Araújo, G.A.: Plataforma Interativa de Análise de Velocidade em Dados Sísmicos usando GPUs. Universidade Federal do Rio Grande do Norte (2018)
2. Bergen, K.J., Johnson, P.A., Maarten, V., Beroza, G.C.: Machine learning for data-driven discovery in solid earth geoscience. Science **363**(6433), eaau0323 (2019)
3. Bianchi, M.B., et al.: The Brazilian seismographic network (RSBR): improving seismic monitoring in Brazil. Seismol. Res. Lett. **89**(2A), 452–457 (2018)
4. Carneiro, C.D.C., Fraser, S.J., Crósta, A.P., Silva, A.M., Barros, C.E.D.M.: Semi-automated geologic mapping using self-organizing maps and airborne geophysics in the Brazilian amazon. Geophysics **77**(4), 17–24 (2012)
5. Fonsêca, J., Ferreira, J., do Nascimen, A., Bezerra, F., Neto, H.L., de Menezes, E.: Intraplate earthquakes in the Potiguar basin, Brazil: evidence for superposition of local and regional stresses and implications for moderate-size earthquake occurrence. J. South Am. Earth Sci. **110**, 103370 (2021)
6. Jia, Y., Ma, J.: What can machine learning do for seismic data processing? An interpolation application. Geophysics **82**(3), V163–V177 (2017)
7. Kong, Q., Trugman, D.T., Ross, Z.E., Bianco, M.J., Meade, B.J., Gerstoft, P.: Machine learning in seismology: turning data into insights. Seismol. Res. Lett. **90**(1), 3–14 (2019)
8. Leandro, W.P., Santana, F.L., Carvalho, B.M., do Nascimento, A.F.: Parallel source scanning algorithm using GPUS. Comput. Geosci. **140**, 104497 (2020)
9. Li, S., Huang, X., Cao, H.: Seismic data prediction lithology sequence model based on machine learning. In: SEG 2018 Workshop: Reservoir Geophysics, Daqing, China, 5–7 August 2018, pp. 249–251. Society of Exploration Geophysicists and the Chinese Geophysical Society (2020)
10. Rouet-Leduc, B., Hulbert, C., Lubbers, N., Barros, K., Humphreys, C.J., Johnson, P.A.: Machine learning predicts laboratory earthquakes. Geophys. Res. Lett. **44**(18), 9276–9282 (2017)
11. Vargas-Solar, G., Farokhnejad, M., Espinosa-Oviedo, J.: Towards human-in-the-loop based query rewriting for exploring datasets. In: Proceedings of the Workshops of the EDBT/ICDT 2021 Joint Conference (2021)
12. Vargas-Solar, G., Kemp, G., Hernández-Gallegos, I., Espinosa-Oviedo, J., Da Silva, C.F., Ghodous, P.: Demonstrating data collections curation and exploration with curare. In: EDBT/ICDT Conference 2019, p. 4 (2019)
13. Vargas-Solar, G., Zechinelli-Martini, J.-L., Espinosa-Oviedo, J.A.: Enacting data science pipelines for exploring graphs: from libraries to studios. In: Bellatreche, L., et al. (eds.) TPDL/ADBIS -2020. CCIS, vol. 1260, pp. 271–280. Springer, Cham (2020). https://doi.org/10.1007/978-3-030-55814-7_23

---

[5] https://adageo.github.io funded by the IEA program of the French CNRS.

PhD Symposium

Design Patterns in an ERP Environment

Aiman Zahid(✉), Sidra Akhtar, and Wafa Basit(✉)

FAST School of Computing, National University of Computer and Emerging Sciences Lahore,
Lahore, Pakistan
aimanzahid007@gmail.com, wafa.basit@lhr.nu.edu.pk

Abstract. Design patterns add quality to a system. The purpose of this research was to identify that whether developers are familiar with the concepts of design patterns in the software houses of Pakistan. As, it is really difficult to include all the software houses in the survey. So, a random sample was taken from three different software houses. This research also explores whether design patterns actually add any value to a system. By value we mean whether they improve the internal quality attributes of the system.

Keywords: ERP · Design patterns · Coupling · Cohesion · UML diagrams · Survey

1 Introduction

The research is based on survey that was held in multiple software houses located in Lahore, Pakistan. The survey is about how different software houses are implementing the design patterns in their daily work practices. Another important matter of concern is that if people are actually familiar with these terms or are these techniques just in practice because they help in creating a more reusable, extendable and flexible software. These concepts are usually not taught in the universities and they are not compulsory as well so it is also a point to ponder that whether the people dealing with these techniques on daily basis are actually familiar with their names and cause and effects as well [1].

The research was performed in three different software houses one of them is dealing with oracle based solutions, other one is working with SRP and the last one with SAP. So, we can include the questions that can help us analyze that which software gets most benefitted with the implementation of design pattern techniques. A comparison can be perform that if people in all three software houses are familiar with these techniques and what do they think about it. Are they implementing it on regular basis because it helps creating a difference or are they not considering it because the things don't really change with or without implementing it? It will be easy to make them understand the concepts and techniques because they are actually connected with this field and they deal with a lot of UML diagrams on daily basis.

The survey form contains questions about the design patterns and some questions about the ERP they are using. This way it can also be analyzed with which ERP people are actually feeling more comfortable with and in which ERP the developers are using

H. Hacid et al. (Eds.): ICSOC 2021 Workshops, LNCS 13236, pp. 255–271, 2022.
https://doi.org/10.1007/978-3-031-14135-5_21

design patterns. The research survey shall enable us to identify if developers are familiar with design pattern concepts or not [4].

Moreover, in order to measure the impact of design patterns on a system. A case study is also performed in SAP environment. A metric of coupling and cohesion was measured in an environment where design patterns were not implemented and later metrics were calculated again so that it can be analyzed that whether adding design patterns to a system increases its quality or not. The system is considered good if it have high cohesion and low coupling. Coupling should not be zero because then it will mean that all the system is implemented in a single class and that is the worst approach to implement a system. High cohesion means the modules of system are not being effected by the change occurring in one part of system. So, in order to calculate the final results it is to be checked that whether the system with design patterns implemented has more cohesion and less coupling then before or not. So, for this purpose two hypotheses were formulated. One hypothesis identifies whether implementing design patterns actually effects the quality of a system or not and second one to analyze if developers working in the software houses are actually familiar with the design patterns or not. The survey is performed in different software houses that are using different ERPs so that the difference of ERPs can be identified and the number of employees in each software house varies. This way it can be identified that if developers in small, medium and large software houses have similar concepts about design patterns or if any of them are better than others [7].

For this research and case study implementation a detailed literature review was performed so that it can be identified that if this work is implemented before or not and if there are any similar projects available then the results could be compared as well.

2 Hypothesis

In this research paper, we will do consider different kind of hypothesis and will evaluate our questionnaire results on the basis of these hypothesis.

H0: The design patterns are not being implemented in most of the software houses in Lahore
H1: The design patterns are being implemented in most of the software houses of Lahore

After getting the results, our hypothesis might be approved or rejected. It solely depends upon the results.

3 Scope

In this paper, we have some findings and limitations as well. As we are covering only Lahore based Enterprise resource Management systems. But we will be studying 4 ERP's (GLUON, GERP, ORACLE and SAP) which are being used worldwide to improve the resource management and a keen analysis of the results shall be performed wgetting from these 4 ERP based Firms.

A questionnaire shall be provided to check that either software developers have any idea about Design patterns and its associated terms. Do they use these in their regular work routine? and if yes, then do they feel any ease after implementing them? We do have some UML Diagram based questions to check how software developers which are working for different ERPs gave the solution for that scenario. The major focus of our paper is how to improve ERP based system. There are various design patterns which we can be applied to improve the quality. But design patterns are mostly not a priority of professionals. There might be several reasons like lack of emphasis at academic level as well as at professional level.

We will be evaluating Developers' knowledge using UML Diagram for their understanding of basic Design patterns.

4 Literature Review

Design patterns provide the ease to mold the software in such a way that it becomes more flexible and less vulnerable [REF]. The major need of introducing design pattern was felt when software becomes very vulnerable to any change happening in any part of software. It is a genuine problem that is still being faced by the developers as they face system crashing down just because of a few very small changes made to any module of system. Design patterns are being used in every single field that is associated with coding/programming. The field chosen for this research was association of design patterns in ERP. Both of these are really vast fields however there is not much research done in this specific area. So, for this purpose a detailed literature was studied so that the potential design patterns have in ERP can be found. Design patterns are related to design heuristics and refactoring [1]. Therefore, papers on these topics were also included in the literature review.

Reusability of code is very important. Refactoring while developing a system and design patterns have made it really easy to let developers use the same code but in a very sophisticated way so that the system don't come crashing down while developer make a change in any part of code. Design patterns are being used in CAD and 3D work as well according to Jing Bai and fellow researchers who worked on how to add design patterns to CAD and how to detect specific patterns in a CAD modeling environment stated that there are specific interfaces in the existing model that contain the key functionality of the system however in order to ensure that completing the model would not make it crash, design patterns are the only solution. However that is not the only use of design patterns. Design patterns are also used to identify the similar structures so that if similar designs are created before than it can help the developer to reuse the code and help save some time and energy of rewriting everything from scratch. Common structures are not difficult to find. In fact, there is an option where developers can find similar design structures from all there designs so, even if the same structure is not in the same file the developer can still copy the similar code from other project and update it accordingly. Such things are performed using clustering. If a design pattern falls into the same cluster then maybe the developer will find it easier and convenient to just update the code rather than redoing everything [2].

Another similar research was performed by Ghulam Rasool and fellow researcher, their research was based on recovery tools being used for design patterns. There are a lot of comparisons done on different design pattern recovery tools. Most of them are open source so other developers can also access them and test the point of view provided by the researchers. It is an important factor to be able to analyze and evaluate the design patterns. The recovery tools just work in the reverse manner. Developers implement the design patterns in the code. However, design pattern recovery tools enable people to detect these design patterns in the code. These recovery tools can be used for multiple purposes such as new developers can use these recovery tools to educate themselves about when, where and which design pattern is to be used however, researchers can use these tools and studies to analyze which tools and practices are providing best possible results. Design patterns are not code or language specific. Design patterns are being used in JAVA, OOP, C++ and python as well. There are all different paradigms and various methods of using the design patters. All languages have specific results and benefits that can be achieved by adding the design patterns to the code. The best possible benefit is that it enables developers to edit a part of system without worrying about rest of the system. Design patterns ensure that the effect of changing a module can be minimized on rest of the system by adding cohesion and reducing coupling. There are different design pattern recovery tools that enables the developer to compare different solutions and patterns before application [3].

A research was performed to understand the importance of cohesion in a system. Software have different metrics to express different aspects of a software. Coupling and cohesion metrics are a part of them and cohesion metrics is really important as it declares the quality of a software in terms of its flexibility and strength against being collapsed. It is really important for a software to have a high cohesion. In this research researchers performed a study where they explained why cohesion metrics is extremely important and secondly they displayed a detailed example of how one can perform the method of developing a cohesion metrics. However in order to perform a comparison on the quality of a software the testing team is required to develop the metrics twice. This way one can find the difference of the quality of software [4].

Saeed Jalili and his fellow researcher performed a research in which they explained how hundreds of design patterns have been declared over the past years and how they have been helpful in designing a more stable and suitable software solution. The most of the part of this research was based on how one can decide between multiple design patterns as it is really difficult to choose from those many design patterns and it is really difficult to identify the need of a design patterns as well. So, in order to overcome this problem a lot of solutions have been offered by multiple programmers as now we have automated design pattern recommendation models. This is a very unique as there are a lot of solution available already but the researchers emphasized on a specific model. The research suggested a two tier strategy. In this strategy there are multiple benefits as now there is no requirement of semi-formal explanation of the design pattern and most importantly in this approach the system suggest the design pattern after analyzing the design problem. So, it kind of works like a customized approach. Every design problem is identified and analyzed before suggesting a valid and more suitable design pattern for the problem [5].

In the recent past papers, several approaches have been indulged to bring an improvement in the ERP System. We have analyzed different case studies and will discuss these papers in detail. In a recent case study, they are trying to highlight the design pattern usage and how it brings the improvement in ERP System and they have tried to answer two of the research questions by considering Dynamics AX ERP. One of them was that how this ERP can get improve by using different design patterns and Are there any techniques in place in existing ERP systems for implementing specialized processes? They have conducted a survey and as a result, they found that improvements can be made in some parts of the application by 17-fold but developers don't use Design patterns because maybe these developers don't come from formal development education. It's entirely feasible that some software engineering principles are sliding under their radar as side effects and secondly they got to know that existing techniques in ERP are mostly not SOLID. For easy and fast maintenance, the SOLID mechanism should be provided [6].

By using an abstract layer on the top of the database access layer through multiple functions and Global methods, ERP Application's structure has improved a lot. A lot of difficulties can be encountered during the analysis of the problem due to the lack of knowledge of abstract concepts of breaking down a major programming problem into a component in the programming language. To bring the improvement, several patterns are being shared, both in coerce and fine grain. Before developing software using ML in a company a few factors needs to be considered like Programming education, Software quality control, Acquiring domain knowledge, and design method. In most cases, the Design of the systems is being started by adopting the MVC design pattern where model, view, and controllers are being handled in a loop [7].

To control the cost of development hours, pattern-based design is an effective way of doing it because through this, we can avoid reinventing, revalidating, and rewriting agnostic software artifacts. It helps to provide reusable solutions to frequently occurring problems. There are multiple composite patterns available for efficient integration of the applications and services. DI design pattern can help to identify the solution for decoupling and integration. Because redevelopment is a costly affair, it is more efficacious to incorporate the applications and services rather than reconstructing to obtain a high level of application and service reusability. Design patterns make us capable to support high-quality software development and reuse the combined design knowledge. DI Pattern has helped to implement loose coupling, service modularity, reusability, and dynamic discoverability of different functions. It helps to bring an abstraction layer between high and low-level modules and that abstraction layer helps out to integration parts to be developed, test and deployed independently [8].

IT Environments have become very complicated as a result of the heterogeneity of existing platforms, making communication between different enterprises more difficult. With the help of Service-oriented architecture, we can improve the interaction and make it simpler. ERP execution and flexibility can be improved by bringing the design patterns. A Design pattern is a best practice or the foundation of a solution that has been documented and is being applied for the problems which occur frequently in specific situations. It's an art to applying the design patterns as it requires a lot of experience so that a bigger picture of the application can be seen rather than just implementing a specific design pattern for the current situation. Developers can be mistaken into applying design patterns because most cases they might rush to bring flexibility and maintainability and unconsciously they overdo it by overdesigning and over engineering. For performance, a singleton design pattern is the better option for stateless services and this design pattern can save CPU processing time as well. For modification, observer pattern and decorator pattern can help out. Decorator design patterns can play a vital role in the major modification. While the Observer design pattern used to synchronize the changes in products and align the things between the customer and the vendor [9].

Literature on ERP systems in past years can be analyzed in 6 different categories like optimization, implementation, deployment, ERP for supply chain, management via ERP, and past research articles. It is being noticed that people started taking interest in the post-deployment phase, customization of development, sociological aspects of ERP, and the interoperation between different ERPs. Many recent case studies shed a light on the poor adaptation and poor maintainability of an ERP while considering the daily work of their users. Because the human factor is an important factor for the good implementation of an ERP. Because every user is handling the ERP according to their own company's strategy and business logic [10].

5 Methodology

5.1 Survey

The link to the survey form is given in references.

The survey was divided into three sections. First section contained the demographics like name of company, number of people working in the company, qualification of the person and the work experience of person in specific field. Second part comprised the questions about the ERP being used at the company so that it can be checked how comfortable people are working in their specific ERP. This was also done to perform a little comparison of different ERPs. Finally the last section contained eight different UML diagrams and people were required to choose a UML out of three or four given options so that it can be checked if they are used to the design patterns and they can identify them in a UML diagram or not. This wasn't totally about the design patterns. It was also a small test to see which specific UML people found more suitable to be used in their program. So, if developers thought it was good approach to include design patterns or not. The data was collected in multiple software houses and as the major target were the developers who are responsible for making and managing the ERPs. So, in total 60 survey forms were filled by three different companies and a few miscellaneous to keep the survey sample well spread and totally generic. It was made sure that people fill in all the questions after reading and understanding the situation so that any missing or wrong values can be avoided.

5.2 Findings

It was found that most of the developers are graduates so, this establishes a fact that may be they never came across the subject of design patterns in their student life however, design patterns are really essential to be used in development so that the system can be flexible and strong.

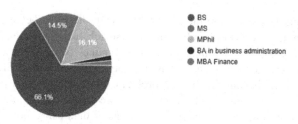

Fig. 1. Educational background

While trying to ensure that people in the survey sample are well spread. Their experience in the development field was asked so that it can be analyzed if increase in experience also increases the knowledge and understanding about design patterns.

Fig. 2. Development experience of developers in years

Even though the distribution of data is not evenly spread but still the sample covers almost all different aspects of experience. In terms of ERP there are three major ERPs GLUON, SAP, there are a lot of other ERPs as well but the one noticeable point is that a few companies are using customized ERP for their organizations. There could be a lot of possibilities to analyze the results of UML diagrams.

The final result of the UML diagrams are to be checked on three aspects. Firstly, the number of employees who got more than four UML diagrams right out of all eight. Work experience of people who got more than four UMLs right. Finally, the ERP of people who got most UMLs right.

According to the results it is determined that it doesn't necessarily depend on the educational background or work experience of developers. Even though the data is extremely spread but still a range can be specified in which most people chose correct answers. Barely, 30% were able to get through the minimum criteria off being considered passing. All people belonged to the work experience of 1 to 3 years so it can be said that recent developers are more interested in implementing the programs while keeping the design pattern under practice.

Even though developers are mostly satisfied with the ERP they are working with however a common problem identified in this research survey was that developers are not very satisfied with the automated testing techniques being suggested by the ERP they are using. Another important fact is that there are customized ERPs being used in some companies. It could've been a problem for developers and people who are managing those systems but people seem to be more comfortable while working in the ERP designed especially for their organization rather than using any other commonly available ERP (Fig. 3).

Gloun ERP	3 years	5
Gluon ERP	4 years	3
Gluon ERP	3 years	3
Gluon ERP	6 months	4
Gluon ERP	2 years	2
SAP	11 Years	3
SAP	2 years	6
Oracle	5 years	3
Microsoft Dynamic:	5 years	3
SAP	4 years	1

Fig. 3. ERP, experience and number of correct answers

Most people who got their answers right were using gluon as an ERP. So, it was required to take a close look at the code from gluon so that the design patterns could be identified. The code was being updated since 2015 so it was difficult to perform a comparison on code with and without design patterns. However, a survey was performed on the code as well to check if the code in the company using gluon. This was done to verify if developers were also familiar and used to the design patterns in their daily practice.

6 Design Pattern Identification

6.1 Design Patterns Familiarity

As per our Survey results, Most of the developers don't have the required knowledge of design patterns. Most of the developers are undergraduates and design pattern is mostly studied during this graduation period but as the companies don't give that much importance during the coding level so considering design patterns during coding is not a priority in most of the cases.

During conversation with different developers in GLUON ERP, we got to know that they were implementing a separate interface or creating an abstract class but these were their coding practices, they didn't know that they are specifically implementing the design patterns at all.

Design patterns are quite important for the clarity and modularity of the code. But unfortunately small companies are not giving that much importance to that side. It helps to enhance the reusability concept that provide a proven solutions to repetitive problems that may arise in specific context. A pattern-based approach helps to avoid costly cycles of revalidation and rediscovery of repetitive software solutions.

6.2   The ERP System

Design pattern applications have long existed in computer science field. It is not possible for ERP System that they don't have a good maintenance score. But there is always a room for an improvement in any application and in this research paper we have tried to examine that which design patterns made an improvement in this application. For this purpose a comparative survey is being conducted and multiple questions are being asked like ERP Support for automation testing for maintenance and the results that around 70% records are average or above average for multiple ERP's like GLUON, SAP B1, ORACLE Dynamics which is used in different companies like Softbeats, ABACUS CONSULTING, AZEEMI TECHNOLOGIES, DIAMOND FABRIC LIMITED, HCC Labs Pvt limited, TekHQ etc. So that 'why we consider to pick one ERP **GLUON** and bring the improvements using different design patterns.

This section discusses how the ERP System works and how it gets updated without interpreting any run-time working.

6.2.1   ERP System Architecture

For ERP based systems, **MVC** compound design pattern is mostly used for maintaining the structure of it but it's more like an architectural pattern. It is related to user interaction layer of ERP System. It can handle large data and provide support in the improvement of ERP. But the code methodology should be good and it should be bad smell free. Testing of a feature is getting easier with the use of **MVC** unit testing feature. But it is not enough for creating an ERP Program, there is still need to use some data access layer, contracts layer, helper layer, service layer. It enables logical functionalities and actions on a controller (Fig. 4).

In ERP System, the ERP system was decomposed into three components (MODEL, VIEW, and CONTROLLER). The figure attached is the MVC architecture for GLUON ERP Software. It gave the margin to developers to work simultaneously on the model, controller and views of ERP. Actually, it works like a loop wherein view is connected with the controller and further controller is connected with the model and it is connected to the view. MVC pattern was first generate for desktop Applications but later on it was used for web apps too. It's more like a modular design having modules, controllers and views rather than a layered structure. Layered structures are best for web applications.

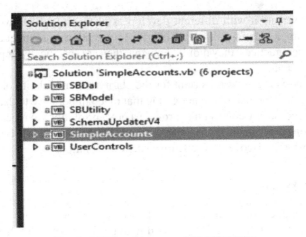

Fig. 4. MVC architecture of GLUON ERP

6.3 Model

This component consists of the classes which will create by considering the data model. In GLUON ERP, it is named as **SBModel**. Data is being stored in a relational database system as SQL SERVER. For storing the data, **data access layer (data logic layer)** is being created which named as **SBDAL. SBDAL** is being utilized for making a connection between data and model. All the data which is being kept in data access layer is being transmitted to model through this layer. It's actually a two way process, we can shift the data from database to model as well as vice versa. We can insert, delete and update the data in SQL using the model by building up a relationship between table from database and class for that table in the code. It contains all the tables, stored procedures, and column names. It holds the business logic of the application. In model, all the master and detail table data are being connected via objects and child classes, so that's how we can create a model for every table.

6.4 View

In this part, the view shows us what a client wants to see. Basically model and controller works together to show the results in view component. Its main purpose is to display data using model class object. It contains the functional requirements of an ERP. For each method, there is a different view, that' why, we keep a sub-folder with the same name, under the view. The view updates itself via the observer pattern. The view is mostly concerned with visual impacts so you can see the strategy pattern in this part. It contains multiple forms to view for user.

6.5 Controller

In the Fig. 1, **SchemaUpdaterV4** and **SBUtility** lies under this section. It contains all the methods, functions, variables behind an ERP. A set of methods and functions are being called in this level to perform a specific action according to the user request. It's more like business logic layer because it will decide how to play with data or manipulate it. The required function is being called from the main interface of the application and perform the functions like get the request and analyze it then call the respective function, then generate a view and verify it and return to the user. It acts as a middleware between the view and the model to control the information exchange. All the logics that are going to be used are being written in this component to play around with the data get from view.

7 Analysis of Existing Patterns in GLUON ERP

In this section, we will discuss the design patterns which already existed in GLUON ERP. We will discuss the definition of those design patterns, implementation and its structure.

7.1 Factory Pattern

7.1.1 Purpose

In this we create the objects without exposing it to the client and create a separate common interface to refer the newly objects.

7.2 Application in GLUON ERP

In the initial stages, the developers have made some functions, methods, Sub Procedures in each class for some common actions just like save, update, delete, new, edit, Refresh and load all. But later on for smooth working, by considering the factory pattern they made an interface with the name of **IGeneral**, which kept all the common functions which are mostly used. So for further development they don't need to write the whole code of function everywhere. They just need to call the function for enhancing the reusability and overall improving the performance of an ERP (Fig. 5).

Just like **SetNavigationButtons** and **GetAllRecords** are the Sub Procedure which can be seen in figure attached. They used to set the navigation button as per mode and show all of the records in master according to the screen respectively. This screen often contain a few functions as well (Fig. 6).

Isvalidate and **Update** are the functions in **IGeneral** interface for **GLUON ERP** Which have the purpose of first end validation and update the record by clicking on update function respectively.

They do have made some Global Variables as well **figure** attached so that they don't have to initialize repeatedly and assign the value whenever they want and improve the reusability of variables and make the code more readable and professional and by keeping the access specifier of these variable, **public,** make it easy for coder to use it in code anywhere (Fig. 7).

The Algorithm for the Global functions are being written in this model figure attached so that we may reuse them later on in the controller section to build a logic (Fig. 8).

Fig. 5. General functions in GLUON ERP

Fig. 6. General functions in Gluon ERP 2

7.3 Strategy Pattern

7.3.1 Purpose

In this technique, an algorithm can be changed at runtime and we create context objects which can varies according to the required strategy. This design pattern lies under the Behavior Category. The objects create in this phase shows the various strategies which can be implemented.

7.4 Application in GLUON ERP

In GLUON ERP, Strategy pattern is being implemented to get the **lastpurchaseprice**. Sometimes the requirements get vary from user to user just like a user want to get the last purchase price through batch no and the other one wants to get it by last production order against that item so to handle this scenario, a separate interface of strategy is being

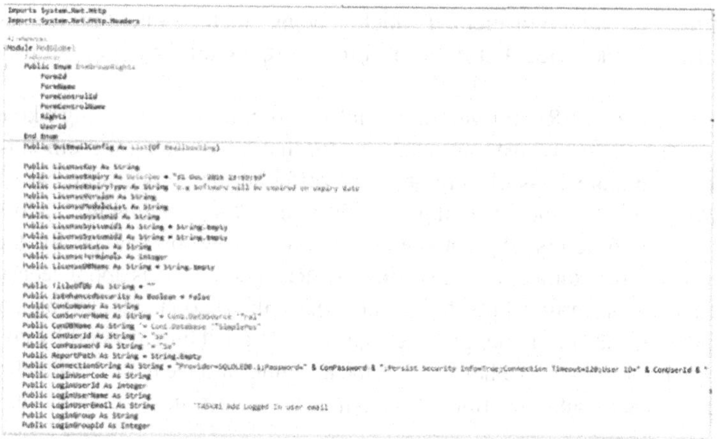

Fig. 7. Global variables in GLUON

Fig. 8. Function of GLUON ERP

created where different runtime functions are being written to get the last purchase price and the one which is required being called.

7.5 Iterator Pattern

7.5.1 Purpose

This pattern is commonly used in coding practice because it is a way to access to elements of a collection in a sequential manner without knowing the primary depiction. It lies under the behavioral category.

7.6 Application in GLUON ERP

In this ERP, The COLLECTION of chart of accounts, is being worked under this pattern because a container of string is being created to get all of the values and an iterator class is being created in this manner that it will work according to the no of rows of the table and an iterator interface have some methods like count and it will just get the data from the string by using a loop or method which is being created in iterator. It doesn't have any concern with the object but it will just loop through all the data until the count of rows.

Same as in Gluon ERP, for getting the information in separate string like inventory items or vendor names or customer name, so for that reason Iterator method is the best by creating a separate class of iterator where different methods can be called without concerning that which object is going to refer. After calling that method, a loop will be executed for n no of terms where **n** is mostly the no of rows of that specific local table.

There is another example of iterator in this ERP and this one is about security rights. To implement the security rights, first of all, get all the accounts, locations, cost center at once and then start implementing the rights where it is being needed by considering the client requirements. If we have got all the accounts at once then the developer don't need to get the accounts, locations, cost center for each time a new security is being added.

7.7 Template Pattern

7.7.1 Purpose

Template method lies under the Behavioral Design pattern. This method have a super-class, mostly named as abstract superclass, and in this number of steps and a skeleton of operations are being defined.

7.8 Application in GLUON ERP

In Aging Receivable report, there is an option of template for showing the data according to the user where the user/client set the template but it will only change the range of data shown but the other columns will remain same.so there is a template setting is being done in the backend which doesn't change the remaining column even the client change the whole structure.

The Fig. 1 is showing that the default layout have 30_60, 60–90, 90+ days range while in Fig. 2 the template1 have 90_120, 120_150 and 150+. Both templates have different layouts but the remaining data remains still same.

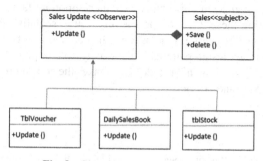

Same as we do have some templates screen where a list of operations are being performed and a skeleton is being created against that record and whenever a customer needs that he will simply load that template and reuse it. The template design pattern is the logic behind this whole scenario to make an ease for the client.

8   Class Diagrams of Design Patterns in ERP

There are certain ways to represent the relationship between different classes and objects in ERP just like UML diagrams, Class diagrams, Sequence Diagrams. A class diagram is being considered a good approach for representing the design pattern as it will cover all the relationships like inheritance, association, aggregation and composition between different classes. Just like in Fig. 1, an Observer design pattern is being observed in Sales module of ERP. Whenever the sales update (observer) updates the sales voucher, it will update that voucher in stock table, ledger table and its local table to which we named as DailySalesBook.

In another example of ERP, a composite pattern is being observer. For reports, there are two ways to run the reports. One is about to get the report directly which have rpt extension from reports folder and there is another way to get that which is, the reports contain the customized reports folder, and it contains the reports with the same name as well. So according to Fig. 10, the reports folder contain the reports which can directly run, and the customized reports is the folder which lies in reports folder having the same name reports (Fig. 9).

Fig. 9. Observer pattern can be added

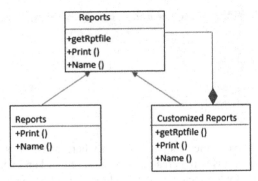

Fig. 10. Composite design pattern in gluon ERP

9 Results/Software Matrics

At initial level around 2005, they didn't consider the design patterns when the start making the different screens of ERP but gradually the code become more professional and the modularity of code getting increase. At initially, they have to write all the basic methods again and again like update or save buttons for each screen then by using design patterns the problem get resolved. At current state, 75% of the code is based on Design pattern consciously and unconsciously as well Because they have made it into their practice to do code in such a way that reusability becomes the main priority. Due to this higher percentage, the application is getting more stable.

10 Conclusion

The survey's results proved that the null hypotheses were correct. Developers are mostly unaware of the design patterns and they are not using it in their daily life practice as much. The study showed that most of the old employees and senior developers are used to the old programming ways. However, the people who are recently graduated and working in software houses are aware of design patterns. It means that either they have studied design pattern in universities. Another important factor is the increase of social communication these days. As people are getting socially aware and they can find suitable solutions for every problem. It can be counted as a factor that developers are trying to implement design patterns in their daily coding practices because it adds cohesion and reduces coupling in the code. This makes the code more flexible, solid and less vulnerable to bugs and crashes.

References

1. Bafandeh Mayvan, B., Rasoolzadegan, A., Ghavidel Yazdi, Z.: The state of the art on design patterns: a systematic mapping of the literature. J. Syst. Softw. **125**, 93–118 (2017). https://doi.org/10.1016/j.jss.2016.11.030
2. Bai, J., Luo, H., Qin, F.: Design pattern modeling and extraction for CAD models. Adv. Eng. Softw. **93**, 30–43 (2016). https://doi.org/10.1016/j.advengsoft.2015.12.005

3. Rasool, G., Maeder, P., Philippow, I.: Evaluation of design pattern recovery tools. Procedia Comput. Sci. **3**, 813–819 (2011). https://doi.org/10.1016/j.procs.2010.12.134
4. Makela, S., Lappanen, V.: Client based cohesion metrics for java programs. Sci. Comput. Program. **74**, 355–378 (2009)
5. Hashminejad, S.M.H., Jalili, S.: Design pattern selection: an automatic two phase method. J. Syst. Softw. **85**, 408–424 (2012)
6. Rajam, S., Cortez, R., Vazhenin, A., Bhalla, S.: Design patterns in enterprise application integration for e-learning arena. In: Proceedings of the 13th International Conference on Humans and Computers, pp. 81–88 (2010)
7. Rooksby, J., Morrison, A., Murray-Rust, D.: Student perspectives on digital phenotyping: the acceptability of using smartphone data to assess mental health. In: Proceedings of the 2019 CHI Conference on Human Factors in Computing Systems, pp. 1–14 (2019)
8. Sağbaş, E.A., Korukoglu, S., Balli, S.: Stress detection via keyboard typing behaviors by using smartphone sensors and machine learning techniques. J. Med. Syst. **44**(4), 1–12 (2020). https://doi.org/10.1007/s10916-020-1530-z
9. Thakur, S.S., Roy, R.B.: Predicting mental health using smart-phone usage and sensor data. J. Ambient Intell. Humaniz. Comput. **12**(10), 9145–9161 (2021). https://doi.org/10.1007/s12 652-020-02616-5
10. Torous, J., et al.: Creating a digital health smartphone app and digital phenotyping platform for mental health and diverse healthcare needs: an interdisciplinary and collaborative approach. J. Technol. Behav. Sci. **4**(2), 73–85 (2019). https://doi.org/10.1007/s41347-019-00095-w
11. Zakaria, Y., Hegazy, O.: Enhancing service design in erp systems using patterns and diverse healthcare needs: an interdisciplinary and collaborative approach. J. Technol. Behav. Sci. **4**(2), 73–85 (2019)
12. Survey form: https://docs.google.com/forms/d/e/1FAIpQLSdQnsEycZWq2wNnb_JQY0ma WPPlZhSg1RRh1bsjMgEU8r89lw/viewform?usp=sf_link

Towards a Semantics-Based Search Engine for Smart Contract Information

Chaochen Shi[1]([✉]), Yong Xiang[1], Jiangshan Yu[2], and Longxiang Gao[1]

[1] Deakin Blockchain InnovationLab, School of Information Technology,
Deakin University, Geelong, Australia
{shicha,yong.xiang,longxiang.gao}@deakin.edu.au
[2] Monash Blockchain Technology Centre, Faculty of Information Technology,
Monash University, Melbourne, Australia
jiangshan.yu@monash.edu

Abstract. Most blockchains are known for transparency since all on-chain records are open and immutable. However, due to the particularity of the blockchain data structure, users cannot perform semantic searches in the blockchain as in traditional search engines, resulting in inefficient access to blockchain data. To solve this problem, we proposed approaches for semantic search in the blockchain. In particular, we focused on the extraction of semantic information and related data for smart contracts. In this paper, we describe the road map, challenges, and preliminary results of our research.

Keywords: Blockchain · Smart contract · Information search

1 Introduction

A blockchain can be considered as a decentralized database that stores encrypted blocks of data. Some blockchains, such as Ethereum, support programmable smart contracts for better scalability. Smart contracts have been applied in many business areas beyond crypto-currencies, such as supply chain management, smart grid, and IoT applications, for trustable transactions. By July 2021, the number of smart contracts created by developers has exceeded two million on Ethereum. The explosive growth of smart contracts has spawned a demand for related data search services. However, mainstream blockchains can merely retrieve a smart contract by transaction ID or contract address. For users, it is inefficient to obtain desired information from complex blockchain data. Thus, it is essential to explore semantic search, such as searching by natural languages. The semantic search engine is to a blockchain network what Google is to the Internet.

Semantic search is a cross-field technology, which mainly relates to Natural Language Processing (NLP), information retrieval, and knowledge graphs. Traditional semantic search is mainly used in web search or database search, and relevant text data is easy to access. However, semantic search in blockchain poses additional challenges:

Supervised by Yong Xiang.

H. Hacid et al. (Eds.): ICSOC 2021 Workshops, LNCS 13236, pp. 272–277, 2022.
https://doi.org/10.1007/978-3-031-14135-5_22

- Smart contracts that implement business logic are written in programming languages, compiled into bytecode, and deployed in the decentralized nodes without any metadata or description. Except for the contract owner, other parties can only treat this contract as meaningless bytecode. How to identify semantic information from smart contract bytecode?
- The format of transaction data related to smart contracts is heterogeneous and has no semantic significance. How to provides these blockchain elements with semantic labels?
- In addition to on-chain data, smart contracts also include vital off-chain data such as source codes, API documentations, and official home pages. How to implement aggregated semantic search across on-chain and off-chain networks?

Therefore, our long-term research objective is to develop a semantic search engine for not only on-chain but also off-chain smart contract data. We can imagine the application scenario: after the user enters the natural language description of a smart contract, the search engine returns the on-chain data (bytecode, hexadecimal address, transaction history, etc.) and the corresponding source code stored off-chain, ranked by similarity. To address the above challenge, we proposed approaches for bytecode-based smart contract classification that can identify semantic features from on-chain bytecodes. We also proposed a semantic search method for off-chain smart contract source code. In this paper, we present these milestones and future works to be completed. We focus on Ethereum since it is the most widely used blockchain platform supporting smart contracts, but the concept of our approaches is applicable to other Turing Complete blockchains as well.

2 Contributions

In this section, we describe our contributions to the semantic search engine for smart contracts. We explain our classification approaches for on-chain contracts and semantic extraction approaches for off-chain contracts, and the corpus collected for experiments.

2.1 Classification Approach for On-Chain Smart Contracts

As the first step to address the semantic identification challenge for on-chain contracts, we need to classify and label smart contracts for further index. Thus, we proposed a novel bytecode-based classification approach [10] to effectively classify smart contracts on Ethereum. Smart contract bytecode is stored as a string of hexadecimal numbers in a Merkle Patricia tree. The numbers and can be translated into equivalent opcodes defined in Ethereum virtual machine via disassemblers such as $evmdis$[1]. We adopted opcode as a feature because it reflects the logic of the smart contract at the stack operation level.

---

[1] https://github.com/Arachnid/evmdis.

According to our statistics, smart contract categories on Ethereum are distributed unevenly—gaming and gambling contracts account for almost 40% among dozens of categories. Therefore, the smart contract classification can be regarded as a multi-classification problem on an imbalanced data set. The objective of the classification is as follows:

The dataset is defined as $\{D_i, y_j\}$, where D_i refers to a smart contract; y_j belongs to Y which is a predefined collection of k categories, $Y = \{y_1, y_2, \ldots, y_k\}$. The objective is to learn a mapping function h which maps the input D_i to the category y_j which it belongs to.

Given the poor performance of traditional classifiers on imbalanced datasets, we proposed a novel BPSO-Adaboost algorithm to solve this problem. The algorithm implements Binary particle swarm optimization [3] as the feature selection method to reduce the noise in the sample space, thereby improving the classification performance of minority classes. We also integrated Adaboost.M1 as the ensemble learning scheme to achieve better accuracy than individual classifiers. Comparative experiments have proved the superiority of each element in our algorithm. Compared with state-of-the-art NLP-based approaches, our bytecode-based approach has two key advantages. First, the bytecode-based approach can classify smart contracts stored as bytecode on Ethereum, thus supporting tag-based search and range query of on-chain smart contracts. Second, our approach can better resist adversarial attacks. Further improvements, experiments, and a comprehensive evaluation of our model will be implemented in the short run.

2.2   Semantic Search for Off-Chain Smart Contract Source Code

As mentioned in Sect. 1, a smart contract search engine is supposed to search for vital off-chain data. Thus, we explore semantic search techniques for smart contract source codes written in Solidity, the most widely used programming language in Ethereum Dapps. Our approach enables users to search Solidity source codes stored in off-chain repositories via natural language, even if there are no shared words between source codes and queries.

To achieve this objective, we mapped source codes and searched queries into a shared vector space where (code, query) pairs with similar semantics are neighbors. Cosine distance is regarded as the metric: The shorter the distance between the code and the query statement means higher the likelihood of returning the code. Since blockchain terms are different from daily usage words in English, we fine-tuned the pre-trained ALBERT model [4] to make our search engine understand the blockchain context and map queries and code comments to 768-dimensional vectors. To extract semantic information from the source code, we proposed Code2doc and Code2vec models, where Code2doc model is the upstream work of Code2vec model. As Fig. 1 shows, Code2doc model is designed to automatically generate summarization from the original source code via Transformer [12]. After training, the encoder of the Code2doc model acquired the mapping method from the source code to the intermediate representation of natural language. In the Code2vec model, we kept the encoder of the Code2vec

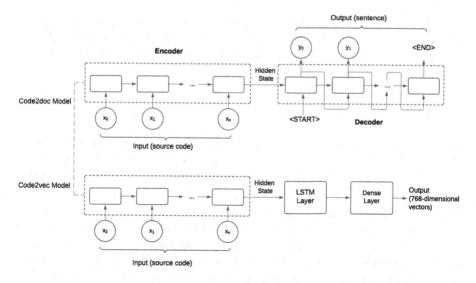

Fig. 1. The frameworks of the Code2doc model and Code2vec model

model and replaced the decoder with LSTM and Dense layers to generate vectors with fixed dimensions. Then we trained this model with (code, comment vector) pairs. Finally, the Code2vec model outputted 768-dimensional vector representations of codes in the shared vector space.

With the trained Code2vec model, we collected and inserted all the code vectors into the off-chain repository. Our search engine can convert the arrived query to the corresponding vector representation and conduct the nearest neighbor search with Hierarchical Navigable Small World (HNSW) graphs [5]. We have developed a prototype and plan to use Precision@k and Mean Reciprocal Rank (MMR) as metrics to evaluate the effectiveness of our approach.

2.3   Datasets and Corpus

Since there are no available public datasets for our training and validation yet, we need to build the required datasets and parallel corpus by ourselves. With our bytecode-based smart contract classification approach, we collected 4,000 smart contracts bytecodes from the top 100 Ethereum Dapps ranked by user activities and manually labeled their categories to train our supervised learning model. For our source code search engine, we collected about 37,000 smart contract code snippets from Etherscan[2] via web crawlers. To build a parallel training corpus, we extracted comments from source codes by regular expressions and then divided the corpus into training and test sets with the 10-fold cross-validation method. In addition, Solidity is a language that involves many reserved terms, specifically designed for blockchain operations. We summarized the blockchain terms and

---

[2] https://etherscan.io/.

special APIs related to blockchain behaviors, such as token transfer functions, which conduces to the fine-tuning process of ALBERT model and Code2doc model capture attention of word embedding. We will release our datasets and corpus to facilitate future research in this area.

3   Related Work

Researches on information search related to smart contract are still in its infancy, and there is no systematic study yet. In this section, we present a review of state-of-the-art techniques for on-chain information search and off-chain code search.

On-Chain Information Search: Peng et al. [7] proposed a verifiable query layer. Transactions are stored in the underlying blockchain system and extracted by the middleware layer, which regroups them into databases to provide block query and transaction query services for everyone. Pratama et al. [11] designed blockchain data retrieval that supports more than one search parameter and leverages multiple criteria to make searching for data easier. Additionally, some system query functionalities with analytic functions were proposed to effectively improve the overall capability of the query layer system. However, unlike our bytecode-based approach, these approaches can only provide keywords or parameter-based search without the ability to identify semantics and categories of on-chain data.

Off-Chain Semantic Code Search: There have been multiple deep learning proposals on semantic code search. The common point is the use of vector distance to measure the semantic correlation between codes and queries. NCS [9] is an unsupervised semantic code search model which uses word and document embeddings only. UNIF [1], an extension to NCS with supervised learning, replaces the sequence-of-words-based networks used in NCS with a less complex bag-of-words-based network. CODEnn [2] is also a supervised neural code search approach that integrates multiple Seq2seq networks. These techniques have difficulty capturing dependencies between long-range word tokens, leading to low accuracy when searching long code snippets. Thus, we use the Transformer to solve the long-range dependency problem. Our customized corpus of blockchain terms and APIs can also make the fine-tuned model more applicable for contexts related to smart contracts.

4   Conclusion and Future Work

In this paper, we have introduced the significance of semantic search for smart contracts and major challenges for on-chain and off-chain search. So far, our contributions include a bytecode-based classification approach for on-chain contracts, a semantic search approach for off-chain contract source codes, and related datasets and corpus.

Our current contributions allow us to generate simple semantic labels for on-chain contracts. In the following work, we will consider the further use of semantic labels to address the challenge of search content heterogeneous mentioned in Sect. 1. Specifically, we can treat the information in blockchain as graph information such as Resource Description Framework (RDF) [6]. Metadata related to smart contracts will be added to the body of the blockchain, just like the current web 3.0 semantic network does. After that, we can obtain the attributes of each element in blockchain and the relationship between the elements in a graph. This RDF information can be stored in a dedicated graph database or registered as assets on the chain for further search. In this way, we can search and analyze the functions of any service deployed to the blockchain and may expose them to the off-chain world for interoperability.

If some data can be modeled in RDF, searches would not be a severe challenge since there is a naturally adaptive query language SPARQL [8]. Natural language can be translated into SPARQL through NLP tools such as NL2Query[3] to realize the semantic search on the blockchain indirectly.

References

1. Cambronero, J., Li, H., Kim, S., Sen, K., Chandra, S.: When deep learning met code search. In: ESEC/SIGSOFT FSE, pp. 964–974. ACM (2019)
2. Gu, X., Zhang, H., Kim, S.: Deep code search. In: ICSE, pp. 933–944. ACM (2018)
3. Khanesar, M.A., Teshnehlab, M., Shoorehdeli, M.A.: A novel binary particle swarm optimization. In: 2007 Mediterranean Conference on Control & Automation, pp. 1–6. IEEE (2007)
4. Lan, Z., Chen, M., Goodman, S., Gimpel, K., Sharma, P., Soricut, R.: ALBERT: a lite BERT for self-supervised learning of language representations. In: ICLR. OpenReview.net (2020)
5. Malkov, Y.A., Yashunin, D.A.: Efficient and robust approximate nearest neighbor search using hierarchical navigable small world graphs. IEEE Trans. Pattern Anal. Mach. Intell. **42**(4), 824–836 (2018)
6. Miller, E.: An introduction to the resource description framework. Bull. Am. Soc. Inf. Sci. Technol. **25**(1), 15–19 (1998)
7. Peng, Z., Wu, H., Xiao, B., Guo, S.: VQL: providing query efficiency and data authenticity in blockchain systems. In: ICDE Workshops, pp. 1–6. IEEE (2019)
8. Pérez, J., Arenas, M., Gutierrez, C.: Semantics and complexity of SPARQL. ACM Trans. Database Syst. (TODS) **34**(3), 1–45 (2009)
9. Sachdev, S., Li, H., Luan, S., Kim, S., Sen, K., Chandra, S.: Retrieval on source code: a neural code search. In: MAPL@PLDI, pp. 31–41. ACM (2018)
10. Shi, C., Xiang, Y., Doss, R.R.M., Yu, J., Sood, K., Gao, L.: A bytecode-based approach for smart contract classification. CoRR abs/2106.15497 (2021)
11. da Silva, F.J.C., et al.: Analysis of blockchain forking on an ethereum network. In: EW, pp. 1–6. VDE/IEEE (2019)
12. Vaswani, A., et al.: Attention is all you need. In: NIPS, pp. 5998–6008 (2017)

---

[3] https://www.lymba.com/nl2query.

A Blockchain-Based Autonomous System for Primary Financial Market

Ji Liu[1,2(✉)] and Shiping Chen[2]

[1] School of Electrical and Information Engineering, The University of Sydney, Sydney, NSW 2000, Australia
jliu3872@uni.sydney.edu.au
[2] CSIRO DATA61, Sydney, NSW 2015, Australia

1 Introduction

The primary financial market has a large size, including both the debt market and equity market. According to Economics of Bloomberg, global debt issuing is 24 trillion US dollars in 2020 [1]. McKinsey Global Private Market Annual Review 2021 [2] reports that the total global private equity market transactions in 2020 have plateaued at 1.1 trillion USD.

However, the ecosystem of the primary financial market is fragmented, and its infrastructure is not as well established as the second financial market. The main reasons for hindering the primary market are: high cost on the discovery of each other correctly, (a) high cost on discovery of each other properly; (b) asymmetric information; (c) high cost and complex process of trading stages; and (d) poor financial derivatives in primary market. Most activities in the traditional financial market are done through centralized intermediatory institutions, i.e., banks, trust, exchanges, etc.

The above issues are mainly caused by the characteristics of the infrastructure of the centralized system. As long as the centralized system is adopted, the above challenges cannot be eliminated. Due to its distributed characteristics, Blockchain technology can be considered a potential solution for the primary market. The object of this PhD-Project is to research and develop a blockchain-based autonomous system for the primary financial market to reform the existing centralized infrastructure of the primary financial market. This revolution will not only improve business automation status but also enhance organizational and ecological autonomy.

2 Solutions and Contributions

The essence of the blockchain is a distributed ledger system. The core technology of the blockchain is mainly to ensure the complete security of the data in the distributed ledger system and the trustworthiness of the entire transaction process. The characteristics of blockchain have different expressions and differences in various papers. However, the following aspects have reached a consensus: decentralization, high transparency, enhanced security, and immutability of information [3].

© Springer Nature Switzerland AG 2022
H. Hacid et al. (Eds.): ICSOC 2021 Workshops, LNCS 13236, pp. 278–282, 2022.
https://doi.org/10.1007/978-3-031-14135-5_23

This project takes the traditional challenges of the primary financial market as the research object and embeds blockchain technology as a practical path for infrastructure optimization. Based on the current related theoretical research results, this project uses the new institutional economics theory as a logical starting point to analyze the financial market infrastructure with organizational evolution as a clue and the practical constraints and problems it faces. On the other hand, this research takes blockchain technology as a software architecture solution by interpreting the "blockchain + securities" characteristics of the blockchain and the institutional economics of blockchain digital governance, paving the way for the application of blockchain in the financial market.

Furthermore, this project combines the "technology logic" of blockchain digital governance with the "system logic" of financial products and analyses the governance mechanism under the dual logic of "technology + system." The proposed infrastructure of the primary financial market will be constructed based on the blockchain architecture. Under this functional framework, the primary governance mechanisms, such as trust, reputation, contracts, etc., on the chain are further discussed. In terms of empirical research, this project will establish a primary bond issuance, regulation, trading system based on blockchain technology to contribute to the autonomy of the primary market in practice.

3 State-of-Art

Many scholars have researched the application of blockchain technology in the financial field. They mainly focus on four scenarios: payment and settlement, asset digitization, innovative securities, and customer identification and credit reporting systems. [4] tries to combine the idea of blockchain into bill market transactions. [5] explains the reason to adopt blockchain into inter-bank payment theoretically. [6] uses a blockchain-based system in practice for three months and concludes with positive results. [7] discusses from the development logic of blockchain, as the underlying technology of payment and clearing, and the basic technology of financial technology. [8–10] from traditional commercial banks process, credit system, and financial risk control aspects, respectively, have concluded that the blockchain is helpful to improve the current systems.

4 Findings and State of Work

Comprehensive literature research, named "Applying Blockchain for Primary Financial Market: A Survey," has been published. According to this survey from academics, we found that the scholars are concentrated on the issue of securities, and most of the research is carried out around technical aspects. Few scholars researched from the perspective of applying technology into the primary financial market.

In addition, there are still challenges of applying blockchain for solving the issues in the financial market: a) regulation and compliance of governments are challenging to build into blockchain; b) technology development and application of blockchain is still at an early stage; c) privacy preservation of blockchain applications is hard to satisfy each party's requirements in such a financial market.

After the direction from the literature review, a requirement collection survey paper from the industry was conducted. A combining method of qualitative and quantitative analysis has been adopted in the research. We interviewed 15 domain experts and handed out 54 questionaries to industry participants. As a result, we figured out that the most concerning issues in the primary financial market are complex due diligence, mismatch, and rugged monitor. The experimental study has been completed, and the outcome is about to submit as another journal paper.

5 Methodology

The primary financial market has a complex mechanism (i.e., involving multi-parties, different organizational governance models, and multiple factors that affect organizational governance models, etc.). The adoption of blockchain to optimize the primary market infrastructure requires the integration of the digital technology theory represented by the blockchain with the economic theory of the primary market and empirical analysis. Therefore, this project will adopt various methods such as interdisciplinary research, literature, typical case analysis, qualitative, game theory, and other methods to research and discuss the related topics.

More specifically, there are five components in my research and categorized into two layers. The first layer is for requirement collection to determine the next layer. The first layer contains two projects: a survey paper from academics and a survey paper from the industry. The second layer has three projects: system designs focusing on 1) bond issuance in the primary market, 2) post-issuance fund monitor, and 3) bond trading system. It is a standard order of managing a bond's lifecycle.

6 Evaluation

In terms of evaluation, I plan to evaluate it from three perspectives. 1) Performance evaluation: it is designed from the perspective of system performance. I will compare it with the corresponding benchmark ratio to test the performance of the system. 2) Usability test: it is one of the evaluation methods of traditional software engineering. It is used to test the usability and experience of the system. 3) Mathematical modeling: it is from a business perspective. I will build a cost model for the proposed blockchain infrastructure system and compare it with the cost of the current central system to perform cost optimization calculations to evaluate whether the system is significant at the business level.

7 Related Work

Scholars have studied a lot around the combination of blockchain and primary market. In terms of securitization issuance, [11] claims that substituting third parties in the primary market with a blockchain-based system can monitor process details more accurately, reduce costs, enhance issuance speed, increase transparency, and therefore increase liquidity. [12] thinks that blockchain technology can help decrease the risk of fraud caused by knowledge asymmetry. [13] presents a Linked Data-based approach that provides both verification and tamper-proof characteristics to avoid collision and efficiently enhance confidence in the financial sector.

Researchers are also debating how blockchain technology might help cut expenses and enhance efficiency. [14] explains how the use of blockchain under solid governance and legal frameworks may enhance corporate transparency and efficiency when the record is generated. explained how blockchain technology has increased automation. [15] shows that by incorporating smart contract terms into an automated programming language, stock transactions may be performed automatically when specific criteria are satisfied. More specifically, it employs machine learning to assess the desire and ability of borrowers and use a blockchain-based database, and it can predict the borrower's future willingness in real-time.

Furthermore, several articles are based on a regulatory perspective, believing that apps are beneficial to departmental oversight. [16] describes the history of employing blockchain in the primary market early and argues why blockchain technology can replace financial institutions. [17] describes the supervision department, which registers nodes on the chain and receives the public key to oversee basic information in real-time. At the same time, the blockchain's immutability ensures the information's validity and traceability. [18] demonstrates that auditing and supervision will no longer be limited to sampling but will acquire and process all information through a database maintained and shared on the blockchain. In contrast to the conventional paradigm, [19] believes that the auditor may validate the data through an internal BCs ecosystem before going external. Once the report is completed, the auditor will place it in the external eco to prevent external institutions and customers from altering the data regularly. It prevents data tampering and leakage during transmission and enhances data authenticity.

References

1. Maki, S.: World's $281 Trillion Debt Pile Is Set to Rise Again in 2021. In: Economics, Bloomberg (2021). https://www.bloomberg.com/news/articles/2021-02-17/global-debt-hits-all-time-high-as-pandemic-boosts-spending-need
2. de Miguel, B.: A year of disruption in the private markets. McKinsey Global Institute (2021). https://www.mckinsey.com/~/media/mckinsey/industries/private%20equity%20and%20p rincipal%20investors/our%20insights/mckinseys%20private%20markets%20annual%20r eview/2021/mckinsey-global-private-markets-review-2021-v3.pdf
3. Nakamoto, S.: Bitcoin: A peer-to-peer electronic cash system. Manubot (2019)
4. Takahashi, K.: Blockchain technology and electronic bills of lading. J. Int. Marit. Law Publ. Lawtext Publ. Ltd. **22**, 202–211 (2016)
5. Yoo, S.: Blockchain based financial case analysis and its implications. Asia Pac. J. Innov. Entrepreneurship (2017)
6. Tsai, W.T., et al.: Blockchain systems for trade clearing. J. Risk Finance (2020)
7. Peters, G.W., Panayi, E.: Understanding modern banking ledgers through blockchain technologies: future of transaction processing and smart contracts on the internet of money. In: Tasca, P., Aste, T., Pelizzon, L., Perony, N. (eds.) Banking beyond banks and money. NEW, pp. 239–278. Springer, Cham (2016). https://doi.org/10.1007/978-3-319-42448-4_13
8. Wu, B., Duan, T.: The advantages of blockchain technology in commercial bank operation and management. In: Proceedings of the 2019 4th International Conference on Machine Learning Technologies, pp. 83–87, June 2019

9. Li, Y., Liang, X., Zhu, X., Wu, B.: A blockchain-based autonomous credit system. In: 2018 IEEE 15th International Conference on e-Business Engineering (ICEBE), pp. 178–186. IEEE, October 2018

10. Bashynska, I., Malanchuk, M., Zhuravel, O., Olinichenko, K.: Smart solutions: risk management of crypto-assets and blockchain technology. Int. J. Civ. Eng. Technol. (IJCIET) 10(2), 1121–1131 (2019)

11. Halevi, T., et al.: Initial Public Offering (IPO) on permissioned blockchain using secure multiparty computation. In: 2019 IEEE International Conference on Blockchain (Blockchain), pp. 91–98. IEEE, July 2019

12. Buterin, V.: A next-generation smart contract and decentralized application platform. White Pap. 3(37), 1–2 (2014)

13. Ahmad, A., Saad, M., Bassiouni, M., Mohaisen, A.: Towards blockchain-driven, secure and transparent audit logs. In: Proceedings of the 15th EAI International Conference on Mobile and Ubiquitous Systems: Computing, Networking and Services, pp. 443–448, November 2018

14. Tinn, K.: Blockchain and the future of optimal financing contracts. Available at SSRN 3061532 (2017)

15. Sutton, A., Samavi, R.: Blockchain enabled privacy audit logs. In: d'Amato, et al. (eds.) ISWC 2017. LNCS, vol. 10587, pp. 645–660. Springer, Cham (2017). https://doi.org/10.1007/978-3-319-68288-4_38

16. Aitzhan, N.Z., Svetinovic, D.: Security and privacy in decentralized energy trading through multi-signatures, blockchain and anonymous messaging streams. IEEE Trans. Dependable Secure Comput. 15(5), 840–852 (2016)

17. Lamarque, M.: The blockchain revolution: new opportunities in equity markets, Doctoral dissertation, Massachusetts Institute of Technology (2016)

18. McCallig, J., Robb, A., Rohde, F.: Establishing the representational faithfulness of financial accounting information using multiparty security, network analysis and a blockchain. Int. J. Account. Inf. Syst. 33, 47–58 (2019)

19. Sheldon, M.D.: A primer for information technology general control considerations on a private and permissioned blockchain audit. Curr. Issues Audit. 13(1), A15–A29 (2019)

Multi-agent Reinforcement Learning for Task Allocation in Cooperative Edge Cloud Computing

Shiyao Ding$^{(\boxtimes)}$

Graduate School of Informatics, Kyoto University, Kyoto, Japan
dingshiyao0217@gmail.com

Abstract. Edge cloud computing has become a fundamental computation infrastructure supporting the resource-limited devices of Internet of Things (IoT). An important problem in edge cloud computing is how to allocate tasks to the servers while minimizing various costs and satisfying task requirements. Studies to date usually assume a self-interested setting where each edge/cloud server is owned by one user who tries to maximize own interests. However, with the strong development of smart communities like smart factory, the servers are usually owned by an organization like an IT corporation. This triggers the necessity for edge/cloud server cooperation to maximize team interests. Thus, in this paper, we consider a new problem called cooperative edge cloud computing where edge/cloud servers cooperate with each other to perform tasks to optimize the interests of the whole system. This problem is difficult due to some features such as 1) the tasks usually have high workloads which cannot be well performed by only one server; 2) the tasks usually have a dependency relationship; 3) edge servers are usually distributed where each server only has a partial observation. Our idea is to formulate the problem as a multiagent system, where each server is regarded as an agent who can learn to execute decision-making for task allocation based on its observation (e.g., current server status and arriving task). Then, we employee multiagent reinforcement learning methods to make agents learn from the environment by themselves without previously designed rules. Our expected impact is that our algorithm can offer significantly better attributes such as low latency and low energy consumption in the cooperative edge cloud computing.

Keywords: Internet of Things (IoT) · Edge cloud computing · Task allocation · Multiagent systems · Reinforcement learning

1 Introduction

Since the number of Internet of Things (IoT) devices such as smart watches, smart phones is exponentially increasing, an enormous volume of data will be generated and need to be processed [18]. However, IoT devices usually have

Supervised by Donghui Lin.

H. Hacid et al. (Eds.): ICSOC 2021 Workshops, LNCS 13236, pp. 283–297, 2022.
https://doi.org/10.1007/978-3-031-14135-5_24

constrained computation resources, which creates reliance on external computing resources [3]. Cloud computing is a classical solution that can provide abundant customizable computation resources, its effectiveness has been verified and many cloud computing providers such as AWS, Google and Azure have proposed various cloud computing services. Edge computing, as a supplement to cloud computing, can offer computing services with lower latency and lower energy than cloud computing as its servers are closer to the users. However, the computation resources of edge servers are not as rich as those of cloud servers. Therefore, edge cloud computing, which combines the advantages of edge and cloud computing is seen as the desirable computing platform for IoT [4].

A fundamental problem with edge cloud computing is how to allocate tasks to the various servers so as to minimize various costs while satisfying the task requirements. Although many studies have tackled this problem, they most often assume a self-interested edge cloud computing environment, where each edge/cloud server tries to maximize its own interests. Accordingly, they usually formulate the problem as a game and apply the Nash equilibrium strategy at each server so each server develops its best response given the other server strategies.

However, with the strong development of IoT devices and services, more and more companies are starting to provide smart communities like Aliyun's city brain. In those scenarios, edge servers are usually owned by an organization rather than a single user [11,17]. The goal is to optimize the overall performance of the edge cloud computing systems rather than each edge server's own interests. Thus, how to make edge and cloud servers cooperate with each other to perform tasks well requires to be studied. In this paper, we consider cooperative edge cloud computing, a new edge cloud computing framework where each edge/cloud server tries to maximize team rewards like total delay costs. Although the methods proposed in a self-interested setting can be applied to this problem, they may not achieve good performance since optimization of each server may not achieve an optimization of the team.

2   Motivating Example

Along with the strong development of information and communication technology, technologies such as object detection algorithms, recommendation algorithms, and virtual reality (VR)/augmented reality (AR) can be used to allow IoT to supply various applications to users. For instance, people in widely separated areas can use VR/AR equipment to realize a remote collaboration by sharing one common virtual space. However, this generates large volumes of data that must be processed efficiently with attributes such as low latency and low energy consumption. This emphasizes the necessity of edge cloud computing, since the devices themselves are usually too resource-constrained to handle the loads.

Let us take the scenario in Fig. 1 as an example; users are distributed in various areas and each area is called a local environment. In each local environment, various devices such as temperature sensors and cameras exist to capture user/environment data. Also, an edge server cluster consisting of several heterogeneous edge servers like RaspberryPi exists. It performs the tasks generated

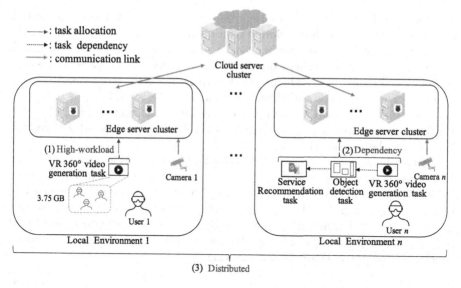

Fig. 1. Three major features of the edge cloud computing system needed to ensure cooperation. (1) High-workload: in order to generate a shared virtual space for user 1, a VR 360° video must be created from based on the camera data collected from the other users (approximately 3.75 GB data must be processed per second). (2) Dependency: in order to provide a service that user 2 may now need, three dependent tasks are needed: VR 360° video generation task, object detection task and service recommendation task. (3) Distributed: all the edge servers spread out over various areas and share one common cloud server cluster. The goal is to optimize the overall performance of the edge cloud computing system rather than a single server.

from the IoT devices to support the various IoT applications offered to the local users. Moreover, all edge server clusters share one cloud server cluster and can choose to offload tasks to it if necessary.

Several features of edge cloud computing require cooperation among the servers. Let us consider the example of remote collaboration by VR equipment shown in Fig. 1. In order to generate a shared virtual environment for users located in different areas, the cameras in each local environment collect the user data and send it to the respective edge server cluster. Consider the instance of user 1 in the left part of Fig. 1, its local edge cluster can generate a VR 360° video of the shared virtual space based on the camera data collected from the users in the other areas. In order to achieve an immersive feeling, high quality real-time videos with low latency are necessary. For instance, a 4K video frame usually has 8 MB data, which demands at least 1.875 Gbps to display the video given the refresh rate 240 Hz. Since VR has two displays and each display requires one 4K video, 3.75 GB of data must be transferred per second. This **high-workload** feature means that just one server is insufficient, which incurs the necessity of cooperation among edge servers.

Moreover, in the virtual shared space, tasks other than VR video generation exist and might have a **dependence** relationship. As for user n shown on right side of Fig. 1, it might be recommended a service based on the current virtual environment. Achieving this recommendation requires three dependent tasks to be performed sequentially. The first task is to generate the current virtual environment. Then, based on the virtual environment, the second task is to execute an object detection algorithm to collect the major elements around the user. Finally, a service recommendation task is invoked to recommend services like Google translation that the user may need.

Although all edge servers are **distributed** across various areas, they can belong to one organization and share one common cloud server cluster, as shown in Fig. 1. Thus, the goal is to optimize the performance of the overall edge cloud computing system rather than a single server. For instance, if many edge servers upload tasks at the same time, network congestion may occur which would hurt overall system quality. Considering the above three major features, this paper considers cooperative edge cloud computing.

3   Research Challenges

As stated in the above section, cooperative edge cloud computing is required to support the truly effective IoT environment. In this section, we first define the problem of cooperative edge cloud computing as follows: "Given an edge cloud computing system with several edge and cloud servers and a certain period such as one day or one month, the goal is to maximize the overall team interest (the whole system) in that period, such as minimizing the total delay/energy cost of all servers." Then, we indicate its three major challenges which are stated as follows.

First, **high-workload** tasks usually cannot be performed well by just one server; adequate performance is assured only if the task can be cooperatively performed by several servers.

Second, the tasks are inter-dependent [20] where some tasks can only be performed after the other tasks have been completed; it corresponds to a **dependence** feature. Then, one server needs to know the situation of its own tasks' dependent tasks allocated to other servers, which corresponds to a cooperation among servers.

Third, the edge servers are usually **distributed** across various areas and share one cloud server cluster [13]. Thus they must cooperate with each other, since one server's performance does not depend on just its own actions, but can be influenced by the other servers' actions.

4   Proposal

The research challenges elucidated above require that task allocation in edge cloud computing be well modeled. Since each server can observe its own status and conduct decision-making for task allocation, it can be regarded as an agent, an intelligent entity that can make decisions based on its observations. Moreover, many servers exist and they will influence each other. Thus, we model the edge cloud computing system as a multi-agent system, which is a classical model to solve agent interaction problems, especially where the multiple agents interact with each other to achieve some goals.

Reinforcement learning (RL) is an efficient method to train agents in an environment without any previously designed rules, and has been applied in many studies on edge cloud computing. However, they usually focus on a single RL method without considering the interaction of agents. In this paper, we desire to make multiple agents learn cooperation by multiagent reinforcement learning (MARL). As shown in Fig. 2, our proposed MARL framework using RL, game theory and other approaches solves the problems of cooperative edge cloud computing. The aim of this paper is to provide a MARL framework to support task allocation in cooperative edge cloud computing. We describe in detail below how to cope with each challenge.

4.1 Reinforcement Learning Based Dynamic Coalition Formation for High-Workload Task Allocation

Objectives and Issues. As for challenge 1, existing studies tackling high-workloads usually employ static distributed computing methods. For instance, MapReduce, a classical distributed algorithm, can divide a high-workload task into several low-workload tasks and then distribute them among several servers [7]. However, they consider just the information of tasks that are independent of the current status of servers. This may incur bad performance in edge cloud computing since the cost of performing the tasks significantly depends on server status; it is dynamically altered by the popping/pushing of tasks. For instance, edge server status in each local environment in Fig. 1, dynamically changes as it performs tasks like VR video generation task. Moreover, with the users coming/leaving the current local environment, some tasks would be invoked/deleted.

Thus, we study the high-workload task allocation problem in edge cloud computing with consideration of the dynamic features. The goal is to identify an optimal policy that can make edge servers perform high-workload tasks cooperatively where the task's workload can be infinitely divided among the various servers. To sum up, solving the high-workload problem has two major problems: 1) it is difficult to divide a high-workload task into several low-workload tasks given the dynamic features. 2) it is difficult to handle the large solution space incurred by infinite task division.

Proposal. To cope with the first issue, we formulate the problem as a Markov decision process (MDP), since MDP is a classical model for modeling discrete-time decision problems. Specifically, the statuses of arriving tasks and servers are regarded as states, and the dynamics can be regarded as state-transitions in MDP. To cope with the second issue, we consider the dividing task workload

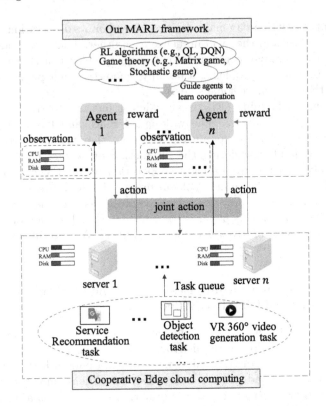

Fig. 2. We propose a MARL framework to guide the servers to cooperate in allocating the tasks. Each server is regarded as an agent which has its own observation such as server status and arriving tasks and can take an action (task allocation) based on the observations. Then, a joint action consisting of the actions from all agents effects to the edge cloud computing system and each agent can obtain a reward. Then, based on the rewards obtained, the agents can learn cooperation by our framework.

problem from the perspective of servers. Specifically, we make the servers form several coalitions, each of which can take a task to perform. The task can be divided among the servers of the same coalition and executed in a distributed manner. Specifically, we propose a dynamic coalition formation algorithm called coalitional R-learning (CR-learning) to guide edge servers in performing tasks by forming coalitions.

Our proposed method can be divided into two phases as shown in Fig. 3. In the first phase, the edge servers can form several coalitions $\{c_1, c_2, ...\}$ where each edge server can join just one coalition c_i. As shown in Fig. 3, at the current step server 1 and server n form a coalition c_1, and server 2 forms a coalition c_2 with the other servers. Each coalition formation way is called a coalition structure cs and each coalition c_i corresponds to an action space defined as A^{c_i} where the action is defined as choosing a task in this paper. Then, in the second phase, each coalition can choose an action from A^{c_i}. As shown in Fig. 3, coalition c_1 chooses task 1 to perform and coalition c_2 chooses task 4 to perform. Then, a high-workload task can be divided among a coalition according to the current server statuses. In the next step, the server statuses are altered by task performance and the task queue is refreshed, which corresponds to a new state. The new state requires the servers to form a new coalition structure that accords the current state.

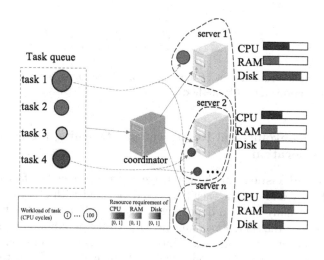

Fig. 3. Illustration of our proposed coalitional R-learning algorithm.

Evaluation. To show the effectiveness of our proposed method, we compare the performance attained by our coalitional-R learning algorithm to those of the baselines of R-learning and a linear programming (LP) algorithm [25]. The results are shown in Fig. 4. Since the LP method yields a deterministic optimal solution of the current step rather considering long-term rewards, its performance is inadequate. Its performance does not improve regardless of iteration number due to its deterministic feature without learning. However, R-learning, as a classical RL algorithm, can well cope with the dynamic feature. It can achieve better performance than the LP method, since it copes with dynamic changes well. However, its performance is inferior to that of our proposed CR-learning algorithm. This is because CR-learning considers both dynamic feature and cooperation feature (coalition formation). We confirmed this by conducting comprehensive experiments with different parameter settings such as task workload, task storage and task number in [9]. Then, that paper identified the problem that the cost of changing coalition structure must be considered in some cases. Thus, we solved this problem by proposing a new theoretical model in [8].

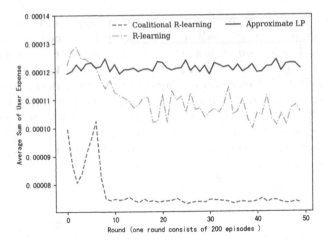

Fig. 4. Performance comparison of our coalitional R-learning algorithm, R-learning algorithm and approximate LP method.

4.2   Graph Convolutional Reinforcement Learning for Dependent Task Allocation

Objectives and Issues. In the challenge 1, we consider the feature of high-workload of tasks and assume each task is independent with other tasks. Besides of that, there usually exists a dependency relationship among the tasks, which is shown in Fig. 1: the service recommendation task can only be performed after finishing the object detection task. Thus, it is necessary to consider the dependency relationship while do a dependent task allocation. Solving this problem has two major issues: 1) how to cope with dependency information for decision-making of task allocation and 2) how to cope with the dynamics whereby server status and arriving tasks change dynamically.

Proposal. To solve these issues, we propose a novel algorithm of graph convolutional reinforcement learning (GCRL) for dependent task allocation: it can deal with the dependency and dynamic issues of the problem effectively. It consists of two parts that tackle the above issues separately: encoding part and decision-making part, as shown in Fig. 5. In the encoding part, we call several dependent tasks as a job where each task is related to at least one another task. Then, we represent a job as directed acyclic graphs (DAG): each task of a job is a node and dependency relationships are represented by directed edges. DAG is handled by a graph convolutional network (GCN), which is an effective tool for dealing with graphical data and can well abstract high dimensional information. Specifically, we can obtain an adjacency matrix and feature matrix for each job based on DAG. We input those matrixes to GCN which outputs the embedding result of the dependency information of tasks. In the decision-making part, we formulate the task allocation problem as a MDP and use the deep reinforcement learning algorithm called deep Q-network (DQN) to cope with the dynamics. The embedding result obtained from GCN is used as a part of the input for decision-making in task allocation. Finally, DQN outputs the state-action value $Q(s, a)$, which is the evaluation for each action (choosing a server for a job allocation). We use the cost generated by task allocation in training the GCRL network.

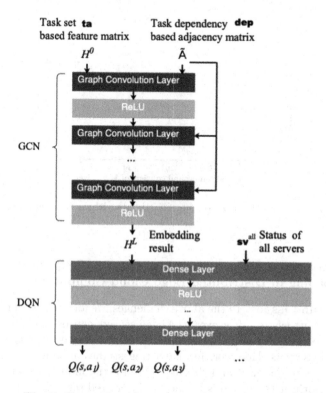

Fig. 5. Illustration of our proposed GCRL algorithm.

Evaluation. To show the effectiveness of our proposed method, we compare the performance of our GCRL algorithm with those of the baselines of Q-learning (QL) algorithm [22] and DQN algorithm [5]. The results are shown in Fig. 6. Since large edge server status incurs a huge state space, QL must maintain a large Q-table to record each Q-value for each state-action pair, which degrades learning effectiveness. Thus, QL has a poor performance which can even learn nothing during iterations. DQN consists of a neural network that can learn the generalization of similar states, and so can handle the large state space well. However, its performance is not better than that of our proposed GCRL algorithm. This is because GCRL well cope both large state space features and dependency features. We performed more comprehensive experiments under different parameter settings such as task workload, task storage and task number in [10].

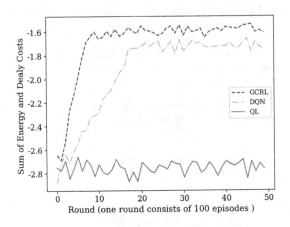

Fig. 6. Performance comparison of our GCRL algorithm, QL algorithm and DQN method.

4.3 Multi-agent Reinforcement Learning for Cooperative Task Offloading in Distributed Edge Cloud Computing

Objectives and Issues. In the above challenges, we focus on a centralized edge cloud computing system. However, the edge servers can be allocated to various areas. Each edge server can only observe its own status and offload its tasks to remote cloud servers. However, most existing studies assume each edge server cluster belongs to one user and they compete for the limited resources like network bandwidth; it corresponds to the self-interested setting. This setting does not suit with the case of all edge cloud servers being owned by one organization. We tackle this cooperation problem by considering this distributed characteristic where each edge server. The goal is to optimize the interest of the whole system rather than a single server. To sum up, solving this problem has two major issues: 1) how to cope with the dynamics whereby server status and arriving tasks change dynamically and 2) how to cope with the partial observation problem since each edge server can only know its own status.

Proposal. As stated in the above section, the problem has two major features: dynamic attributes and partial observation. Thus, we formulate it as a decentralized partially observable Markov decision process (Dec-POMDP) which is a classical model for discrete-time decision problems under partial observation. Although traditional algorithms like DQN can also be applied to solve

Dec-POMDP by making each agent maintain a DQN, it is difficult to achieve a cooperation. Thus, based on the cooperative MARL algorithm called value decomposition network (VDN) [19], we propose a cooperative task offloading algorithm called VDN based task offloading (VDN-TO). As shown in Fig. 7, in VDN-TO there are two types of Q-values: total Q-value and individual Q-value. The total Q-value is used to evaluate the team interests which is trained using team rewards (in this paper, we define team rewards as the total summation of all individual rewards). Then, each individual Q-value is updated to the direction that can maximize the total Q-value rather than its own Q-value.

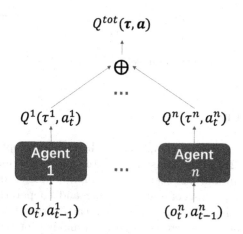

Fig. 7. Illustration of our proposed VDN-TO algorithm.

Evaluation. In this section, we apply the algorithm in [15] called IDQL-TO and random policy as the baseline algorithms, and compare their performance with that of our VDN-TO algorithm. Since the tasks include some random elements which make them different at each episode, we take 5 episodes as one round and use the average reward values in one round in the performance comparison. We ran the experiments for 500 episodes, which corresponds to 100 rounds, and the results are shown in Fig. 8. Although IDQL-TO can sometimes match the performance of VDN-TO, its performance is unstable. This is because each edge server considers just its own interests and they can conflict if they aim to maximize its own interests at the same time. On the other hand, our proposed VDN- TO algorithm utilizes a total Q and each edge server tries to optimize it in a cooperation way, which yields good and stable performance.

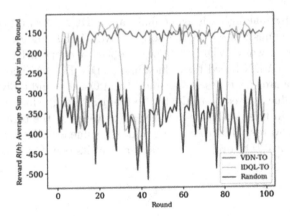

Fig. 8. Performance comparison of our VDN-TO algorithm, IDQL-TO algorithm and random policy.

5   Related Work

Task allocation in edge cloud computing with the aim of optimizing costs such as offloading cost [21], data transfer cost [14] and deployment cost [12], is a fundamental problem and has been studied often. These studies can be divided into two main types: non-cooperative setting and cooperative setting. In the non-cooperative setting, it is assumed that each server is self-interested and they only care about own interests in decision-making for task allocation. This kind of problem is usually formulated as a game whose solution usually involves a Nash equilibrium. In the cooperative setting, it is assumed that each server has a common team interest called social welfare to optimize rather than merely considering its own interest. We review here the studies that consider the non-cooperative and cooperative settings.

Non-cooperative Task Allocation. In the non-cooperative setting, each edge server is self-interested and might trigger a competitive relationship with other edge servers since the common resources might be limited. Chen et al. [6] studied a multi-user task offloading problem in edge cloud computing in the self-interested setting. Specifically, since the channel resources are limited, each edge server tries to compete for access, which corresponds to a competition relationship. The problem is formulated as a game theoretic model and the goal is to solve a Nash equilibrium strategy for task offloading. Liu et al. [16] studied an edge cloud computing network involving of cloud layer, edge layer and user layer, where the users choose to offload tasks to edge or cloud layers. The users are divided into different groups by a clustering method. Then, each group is assigned a task offloading priority and whether tasks are offloaded depends in part on the priority. Chen et al. [5] also considered a non-cooperative environment in the distributed setting. Specifically, each end user can observe its local environment and make a decision on the use of either local computing or edge computing. The problem is formulated as a stochastic game and each end user is regarded as an agent. Then, each agent has its independent policy and tries to optimize its own interest.

Cooperative Task Allocation. Although some studies tackle the task allocation in the cooperative setting, they usually take static cooperation approaches. Yu et al. [23] considered a joint computation and communication user cooperation problem in edge computing, where one user can share its computation resources with the others to improve overall user performance. Barbarossa et al. [1] proposed a cooperative task offloading algorithm. It divides edge servers into two groups according to the amount of data that they need to offload. The edge servers in the first group are allowed to access the cloud server, while the second group can process tasks only locally. Cao et al. [2] considered mobile edge computing systems with the aim of improving the energy efficiency while satisfying latency-constraints. They proposed a joint computation and communication cooperation method by setting a server as a helper and allowing other edge servers to offload computation tasks to the helper which cooperatively computes these tasks. Yuan et al. [24] considered a cooperative edge computing platform shared by different stakeholders. Due to the constrained computing resources of edge servers, it requires them to cooperate in a distrusted environment. Since the edge servers belong to different stakeholders, the key is how to resolve trust and incentive issues. Their solution is a blockchain based method. Each edge server can publish its own tasks for other candidate edge servers to contend for. Then, a winner is selected from candidate edge servers based on their reputations and a consensus is reached by using blockchain to record the performance of task execution.

The above studies do not consider dynamic cooperation so server status remains unchanged and the cooperation is usually based on a static approach. This setting is unlikely to satisfy the dynamic attributes of edge cloud computing and so we studied dynamic cooperation methods in this paper.

6 Conclusion and Future Work

In a departure from most existing research into edge cloud computing with the emphasis on the self-interested setting, this paper introduced a new framework called cooperative edge cloud computing wherein each edge/cloud server tries to optimize a team reward. We considered three major challenges raised by cooperative edge cloud computing as three research topics: high-workloads, task dependency and distributed features. To resolve these challenges, we formulated the problem as a multiagent system and applied multiagent reinforcement learning methods to solve. Extensive experiments confirmed the effectiveness of our proposed methods by comparison with existing methods. However, there remain some gaps from real-world deployment. Thus, in future work, we intend to run the proposed algorithms on real physical edge and cloud servers. We will improve our algorithms based on the feedback from the real deployment.

Acknowledgment. This research was partially supported by a Grant-in-Aid for Scientific Research (B) (21H03556, 2021–2024), and a Grant-in-Aid for Challenging Exploratory Research (20K21833, 2020–2023) from the Japan Society for the Promotion of Science (JSPS).

References

1. Barbarossa, S., Sardellitti, S., Di Lorenzo, P.: Joint allocation of computation and communication resources in multiuser mobile cloud computing. In: 2013 IEEE 14th Workshop on Signal Processing Advances in Wireless Communications (SPAWC), pp. 26–30. IEEE (2013)
2. Cao, X., Wang, F., Xu, J., Zhang, R., Cui, S.: Joint computation and communication cooperation for mobile edge computing. In: 2018 16th International Symposium on Modeling and Optimization in Mobile, Ad Hoc, and Wireless Networks (WiOpt), pp. 1–6. IEEE (2018)
3. Chang, C., Srirama, S.N., Buyya, R.: Internet of Things (IoT) and new computing paradigms. In: Fog and Edge Computing: Principles and Paradigms, pp. 1–23 (2019)
4. Chang, H., Hari, A., Mukherjee, S., Lakshman, T.: Bringing the cloud to the edge. In: 2014 IEEE Conference on Computer Communications Workshops (INFOCOM WKSHPS), pp. 346–351. IEEE (2014)
5. Chen, X., Zhang, H., Wu, C., Mao, S., Ji, Y., Bennis, M.: Optimized computation offloading performance in virtual edge computing systems via deep reinforcement learning. IEEE Internet Things J. **6**(3), 4005–4018 (2018)
6. Chen, X., Jiao, L., Li, W., Fu, X.: Efficient multi-user computation offloading for mobile-edge cloud computing. IEEE/ACM Trans. Network. **24**(5), 2795–2808 (2015)
7. Dean, J., Ghemawat, S.: MapReduce: simplified data processing on large clusters. Commun. ACM **51**(1), 107–113 (2008)
8. Ding, S., Lin, D.: A coalitional Markov decision process model for dynamic coalition formation among agents. In: 2020 IEEE/WIC/ACM International Joint Conference on Web Intelligence and Intelligent Agent Technology (WI-IAT), pp. 308–315. IEEE (2020)
9. Ding, S., Lin, D.: Dynamic task allocation for cost-efficient edge cloud computing. In: 2020 IEEE International Conference on Services Computing (SCC), pp. 218–225. IEEE (2020)
10. Ding, S., Lin, D., Zhou, X.: Graph convolutional reinforcement learning for dependent task allocation in edge computing. In: 2021 IEEE International Conference on Agents (ICA) (2021)
11. Donovan, S., Chung, J., Sanders, M., Clark, R.: MetroSDX: a resilient edge network for the smart community. In: 2017 IEEE International Conference on Pervasive Computing and Communications Workshops (PerCom Workshops), pp. 575–580. IEEE (2017)
12. Gu, L., Zeng, D., Guo, S., Barnawi, A., Xiang, Y.: Cost efficient resource management in fog computing supported medical cyber-physical system. IEEE Trans. Emerg. Top. Comput. **5**(1), 108–119 (2015)
13. Jošilo, S., Dán, G.: Selfish decentralized computation offloading for mobile cloud computing in dense wireless networks. IEEE Trans. Mob. Comput. **18**(1), 207–220 (2018)

14. Li, S., Huang, J.: Energy efficient resource management and task scheduling for IoT services in edge computing paradigm. In: 2017 IEEE International Symposium on Parallel and Distributed Processing with Applications and 2017 IEEE International Conference on Ubiquitous Computing and Communications (ISPA/IUCC), pp. 846–851. IEEE (2017)

15. Liu, X., Yu, J., Feng, Z., Gao, Y.: Multi-agent reinforcement learning for resource allocation in IoT networks with edge computing. China Commun. **17**(9), 220–236 (2020)

16. Liu, X., Yu, J., Wang, J., Gao, Y.: Resource allocation with edge computing in IoT networks via machine learning. IEEE Internet Things J. **7**(4), 3415–3426 (2020)

17. Nishi, H.: Information and communication platform for providing smart community services: system implementation and use case in Saitama city. In: 2018 IEEE International Conference on Industrial Technology (ICIT), pp. 1375–1380. IEEE (2018)

18. Pan, J., McElhannon, J.: Future edge cloud and edge computing for Internet of Things applications. IEEE Internet Things J. **5**(1), 439–449 (2017)

19. Sunehag, P., et al.: Value-decomposition networks for cooperative multi-agent learning. arXiv preprint arXiv:1706.05296 (2017)

20. Tang, Z., Lou, J., Zhang, F., Jia, W.: Dependent task offloading for multiple jobs in edge computing. In: 2020 29th International Conference on Computer Communications and Networks (ICCCN), pp. 1–9. IEEE (2020)

21. Tao, X., Ota, K., Dong, M., Qi, H., Li, K.: Performance guaranteed computation offloading for mobile-edge cloud computing. IEEE Wirel. Commun. Lett. **6**(6), 774–777 (2017)

22. Watkins, C.J., Dayan, P.: Q-learning. Mach. Learn. **8**(3–4), 279–292 (1992). https://doi.org/10.1007/BF00992698

23. Yu, Y., Zhang, J., Letaief, K.B.: Joint subcarrier and CPU time allocation for mobile edge computing. In: 2016 IEEE Global Communications Conference (GLOBECOM), pp. 1–6. IEEE (2016)

24. Yuan, L., et al.: CoopEdge: a decentralized blockchain-based platform for cooperative edge computing. In: Proceedings of the Web Conference 2021, pp. 2245–2257 (2021)

25. Zhang, Y., Chen, X., Chen, Y., Li, Z., Huang, J.: Cost efficient scheduling for delay-sensitive tasks in edge computing system. In: 2018 IEEE International Conference on Services Computing (SCC), pp. 73–80. IEEE (2018)

Demonstrations

Occupancy Estimation from Thermal Images

Zishan Qin[1]([⊠]), Dipankar Chaki[2], Abdallah Lakhdari[2], Amani Abusafia[2],
and Athman Bouguettaya[2]

[1] School of Computing, Australian National University, Canberra, Australia
taylor.qin2@anu.edu.au
[2] School of Computer Science, University of Sydney, Sydney, Australia
{dipankar.chaki,abdallah.lakhdari,amani.abusafia,
athman.bouguettaya}@sydney.edu.au

Abstract. We propose a non-intrusive, and privacy-preserving occupancy estimation system for smart environments. The proposed scheme uses thermal images to detect the number of people in a given area. The occupancy estimation model is designed using the concepts of intensity-based and motion-based human segmentation. The notion of difference catcher, connected component labeling, noise filter, and memory propagation are utilized to estimate the occupancy number. We use a real dataset to demonstrate the effectiveness of the proposed system.

Keywords: Smart home · Occupancy estimation · Thermal sensor · Human segmentation · K-means algorithm · Connected component labeling

1 Introduction

The emergence of intelligent technologies enables *smart services* in the home environment to provide the residents with *convenience* and *efficiency* in our daily life [5]. Many research are based on a *single occupant* environment [3]. In reality, *multiple occupants* live in a dwelling. Therefore, a new functional model is needed to determine the number of people in a space, referred to as *occupancy estimation*. In application, it can help with heating, ventilation, and air conditioning systems control and even security monitoring in the smart buildings [5].

This demo focuses on occupancy estimation using thermal images. Thermal imaging cameras are chosen because they offer advantages over intrusive sensors (such as RGB cameras) in terms of privacy protection and better than non-intrusive sensors (such as passive infrared (PIR) sensors) in terms of estimation results [1]. There are many challenges in this task. One challenge is that many objects, such as the CPU of a computer, are similar in temperature to a human [2]. Another challenge is that indoor environments may have excessive illustrations and far too much thermal noise [5]. We propose a computationally efficient, non-intrusive, and privacy-preserving occupancy estimation system for

© Springer Nature Switzerland AG 2022
H. Hacid et al. (Eds.): ICSOC 2021 Workshops, LNCS 13236, pp. 301–305, 2022.
https://doi.org/10.1007/978-3-031-14135-5_25

the smart environment. Our proposed model predicts the number of occupants with an average accuracy of 71.6% in six experiments by combining intensity-based and motion-based human segmentation.

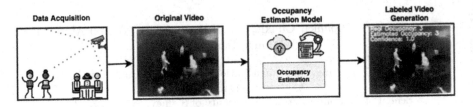

Fig. 1. The architecture of occupancy estimation system

2   System Architecture

The architecture of the system is shown in Fig. 1. The system has three major components: *Data Acquisition, Occupancy Estimation Model,* and *Labeled Video Generation.* Thermal videos from an overhead view can be used as an input to our system. To guarantee the quality of the estimation results, the resolution of the video should be at least 200 * 100. Furthermore, people should not wear too thick clothes under the camera. In the second component, human segmentation and classification are performed in the occupancy estimation model. Filters are used to determine the classes for the segments, and the number of selected classes determines the occupancy number. As output of our system, a labeled video is generated with predicted values indicating the number of people in the scene. From the video, each frame can be visually inspected by the user.

3   Occupancy Estimation Model

The proposed estimation model consists of two phases: *Preliminary Parameter Configuration* and *Occupancy Estimation.* A process flow that describes the process from the parameter configuration through the occupancy estimation is shown in Fig. 2. We discuss each phase in the following subsections:

3.1   Preliminary Parameter Configuration

In this phase, the main focus is to test the domain-specific parameters used in the estimation model via a binary search to get the suitable parameter combination. Several parameters include the mask updating frequency, lighting threshold, lower and upper bounds for the noise filter, and the memory preserving number.

3.2   Occupancy Estimation

This phase aims to use an efficient strategy to separate humans from other objects and then to calculate how many classes remain in the human layer through a classification algorithm. The main challenges include a large volume

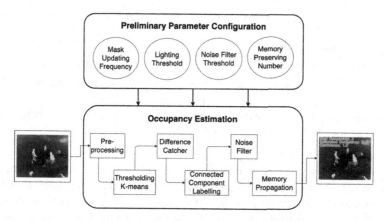

Fig. 2. Process flow diagram of the occupancy estimation model

of thermal noises, either caused by the potential over-lighting or by other objects in a room with a similar temperature to humans. We perform the following steps to deal with these challenges for the occupancy estimation.

Pre-processing of the Input Video: We divide the video into a sequence of images with an interval of two seconds for subsequent operations. Then, all pictures are cropped into a uniform resolution of 200 * 100 for faster runtime while maintaining important information. The first frame in the image sequence is stored into the model as an initialization mask, and this mask is updated according to the update frequency. The mask is used to reduce the effect of lighting variations between frames on the estimation results.

Thresholding K-Means: K-means allocates each point to the cluster with the nearest mean. This step first applies K-means to two consecutive frames to obtain two approximate segmentation results, respectively. Isolated thermal noises from over-segmentation are removed based on a lighting threshold from the preliminary configuration, followed by a Gaussian filter to blur the results. This reduces the side effects of hard thresholding. Then, we apply the mask from the previous step to eliminate the lighting issue in every two consecutive frames.

Difference Catcher: This step considers the motion between two consecutive frames. The difference is taken between two images generated from the prior step into a difference map. It is used to represent the dynamic change in the two consecutive frames which is caused by the movement of people.

Connected Component Labeling: Connected component labeling is a commonly used method to detect connected regions in binary images [4]. We apply this on the difference graph we get from the last step. Thus, we get a simulation of the dynamic differences in the actual movement of people in the room.

Table 1. Experimental result

Experiment	1	2	3	4	5	6	Average
Accuracy (%)	66.7	71.4	66.7	74.3	80.0	70.6	71.6
Confidence	0.833	1.258	0.918	0.863	1.000	1.118	0.998
Environment	OL	OL	MP	LL, TN	OL	TN	

OL: Over Lighting **MP**: Multiple People **LL**: Local Lighting **TN**: Thermal Noises

Noise Filter: The output from the connected component may contain many small connected components, which are likely to be caused by thermal noises or over-illumination. That's why we need this step to eliminate the thermal noises. We use the thresholds from the parameter configuration phase to eliminate these small parts. Usually, we would not have a person occupying a large part of the overhead vision frame. In general, the low threshold is set relatively high when there are more other objects in the scene with similar temperatures to the human body. The more significant the lighting changes between these two images in the scene, the higher the high threshold will be.

Memory Propagation: This step prevents some outrageous predictions due to external factors. For example, the proposed model may over-segment when a person wears thicker clothes in the picture. The more complicated the indoor environment is, the smaller this memory propagation number will be.

4   Demo Setup

We use the Flir FB-Series thermal camera in our experimental setup, which has a relatively high resolution. The experiments have been conducted on a Mac operating system with a Core i3 processor and an 8 GB RAM. Six experiments under different scenarios are done to test the effectiveness of our system, shown in Table 1. The output is a video of all frames of size 200 * 400, marked with the corresponding predicted occupancy, actual occupancy, and confidence scores calculated by $confidence_i = 1 - \left(\frac{estimation_i - real_i}{real_i} \right)$. We estimate the occupancy every two seconds, and the predicted occupancy number is shown in the display's top left corner. A video demonstrating the system can be found in the following link: https://youtu.be/Av9BkB_ZZJc.

Acknowledgement. This research was partly made possible by DP160103595 and LE180100158 grants from the Australian Research Council. The statements made herein are solely the responsibility of the authors.

References

1. Beltran, A., Erickson, V.L., Cerpa, A.E.: ThermoSense: occupancy thermal based sensing for HVAC control. In: 5th ACM Workshop on ESEEB, pp. 1–8 (2013)

2. Chidurala, V., Li, X.: Occupancy estimation using thermal imaging sensors and machine learning algorithms. IEEE Sens. J. **21**(6), 8627–8638 (2021)
3. Du, Y., Lim, Y., Tan, Y.: A novel human activity recognition and prediction in smart home based on interaction. Sensors **19**(20), 4474 (2019)
4. He, L., Ren, X., Gao, Q., Zhao, X., et al.: The connected-component labeling problem: a review of state-of-the-art algorithms. Pattern Recogn. **70**, 25–43 (2017)
5. Naser, A., et al.: Adaptive thermal sensor array placement for human segmentation and occupancy estimation. IEEE Sens. J. **21**(2), 1993–2002 (2020)

Constraint-Aware Trajectory for Drone Delivery Services

Jermaine Janszen[✉], Babar Shahzaad, Balsam Alkouz,
and Athman Bouguettaya

University of Sydney, Sydney, Australia
jjan3640@uni.sydney.edu.au,
{babar.shahzaad,balsam.alkouz,athman.bouguettaya}@sydney.edu.au

Abstract. Drones are becoming a novel means for delivery services. We present a demonstration of drone delivery services in a skyway network that uses the service paradigm. A set of experiments is conducted using Crazyflie drones to collect the data on various positions of drones, wind speed, wind direction, and battery consumption. We run the experiments for a range of flight patterns including linear, rectangular, and triangular shapes. Demo: https://youtu.be/tlXnUSIrRp0.

Keywords: Drone delivery · Delivery service · Flight trajectory · Intrinsic and extrinsic constraints

1 Introduction

Drones are autonomous aircraft that offer potential benefits for a multitude of civilian applications [8]. Drones enable new services in various domains such as surveillance, agriculture, and delivery of goods [3]. Companies such as Amazon and Google are investing in the use of drones for delivery services [4]. The targeted *beneficiaries* of drone delivery services include consumers, transport companies, and suppliers of goods (e.g., medical suppliers, retailers, etc.) [5].

Current research focuses on developing techniques for fast and cost-efficient deliveries using drones [6]. However, existing works rely on simulation analysis of the proposed techniques using synthetic datasets. There is a paucity of real datasets especially those that include intrinsic and extrinsic factors affecting the drone services [1]. Examples of intrinsic factors include flight range, battery, and payload capacity of a drone [2]. Examples of extrinsic factors include number of recharging stations and weather conditions (e.g., wind speed and direction).

We leverage the *service paradigm* to abstract a drone's capabilities as *drone services*. Drone services usually operate in a skyway network taking into account no-flight and restricted zones [7]. A skyway network is a set of predefined *line-of-sight skyway segments*. A skyway segment is a straight line between two particular nodes. *Each segment represents a service that is served by a drone.*

This demonstration focuses on drone delivery services in a skyway network to collect a real dataset that records the impact of different payloads and wind

© Springer Nature Switzerland AG 2022
H. Hacid et al. (Eds.): ICSOC 2021 Workshops, LNCS 13236, pp. 306–310, 2022.
https://doi.org/10.1007/978-3-031-14135-5_26

Fig. 1. 3D model of Sydney CBD

Fig. 2. Fan for wind speed control

Fig. 3. Crazyflie carrying payload

conditions over a set of trajectory patterns. We measure the battery consumption of a drone while carrying different payloads under varying wind conditions. We autonomously fly a drone through various flight patterns including *triangular, rectangular, linear,* and *hovering.* The collected dataset includes the drone's XY positions, altitude, battery consumption, and wind for each flight. Finally, we plot the data collected on each flight to visualize and assess the impact of varying payloads and wind conditions on the battery consumption rate of the drone.

2   Trajectory Tracking and Data Collection

We gather data on various drone parameters under the considered intrinsic and extrinsic factors with a *focus on measuring battery consumption rate.* The experiments were performed to empirically measure the energy use of small drones. To run the experiment, the Crazyflie 2.1 drone by Bitcraze was used as it provides a modular setup with a python API. We setup a 3D model of Sydney CBD as an indoor testbed to mimic a skyway network (Fig. 1). The drone locates its precise position during its flight with the aid of HTC Vive base stations, fitted at the corners of the lab. A fan with different speed settings is used to simulate extrinsic constraints (Fig. 2). The drone is fitted with a payload to simulate intrinsic constraints (Fig. 3). Two main sets of trajectories were collected including hovering flights and predefined flight paths trajectories.

2.1   Hovering Flight with Different Conditions

Hovering with Extended Flight Time. This test measures how the battery voltage changes over time. The drone hovered in a fixed position 50cm off the ground with no payload and no wind. The drone maintained its position for 5 min as various drone parameters were logged in 100ms intervals.

Hovering with Different Wind Speeds. This test measures how the battery voltage changes over time under different wind speeds. The drone hovered in a fixed position 60 cm off the ground and 3 m away from a fan with the same height. The drone would hover for 2 min while parameters were logged in 50 ms intervals. To procure a robust set of data, 4 different wind speeds (1.8 km/h,

2.2 km/h, 2.9 km/h, and 3.6 km/h) were used, with each speed being tested twice. Greater speeds weren't used as the drone's stability was greatly compromised above 4 km/h when hovering.

Hovering with Different Payload Weights. This test measures how the battery voltage changes over time with varying payload weights. The drone hovered in a fixed position 50 cm off the ground while carrying different payload weights. The drone would hover for 2 min while parameters were logged in 50 ms intervals. Four different payload weights (2 g, 4 g, 6 g, and 8 g) were used to procure a robust set of data, with each payload being tested 4 times.

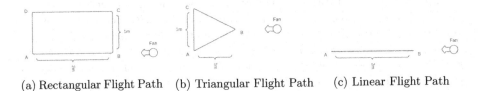

(a) Rectangular Flight Path   (b) Triangular Flight Path   (c) Linear Flight Path

Fig. 4. Flight paths with fan placement and direction

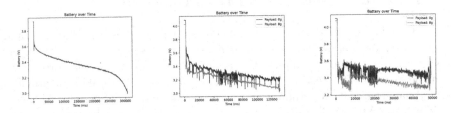

Fig. 5. Battery consumption when hovering

Fig. 6. Battery consumption in hovering state

Fig. 7. Battery consumption in rectangular flight

2.2 Fixed Flight Paths with Different Conditions

In addition to the hovering tests, the drone parameters were also logged during the drone flights over predefined paths. The selected paths include a rectangle, triangle, and line which allow the drone to fly forwards, backwards, side to side, and diagonally. These directions are the typical directions that drones follow traversing the nodes in a skyway network. Figure 4 illustrates different flight paths collected in the dataset. Throughout the flights, the drone travelled at a fixed speed of 0.3 m/s at 0.5 m height and hovered for 5 s at each endpoint along the path. To generate control data points with drone only settings, the paths were flown without any payload or wind. We again run the tests with the same payloads as used for the hovering tests. We used a fan to generate artificial wind. For the rectangle and line paths, the fan was placed 1m away from point B facing back at point A while the fan was placed 1m away from point B facing the midpoint between points A and C for the triangle path.

3   Data

We performed 72 flights under a number of operational parameters (payloads, wind speeds). In addition, 34 recordings were performed to assess the drone's ancillary power and hover conditions. The data collected from the experiments are organized in CSV files where each sheet contains a log under different settings.

4   Results

In what follows, we present an analysis of the collected dataset in terms of battery consumption with different payloads under various wind conditions. Figure 5 shows the battery consumption trend over time without payload in the hovering state until the battery was drained. We observe that the battery consumption of a drone increases as the payload weight increases (Fig. 6 and 7). In addition, we observe that the battery consumption behaviour shows a fluctuating behaviour under different wind conditions (Fig. 8 and 9). This fluctuation is due to the changing wind directions during the drone flight. Flying a drone with a headwind is more energy-efficient [9]. Therefore, when the drone flies in a headwind direction, it consumes less battery compared to the tailwind direction. Figure 10 presents the battery consumption behaviour while flying in a rectangular path

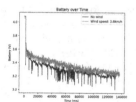
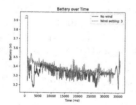
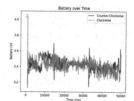

Fig. 8. Battery consumption in hovering state

Fig. 9. Battery consumption in triangular

Fig. 10. Battery consumption in rectangular

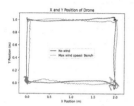
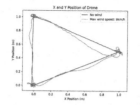
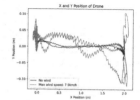

Fig. 11. Stability in rectangular path

Fig. 12. Stability in triangular path

Fig. 13. Stability in linear path

in clockwise and anti-clockwise directions. Figure 11, 12, and 13 shows the stability of the drone under various wind conditions in different flight patterns. The drone's stability is highly affected when flying under strong wind conditions.

Acknowledgment. This research was partly made possible by DP160103595 and LE180100158 grants from the Australian Research Council. The statements made herein are solely the responsibility of the authors.

References

1. Alkouz, B., Bouguettaya, A.: Formation-based selection of drone swarm services. In: EAI Mobiquitous Conference (2020)
2. Alkouz, B., Bouguettaya, A.: Provider-centric allocation of drone swarm services. In: IEEE ICWS (2021)
3. Lee, W., et al.: Package delivery using autonomous drones in skyways. In: Proceedings of the UbiComp/ISWC, pp. 48–50 (2021)
4. Shahzaad, B., Bouguettaya, A., Mistry, S.: A game-theoretic drone-as-a-service composition for delivery. In: IEEE ICWS, pp. 449–453 (2020)
5. Shahzaad, B., Bouguettaya, A., Mistry, S.: Robust composition of drone delivery services under uncertainty. In: IEEE ICWS (2021)
6. Shahzaad, B., Bouguettaya, A., Mistry, S., Neiat, A.G.: Constraint-aware drone-as-a-service composition. In: Yangui, S., Bouassida Rodriguez, I., Drira, K., Tari, Z. (eds.) ICSOC 2019. LNCS, vol. 11895, pp. 369–382. Springer, Cham (2019). https://doi.org/10.1007/978-3-030-33702-5_28
7. Shahzaad, B., et al.: Resilient composition of drone services for delivery. Future Gener. Comput. Syst. **115**, 335–350 (2021)
8. Shakhatreh, H., et al.: Unmanned aerial vehicles (UAVs): a survey on civil applications and key research challenges. IEEE Access **7**, 48572–48634 (2019)
9. Tseng, C.M., et al.: Autonomous recharging and flight mission planning for battery-operated autonomous drones. arXiv preprint arXiv:1703.10049 (2017)

Seamless Synchronization
for Collaborative Web Services

Kristof Jannes[✉], Bert Lagaisse, and Wouter Joosen

imec-DistriNet, KU Leuven, Leuven, Belgium
{kristof.jannes,bert.lagaisse,wouter.joosen}@cs.kuleuven.be

Abstract. Collaborative web services, which allow multiple people to work together on the same data, are becoming increasingly popular. However, current state-of-the-art frameworks for interactive client-side replication cannot handle network disruptions well, or suffer from large metadata overhead when clients are short-lived. This demonstration will show OWebSync, a generic web middleware for data synchronization in browser-based applications and interactive groupware. It offers a fine-grained data synchronization model, using state-based Conflict-free Replicated Data Types, and leverages Merkle-trees in the data model for efficient synchronization. We provide an interactive demonstration of a drawing application that workshop attendees can experiment with. We will also demonstrate the robustness in disconnected and offline settings.

Keywords: CRDTs · Online collaboration · Eventual consistency

1 Introduction and Motivation

The use of online software services to collaborate remotely has been increasing in the last decade, especially in the last year due to the COVID-19 pandemic. Collaborative groupware applications, such as Google Docs or Microsoft Whiteboard, allow people to work together on the same document, without them being present in the same geographic location. People can work from anywhere they want, even in unstable network conditions, or while being offline. When a connection is available, changes should be replicated to all other client replicas within 1–2 s to keep the user experience interactive. Five seconds is the absolute maximum before users are becoming annoyed [9]. When offline, the user should be able to work further on the local copy of the data. Once the user comes back online, any changes should be replicated as fast as possible. This is especially important in unstable network conditions, where there is a limited time frame available to replicate all updates. The requirement for offline support implies the evolution to a more client-centric architecture, in which the different clients all become the authoritative data replicas [2]. This is in contrast to the classical client-server model, where the server is responsible for both data and business logic, typically organized in a data-tier and a business-tier. While this gives reasonable good performance when online, it comes at a cost of higher latency for clients located geographically far from the main server.

© Springer Nature Switzerland AG 2022
H. Hacid et al. (Eds.): ICSOC 2021 Workshops, LNCS 13236, pp. 311–314, 2022.
https://doi.org/10.1007/978-3-031-14135-5_27

The most used client-server technology for collaborative groupware is Operational Transformation (OT) [1]. OT is used in Google Docs. It uses a central server that transforms the conflicting operations for each replica to allow them to be applied in a different order on the other replicas. However, these transformations are rather complex and resource-intensive on the server, limiting the scalability of this technique. Moreover, OT only works for short-time disconnections and cannot be used when the client is offline for a longer time.

Several client-centric frameworks exist for collaborative web services. They rely on Conflict-free Replicated Data Types (CRDTs) [10] to automatically resolve any conflicts that would arise from multiple people editing the same data. There are several kinds of CRDTs. Operation-based CRDTs (CmRDTs) must still send all operations between the replicas using a reliable, exactly-once, message channel, similar to OT. However, no central component to transform these operations is required, as all operations are commutative. CmRDTs are used in Yjs [8] and Automerge [4,5]. State-based CRDTs (CvRDTs) do not use operations, but instead, they send the full state to other replicas, who will merge that state with their local state. CvRDTs are not suitable for client-centric interactive applications, as the full state is too expensive to send every time. It can however be used to replicate data between backend servers. Delta-state-based CRDTs [7] use vector clocks to calculate which part of the data needs to be sent to other replicas. They require much less of the message channel compared to operation-based CRDTs, however, the total size of the metadata will grow with every client that makes an edit. Especially in a web-based environment, where clients are often short-lived, the metadata will become larger over time, reducing the interactive performance. Delta-state-based CRDTs are used in Legion [6].

This demonstration shows OWebSync[1] [3], a generic web middleware for data synchronization in the context of web-based services and interactive groupware. OWebSync leverages nested state-based CRDTs and Merkle-trees to efficiently replicate changes. Compared to state-of-the-art frameworks, OWebSync offers:

- continuous and interactive synchronization of online web clients,
- prompt resynchronization of offline clients when they come back online,
- no meta-data explosion.

Application developers can leverage OWebSync to create collaborative services that are resilient against network failures. OWebSync offers a flexible data model, with fine-grained synchronization and automatic conflict resolution. Online web clients achieve interactive synchronization, making it possible for several people to work fluently on the same document. However, if no internet connection is available, such as in a tunnel or an airplane, clients can continue on their local copy. OWebSync is especially robust against these offline periods, and is able to quickly replicate all missed updates, and achieve the same interactive performance as before within seconds. This robustness also makes OWebSync interesting for the field-services industry, where technicians are often on the road going from customer to customer for technical interventions. A stable internet

---

[1] https://distrinet.cs.kuleuven.be/software/owebsync/.

connection is not always available on their location, however, writing off all used materials is important for correct billing and inventory. Using OWebSync, those offline reports will be synchronized quickly when an internet connection is available again, even when multiple technicians are working on the same job.

2   Overview of the OWebSync Framework

OWebSync is a JavaScript framework for application developers to synchronize data between browser-clients. OWebSync provides Strong Eventual Consistency out-of-the-box, without letting the developer worry about it. Conflicts are solved automatically by the framework.

Data Model. OWebSync can be used to replicate JSON data structures containing strings, numbers, booleans, and objects; the latter can include any of those recursively. OWebSync uses this tree-structure of the JSON data to create a Merkle-tree internally, which is used for efficient synchronization. State-based CRDTs are used to resolve conflicts under-the-hood. Application developers do not need to concern themselves with these internals. However, they need to be aware that data is only eventually consistent. Since we are using state-based CRDTs, there is little required from the message channel, compared to existing operation-based approaches. There is also no need to keep track of clients or client-specific metadata such as vector clocks.

Architecture and API. The deployment architecture of OWebSync is depicted in Fig. 1. OWebSync provides a JavaScript API for web applications to read and modify the tree-structured data. All data is stored locally in the browser using the IndexedDB key-value store, which is built-in in every modern browser. Data is replicated to a server running on NodeJS using a direct WebSocket connection. This WebSocket connection and the server are also used as a signaling channel to set up peer-to-peer WebRTC connections between the other browser instances. Once a WebRTC connection is initialized, the different OWebSync replicas can replicate the changes directly with each other. This reduces the latency to replicate changes to other browser instances, and also improves the scalability, as the central server is no longer a bottleneck. Figure 2 shows an example code snippet using the public API of OWebSync. It connects to the WebSocket endpoint of the NodeJS server. Developers can then use the CRUD operations get, set and del, and the path in the tree, to retrieve and modify data. The full data is immediately stored locally, and in the background OWebSync will replicate the changes to the server and to any other connected browser clients.

Internal Synchronization Protocol. Internally, the synchronization protocol always runs directly between two different replicas, either browser-to-browser or browser-to-server. The protocol uses the Merkle-tree to find out which part of the tree needs to be sent to the other replica. If the local hash does not match the hash on the remote replica, then the corresponding state-based CRDT will be used to merge the remote state with the local state. We refer to Jannes et al. [3] for a detailed specification of this CRDT merge operation.

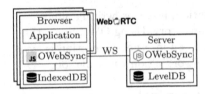

```
<script src="owebsync-browser.js"></script>
<script>
OWebSync("ws://localhost:8080").then(
  async (owebsync) => {
    await owebsync.set("obj11.color", "#f00");
  }
)
</script>
```

Fig. 1. Deployment architecture   **Fig. 2.** Public API example

3 Interactive Demonstration

We demonstrate OWebSync with an interactive web-based drawing application for all the demo attendees world-wide. The drawing can be edited by multiple users simultaneously, and any conflicts that might arise will be solved automatically by the underlying CRDTs. We will demonstrates the interactive latency with worldwide collaboration when everyone is online. OWebSync is especially robust against network failures, and we demonstrate this with two scenario's. First, the server is stopped, yet, all browser clients can continue to work together by using the peer-to-peer network between them. Second, one browser instance loses its internet connection temporarily. We also demonstrate that both the online clients, as well as the offline client, can continue to work on their copies of the data. When the internet connection is restored, we demonstrate that the changes are merged quickly, and interactive performance is resumed.

References

1. Ellis, C.A., Gibbs, S.J.: Concurrency control in groupware systems. In: SIGMOD REC (1989)
2. Jannes, K., Lagaisse, B., Joosen, W.: The web browser as distributed application server: towards decentralized web applications in the edge. In: EdgeSys 2019 (2019)
3. Jannes, K., Lagaisse, B., Joosen, W.: OwebSync: Seamless synchronization of distributed web clients. IEEE Trans. Parallel Distrib. Syst. **32**, 2338–2351 (2021)
4. Kleppmann, M., Beresford, A.R.: A conflict-free replicated JSON datatype. IEEE Trans. Parallel Distrib. Syst. **28**, 2733–2746 (2017)
5. Kleppmann, M., Beresford, A.R.: Automerge: real-time data sync between edge devices. In: MobiUK 2018 (2018)
6. van der Linde, A., Fouto, P., Leitão, J.A., Preguiça, N., Castiñeira, S., Bieniusa, A.: Legion: enriching internet services with peer-to-peer interactions. In: WWW 2017 (2017)
7. van der Linde, A., Leitão, J.A., Preguiça, N.: Δ-crdts: making δ-crdts delta-based. In: PaPoC 2016 (2016)
8. Nicolaescu, P., Jahns, K., Derntl, M., Klamma, R.: Near real-time peer-to-peer shared editing on extensible data types. In: GROUP 2016 (2016)
9. Nielsen, J.: Usability Engineering. Nielsen Norman Group (1993)
10. Shapiro, M., Preguiça, N., Baquero, C., Zawirski, M.: Conflict-free replicated data types. In: Défago, X., Petit, F., Villain, V. (eds.) SSS 2011. LNCS, vol. 6976, pp. 386–400. Springer, Heidelberg (2011). https://doi.org/10.1007/978-3-642-24550-3_29

Crowd4ME: A Crowdsourcing-Based Micro-expression Collection Platform

Xun Wang, Liping Lv, Yili Fang, Xinyi Ding, and Tao Han[✉]

Zhejiang Gongshang University, Hangzhou, China
{wx,19020100029,fangyili,xding,hantao}@zjgsu.edu.cn

Abstract. Machine learning-based approaches have greatly accelerated the progresses in the field of micro-expression detection and recognition. However, many models, especially those based on deep learning require very large databases of hand labeled data for training. Existing micro-expression data collection processes usually cost a lot and are time-consuming. The labeling of very large datasets is becoming a bottleneck of building robust models for micro-expression detection and recognition. With the wide success of crowdsourcing platforms in creating large datasets like ImageNet, in this paper, we present Crowd4ME. Crowd4ME is a crowdsourcing-based platform for collecting large scale micro-expression data. We demonstrate that using Crowd4ME can help collect and manage micro expression samples more easily and efficiently.

Keywords: Datasets · Crowdsourcing · Micro expression

1 Introduction

Micro expressions are brief and subtle facial expressions revealed when people try to hide their true emotions. The detect and recognition of micro-expressions has wide applications in fields such as polygraph detection, mental health, and national security. However, due to the lack of large scale micro-expression data sets, the current supervised learning methods, especially those based on deep neural networks cannot fully exert its due power on micro-expression problems.

Table 1. Available datasets containing micro-facial expressions

	Polikovsky's	USF-HD	SMIC	CASME	CASME II	SAMM
Samples	42	100	164	195	247	159
Participants	10	N/A	16	35	35	32
Spontaneous/Posed	Posed	Posed	Spon.	Spon.	Spon.	Spon.
Emotion classes	6	6	3	7	5	7
Environment	Laboratory	Laboratory	Laboratory	Laboratory	Laboratory	Laboratory

© Springer Nature Switzerland AG 2022
H. Hacid et al. (Eds.): ICSOC 2021 Workshops, LNCS 13236, pp. 315–318, 2022.
https://doi.org/10.1007/978-3-031-14135-5_28

Currently, most existing micro-expression datasets in Table 1 are collected in a controlled laboratory environment. Thus, the number of participants is often limited and it is often impossible to collect large scale samples required for deep neural networks training. Besides, the administration process might result in extra cost and bring in human errors. More importantly, data collected in a controlled environment might differ from those collected in the wild.

Many data sets in other fields (non-micro expression field) have been collected by crowdsourcing methods and have achieved great success [1]. Crowdsourcing has many advantages like open innovation, scalability, and cost-efficiency [2]. In this demo, we present a crowdsourcing-based platform Crowd4ME for the collection of micro-expression data. By using a crowdsourcing-based platform, we will be able to collect large scale samples from different sources. The entire micro-expression collection process is shown in Fig. 1.

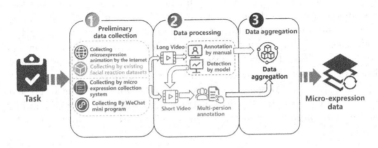

Fig. 1. The workflow of the crowdsourcing-based micro-expression.

2   Crowd4ME Collection and Processing Flow

2.1   Preliminary Micro-expression Samples Collection

We use a variety of ways to initially collect data samples that may contain micro-expressions. As shown in Fig. 2, we collect posed and spontaneous micro-expression data in various ways from the Internet:

Collecting micro expression animation through the Internet. In this way, we collect micro-expression samples from existing "wild" environments on the Internet, which increases the diversity of our data.

Collecting from existing facial reaction video databases. Since micro expressions exist in different scenes, we may find facial micro expressions in those face databases that are not specifically designed for micro expressions.

Collecting using micro-expression collection system. The system gives the participant seven categories of micro-expressions to imitate.

Collecting by WeChat mini program. The WeChat mini program contains two modules. In the imitating micro-expression module, the participant will be given seven categories micro-expressions to imitate. In the second module, we use emotional videos to induce micro-expressions. We select videos that can cause mood swings for the participants and ask the participants to maintain a neutral expression (no expression) when watching the video, not showing their true emotions. Using the WeChat mini program, we can obtain both posed and naturally induced micro-expression samples.

2.2 Data Processing

Through the preliminary micro-expression samples collection, we can obtain various micro expression samples as shown in Fig. 2. Among them, short videos refer to those with micro expressions that can be marked directly. For the long videos, we spot micro-expression clips using two methods: manual detection and machine detection. The manual detection method is to put a long video containing micro-expression into the micro-expression collection system. Participants will watch the video and annotate short video clips containing micro-expression. Machine inspection is by employing the method of Main Directional Maximal Difference Analysis (MDMD) [3]. For the short videos, we put them into micro expression collection system for manual labeling. Through data processing, the various preliminary micro-expression samples are processed into short videos that can be directly annotated.

2.3 Data Aggregation

In order to obtain accurate labeling results with maximum benefit, we calculate the information entropy to make labeling judgments. The calculation formula of the information entropy H(X) is given below: $H(X) = -\sum_{i=1}^{n} P_i \log(P_i)$. P_i is the probability that the true label of the current video clip is the i-th micro-expression. We have 5 taggers for each video segment. If H(X) is less than some threshold t, the majority voting algorithm will be used to tag the micro-expression video; otherwise, the video segment is considered to be one that is difficult to tag.

3 Data Samples

Using the above crowdsourcing-based micro-expression collection and processing flow, we can obtain three types of micro expression samples. We have deployed the crowdsourcing-based platform to collect large scale micro-expression data. Figure 3 shows some micro-expression samples we have collected so far. We have gathered 1062 facial reaction animation through the Internet. After went through the micro expression system and aggregating process, we gathered 483 micro expression. In addition, we have collected 42 posed and 23 spontaneous data by WeChat mini program and micro expression collection system web, among which there are 30 valid micro expression samples. Our system is actively running and we expect more data come in the near future.

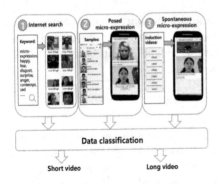

Fig. 2. Preliminary micro-expression samples collection

Fig. 3. The micro-expression collection samples

4   Conclusion and Expected Contributions

In this demo, we present Crowd4ME, a crowdsourcing-based platform for collecting large scale micro-expression data. Compared with other methods, our platform allows us to collect both posed and spontaneous micro expressions in a large scale while at the same time saving labor and administration cost. The URL linking to demonstration video is https://youtu.be/jfyuDOXTtZo. We believe data collection through crowdsourcing will attract more researchers in the near future considering all these benefits and our work provide one good example. The micro expression data collected through our platform will also benefit researchers in this field.

Acknowledgements. This work was supported in part Supported by the National Key Research and Development Program of China (2018YFB1404102), the National Natural Science Foundation of China (61976187, 92046002 and 61976188) and the Research Program of Zhejiang Lab (2019KD0AC02).

References

1. Shan, C., Mamoulis, N., Li, G., et al.: A crowdsourcing framework for collecting tabular data. IEEE Trans. Knowl. Data Eng. **32**(11), 2060–2074 (2019)
2. Felizardo, K.R., de Souza É.F., Lopes, R., et al.: Crowdsourcing in systematic reviews: a systematic mapping and survey. In: 2020 46th Euromicro Conference on Software Engineering and Advanced Applications (SEAA), pp. 404–412. IEEE (2020)
3. He, Y., Wang, S.J., Li, J., et al.: Spotting macro-and micro-expression intervals in long video sequences. In: 2020 15th IEEE International Conference on Automatic Face and Gesture Recognition (FG 2020), pp. 742–748. IEEE (2020)

Tutorials

AI-Enabled Processes: The Age of Artificial Intelligence and Big Data

Amin Beheshti[1(✉)], Boualem Benatallah[2], Quan Z. Sheng[1], Fabio Casati[3],
Hamid-Reza Motahari Nezhad[1], Jian Yang[1], and Aditya Ghose[4]

[1] Macquarie University, Sydney, Australia
{amin.beheshti,michael.sheng}@mq.edu.au,
{hamidreza.motaharinezhad,jian.yang}@mq.edu.au
[2] University of New South Wales, Sydney, Australia
boualem@cse.unsw.edu.au
[3] Servicenow, Santa Clara, CA, USA
fabio.casati@servicenow.com
[4] University of Wollongong, Wollongong, Australia
aditya@uow.edu.au

Abstract. Business processes, i.e., a set of coordinated tasks and activities carried out manually/automatically to achieve a business objective or goal, are central to the operation of public and private enterprises. Modern processes are often highly complex, data-driven, and knowledge-intensive. In such processes, it is not sufficient to focus on data storage/analysis; and the knowledge workers will need to collect, understand, and relate the big data (from open, private, social, and IoT data islands) to process analysis. Today, the advancement in Artificial Intelligence (AI) and Data Science can transform business processes in fundamental ways; by assisting knowledge workers in communicating analysis findings, supporting evidence, and making decisions. This tutorial gives an overview of services in organizations, businesses, and society. We introduce notions of Data Lake as a Service and Knowledge Lake as a Service and discuss their role in analyzing data-centric and knowledge-intensive processes in the age of Artificial Intelligence and Big Data. We introduce the novel notion of AI-enabled Processes and discuss methods for building intelligent Data Lakes and Knowledge Lakes as the foundation for Process Automation and Cognitive Augmentation in Business Process Management. The tutorial also points out challenges and research opportunities.

Keywords: Business process management · Process data science · AI-enabled processes · Artificial intelligence

1 Introduction

Information processing using knowledge-, service-, and cloud-based systems have become the foundation of 21th century [14]. These systems run and support processes in our governments, industries, transportation, hospital, and even our social life. In this context, business processes (BPs) and their continuous improvements are critical to the operation of any system supporting our lives and enterprises. A typical example of BPs supported by systems includes those that automate the process of commercial

© Springer Nature Switzerland AG 2022
H. Hacid et al. (Eds.): ICSOC 2021 Workshops, LNCS 13236, pp. 321–335, 2022.
https://doi.org/10.1007/978-3-031-14135-5_29

enterprises, such as banking and financial transaction processing systems. Over the last decade, many BPs across and beyond the enterprise boundaries have been integrated. Process data is stored across different systems, applications, and services in the enterprise and sometimes shared between different enterprises to provide the foundation for business collaborations [3]. These systems are distributed over various networks, but when viewed at a macro level, multiple organizations and systems are components of a more extensive, logically-coherent system. To understand businesses, one needs to perform analytics over extensive hybrid collections of heterogeneous and partially unstructured process-related execution data [14]. This data increasingly come to show all typical properties of the *big data*: wide physical distribution, diversity of formats, non-standard data models, independently-managed and heterogeneous semantics.

With data science continuing to emerge as a powerful differentiator across industries, almost every organization is now focused on understanding their business and transforming data into actionable insights [9]. For example, governments derive insights from vastly growing private, open, and social data for improving government services, such as personalizing the advertisements in elections, improving government services, predicting intelligence activities, and improving national security and public health. In this context, organizing a vast amount of data gathered from various private/open data islands, i.e., Data Lake, will facilitate dealing with a collection of independently-managed datasets (from relational to NoSQL), diverse formats, and non-standard data models. The notion of a Data Lake was coined to address this challenge and convey the concept of a centralized repository containing limitless amounts of raw (or minimally curated) data stored in various data islands. In our previous work, we introduced the notion of *Data Lake as a Service* [5] which offers researchers and developers a single REST API to organize, index and query their data and metadata in a range of database management systems, from relational to NoSQL, in an easy way.

The rationale behind a Data Lake is to store raw data and let the data analyst decide how to cook/curate them later. While Data Lakes do a great job organizing big data and providing answers to general questions, the main challenges are understanding the potentially interconnected data stored in various data islands and preparing them for analytics. To address this challenge, previously, we presented the notion of Knowledge Lake [6], i.e., a contextualized Data Lake. The term Knowledge refers to a set of facts, information, and insights extracted from the raw data using data curation [8, 16] techniques such as extraction, linking, summarization, annotation, enrichment, classification, and more. The Knowledge Lake will provide the foundation for process data analytics by automatically curating the Data Lake's raw process data and preparing them for deriving insights.

This tutorial gives an overview of Services in Organizations, Business, and Society (Sect. 2). We present novel techniques for organizing process data (Sect. 3) as well as contextualizing process data (Sect. 4). We present the novel notion of AI-enabled Processes (Sect. 5) and discuss methods for building intelligent Data Lakes and Knowledge Lakes as the foundation for Process Automation and Cognitive Augmentation in Processes. The tutorial also points out challenges and research opportunities (Sect. 7).

2 Overview: Services in Organizations, Business, and Society

Web services play an essential role in service-oriented architectures (SOA) in the modern world. For example, when an enterprise wishes to controllably share data (e.g., structured data such as relational tables, semi-structured information such as XML documents, and unstructured information such as commercial data from online business sources) with its business partners, via the Internet, it can use data services to provide mechanisms to find out which data can be accessed, what are the semantics of the data, and how the data can be integrated from multiple enterprises. In particular, a service is a software component that provides rich metadata, expressive languages, and APIs for service consumers to facilitate outsourcing the processes [2].

A Web service, i.e., a method of communication between two electronic devices over the Web [2], can be specialized for each tier[1] of an application: from the functional process logic, data access, to computer data storage and user interface. For example, a data service can encapsulate a wide range of data-centric operations, where these operations need to offer a semantically richer view of their underlying data to use or integrate entities returned by different data services [2]. Web services can be leveraged to reduce the effort required to set up an integration system and to improve the system in a 'pay-as-you-go' fashion as it is used. In this context, composition and integration approaches require semantic techniques before any services can be used. It is crucial as business processes are scattered across several systems and sources, and there is no single schema to which all the process-related data conforms. To address this challenge, *Process Spaces* [20] proposed to overcome some of the problems encountered in the process integration system and to promote awareness of the processes.

Besides the need to extend decision support in process analysis scenarios, the other challenge is the need for scalable analysis techniques. Similar to scalable data processing platforms [13], such analysis and querying methods should offer automatic parallelization and distribution of large-scale computations, combined with techniques that achieve high performance on large clusters, e.g., cloud-based infrastructure, and be designed to meet the challenges of process representation. In particular, there is a need for new scalable and process-aware methods for querying, exploration, and analysis of processes and their data in the enterprise because [4]: (i) process analysis methods should be capable of processing and querying a large amount of data effectively and efficiently, and therefore have to be able to scale well with the infrastructure's scale; and (ii) the querying methods need to enable users to express their data analysis and querying needs using process-aware abstractions rather than other lower-level abstractions. Recently some vendors provided BPM in the Cloud solutions to offer visibility and management of business processes, low start-up costs, and fast return on investment. These solutions can drive new growth opportunities, increase profit margins for the private sector, and achieve more efficient and effective missions for organizations.

---

[1] Three-tier architecture is a well-established software application architecture that organizes applications into three logical and physical computing tiers: the presentation tier, or user interface; the application tier, where data is processed; and the data tier, where the data associated with the application is stored and managed.

3   Data Lake as a Service: Organizing Process Data

Improving business processes is critical to any corporation. Process improvement requires analysis as its first primary step. Before analyzing the process data, there is a need to capture and organize the process data. This is important as executions of process steps in modern enterprises leave temporary/permanent traces in various systems and organizations. To analyze process data, it is possible to collect the data into a data warehouse, using extract, transform, load (ETL) tools, and then leverage an OLAP tool to slice and dice data along different dimensions [14].

In this context, the process data warehousing presents interesting challenges [14]: (i) outsourcing: developing ad-hoc and process-specific solutions for warehousing and reporting on process data is not a sustainable model; (ii) process data abstraction: the typical process executed in the IT system is very detailed and consists of dozens of steps, including manual operations (e.g., scanning invoices), database transactions, and application invocations; and (iii) data evolution: the business process automation/analysis application are co-developed, which means that, during development, changes to the data sources and even to the reporting requirements are relatively frequent.

Considering the above-mentioned challenges, and in the domain of business processes, it is essential to devise a method for minimizing the impact of changes and be able to quickly modify and re-test the ETL (extract, transform, and load) procedures, the warehouse model, and the reports. For example, consider an analyst interested in analyzing the Government Budget through engaging the public's thoughts and opinions on social networks. To achieve this, the analyst may need to deal with a wealth of digital information generated through social networks, blogs, online communities, and mobile applications, which forms a complex data lake [5, 8]: a collection of datasets that holds a vast amount of data gathered from various private/open data islands. Organizing and indexing the growing volume of internal data and metadata in the data lake is challenging. It requires a vast amount of knowledge to deal with dozens of new databases and indexing technologies. To address these challenges, in our recent work [5], we introduced the notion of Data Lake as a Service, which offers researchers/developers a single REST API to organize, index and query their data and metadata. The Data Lake as a Service manages multiple database technologies (from Relational to NoSQL databases), exposes the power of Elasticsearch[2] and weave them together at the application layer. The Data Lake as a Service offers a built-in design to support: (i) Security and Access Control: to provide a database security threat including authentication, access control, and data encryption; and (ii) Tracing and Provenance: to collect and aggregate tracing metadata, including descriptive, administrative, and temporal metadata, and build a provenance graph. Provenance refers to the documentation of an object's lifecycle [15]. This documentation (often represented as a graph) should include all the information necessary to reproduce a particular piece of data or the process that led to it. Figure 1 illustrates the architecture and the main components of the proposed Data Lake as a Service, called CoreDB.

---

[2] https://www.elastic.co/elasticsearch/.

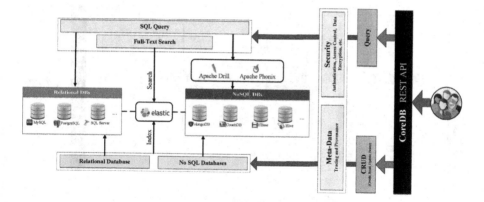

Fig. 1. Data Lake as a service architecture [5].

4   Knowledge Lake as a Service: Contextualizing Process Data

To understand available data (events, business artifacts, data records in databases, etc.) in the context of process execution, we need to contextualize the raw data stored in the data lake, understand the relationships among the information items, and enable the analysis of those relationships from the process execution perspective. It is possible to represent process-related data as entities and any relationships among them (e.g., events relationships in process logs with artifacts, etc.) in an entity-relationship graph. In this context, business analytics can facilitate the analysis of process graphs straightforwardly and intelligently through describing the applications of analysis, data, and systematic reasoning [11,13]. Consequently, an analyst can gather more complete insights using modeling, summarizing, and filtering techniques.

Applications of business analytics extend to nearly all managerial functions in organizations. For example, considering financial services, applying business analytics on customer dossiers, and financial reports can specify the company's performance over periods. As another example, consider the collaborative relationship between researchers, affiliated with various organizations, in the process of writing scientific papers, where it would be interesting to analyze the collaboration-patterns [13] (e.g., frequency of collaboration, degree of collaboration, mutual impact, and degree of contribution) among authors or analyze the reputation of a book, an author, or a publisher in a specific year. Such operations require supporting n-dimensional computations on process graphs, providing multiple views at different granularities, and analyzing set of dimensions coming from the entities and the relationship among them.

While existing analytics solutions, e.g., OLAP techniques and tools, do a great job in contextualizing the process-related data and providing answers to known questions, key business insights remain hidden in the interactions among objects and data: most objects and data in the process graphs are interconnected, forming complex, heterogeneous but often semi-structured networks. Traditional OLAP technologies were conceived to support multidimensional analysis. However, they cannot turn the raw data (stored in Data Lakes) into contextualized data and knowledge. To address these challenges, in

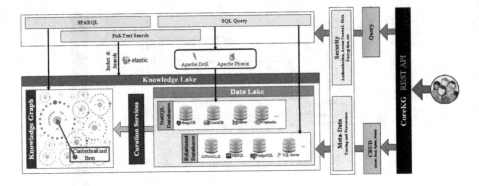

Fig. 2. Knowledge Lake architecture [6].

our recent work [6], we introduced the notion of Knowledge Lake, i.e., a contextualized Data Lake. The term *Knowledge* here refers to a set of facts, information, and insights extracted from the raw data using data curation techniques such as extraction, linking, summarization, annotation, enrichment, classification, and more.

In particular, a Knowledge Lake is a centralized repository containing virtually inexhaustible amounts of both raw data and *contextualized data* that is readily made available anytime to anyone authorized to perform analytical activities. The Knowledge Lake has the potential to provide the foundation for process analytics, process automation, and process augmentation (see Sect. 5). For the basic information items stored in the Knowledge Lake, we provide services to automatically [16]: (a) Extract features such as keyword, part of speech, and named entities such as Persons, Locations, Organizations, Companies, Products, and more; (b) Enrich the extracted features by providing synonyms and stems leveraging lexical knowledge bases for the English language such as WordNet; (c) Link the extracted enriched features to external knowledge bases (such as Google Knowledge Graph[3] and Wikidata[4]) as well as the contextualized data islands; and (d) Annotate the items in a data island by information about the similarity among the extracted information items, classifying and categorizing items into various types, forms or any other distinct class.

To transform the Data Lake into a Knowledge Lake (i.e., a contextualized Data Lake), we propose a framework for data curation feature engineering: characterizing variables that grasp and encode information from raw or curated data, thereby enabling to derive meaningful inferences from data. To facilitate the data curation process and enhance the productivity of researchers and developers, we implemented the above set of basic data curation APIs and made them available as services to researchers and developers to assist them in transforming their raw data into curated data [16]. Figure 2 illustrates the architecture and the main components of the proposed Knowledge Lake, called CoreKG.

---

[3] https://developers.google.com/knowledge-graph/.
[4] https://www.wikidata.org/.

5 AI-Enabled Processes

Today, the advancement in Artificial Intelligence (AI) and Data Science has the potential to transform business processes in fundamental ways; by assisting knowledge workers in communicating analysis findings, supporting evidence, and making decisions. In particular, intelligence is the ability to learn from experience and use domain experts' knowledge to adapt to new situations [10]. In this context, intelligent processes should be able to learn from domain experts' knowledge and experience. Understanding, analyzing, and ultimately improving business processes is a goal of enterprises today. Most related work in analyzing business process execution assumes well-defined processes; however, the business world is getting increasingly dynamic. There are cases where the process execution path can change dynamically and ad-hoc manner. Understanding business processes and analyzing BP execution data is difficult due to the lack of documentation, significantly as the process scope and how to process events across these systems are correlated into process instances are subjective: depending on the perspective of the process analyst.

Consequently, there is a need for Artificial Intelligence-enabled (AI-enabled) approaches that enable analysts to analyze the process events from their perspectives, for the specific goal they have in mind, and in an explorative manner. Understanding modern business processes entails identifying the relationships among entities in process graphs. Viewing process logs as a network, process graphs, and studying systematically the methods for mining such networks, of events, actors, and process artifacts is a promising frontier in database and data mining research: process mining provides an important bridge between data mining and business process modeling and analysis [1]. Applications of AI-enabled processes can be extended to nearly all organizational, managerial functions. For example, considering financial services, applying analytics on customer dossiers, and financial reports can specify the company's performance over periods. Such operations require various open source and commercial software for process analytics.

5.1 Data-Centric Processes

The problem of understanding the behavior of information systems and the processes and services they support has become a priority in medium and large enterprises. This is demonstrated by the proliferation of tools for the analysis of process executions, system interactions, and system dependencies and by recent research work in process data warehousing, discovery, and mining [1]. Accordingly, identifying business needs and determining solutions to business problems requires the analysis of business process data, which will help discover helpful information and support decision-making for enterprises.

The state-of-the-art in process data analytics discussed in our recent book [14], that focus on various topics such as Warehousing Business Process Data, Data Services and DataSpaces, Supporting Big Data Analytics Over Process Execution Data, Process Spaces, Process Mining and Analyzing Cross-cutting Aspects (e.g., provenance) in Processes' Data. This book provides defrayals on: (i) technologies, applications, and practices used to provide process analytics from querying to analyzing process data;

(ii) a broad spectrum of business process paradigms that have been presented in the literature from structured to unstructured processes; (iii) the state-of-the-art technologies and the concepts, abstractions and methods in structured and unstructured BPM including activity-based, rule-based, artifact-based, and case-based processes; and (iv) the emerging trend in the business process management area such as process spaces, big-data for processes, crowdsourcing, social BPM, and process management on the cloud.

5.2   Knowledge-Intensive Processes and Cognitive Assistants

Knowledge-Intensive Processes involve operations that are heavily reliant on professional knowledge. For these reasons, it is considered that human knowledge workers are responsible for driving the process, which cannot otherwise be automated as in workflow systems [3]. Knowledge-intensive processes almost always involve collecting and presenting a diverse set of artifacts and capturing the human activities around artifacts. This emphasizes the artifact-centric nature of such processes. In our previous work [7], we introduced Intelligent Knowledge Lakes to facilitate linking Artificial Intelligence (AI) and Data Analytics. The goal is to enable AI applications to learn from contextualized data, automate business processes, and develop cognitive assistance to facilitate knowledge-intensive processes or generate new rules for future business analytics.

Case-managed processes are primarily referred to as semistructured processes since they often require the ongoing intervention of skilled and knowledgeable workers [14]. Such Knowledge-Intensive processes involve operations that are heavily reliant on professional knowledge. For these reasons, it is considered that human knowledge workers are responsible for driving the process, which cannot otherwise be automated as in workflow systems [3]. Knowledge-intensive processes almost always involve collecting and presenting a diverse set of artifacts and capturing the human activities around artifacts. This emphasizes the artifact-centric nature of such processes. Many approaches used business artifacts that combine data and process in a holistic manner and as the basic building block [12, 14]. An important fact is the evolution of process artifacts over time as they are touched by different people in the context of a knowledge-intensive process. This highlights the need for tracking process artifacts in order to find out their history (artifact versioning) and also provenance (where they come from, and who touched and did what on them) [12].

Cognitive Assistants. Intelligent Knowledge Lakes provide the knowledge and experience of domain experts in a specific field, e.g., Banking or Education. This source of knowledge can be used as the core of AI-enabled Processes, for example, to facilitate understanding the intentions of end-users better (e.g., understanding customer journey in a banking scenario), or providing a comprehensive method to retrieve useful information from the enterprise platform (e.g., assisting a police investigator in investigation processes, to identify important facts/evidence and choosing the best next step). Natural Language Processing systems based on deep learning can be used to translate the conversations between the analyst and the Intelligent Knowledge Lakes. In Sect. 6, we will present a motivating scenario in Policing, where police investigators can use a cognitive assistant to analyze the process data and link that to process analysis in an intelligent way.

5.3   Goal-Oriented BPM

There is a growing realization of the need for *flexible process management*. Real-life processes rarely follow a fixed pre-defined schema. Human operators often leverage their own insights to create *variants* of the mandated process design to suit the exigencies of individual cases. This raises important questions about what qualifies a given instance to be described as a variant of a process design. The capability to generate context-sensitive variants on the fly is also fundamental to any attempt to achieve true process automation (for very similar reasons - humans are good at doing this but present-day automated processes can only follow pre-defined process schemas).

The notion of a *goal* sits at the heart of any attempt to solve these problems. A goal as a useful abstraction to understand user requirements has been well-recognized for the past several decades. A goal is a statement of intent, representing an aspiration to achieve a given state of affairs (or a set of states of affairs. A number of influential approaches to reasoning with goals have been proposed (mainly in the context of requirements engineering), including KAOS [19], i* [29], and TROPOS [17].

The emerging field of *goal-oriented BPM* seeks to place goals as the centerpiece in any account of flexible process management. Goals provide an answer [23] to the question: what qualifies a process instance to be described as a variant of a process design? The simple answer is that the achievement of the same goals as the original process enables us to describe a process instance as a variant of the original process. Variants are thus alternative means of achieving the same goals. Goals have been used as first-class process modeling constructs in *goal orchestrations* [24]. In a goal orchestration, a process task is replaced by a goal. While process tasks admit very little variation in how they are executed, the use of a goal permits us to explore a range of possible realizations.

Goal-oriented processes are data-driven constructs. Recent work has shown how post-condition annotations of process tasks (a post-condition can be viewed as an answer to the question: what will have happened if the process executes upto this point?) can be mined from a combination of process (activity) logs and event logs [25]. The final set of post-conditions achieved by a process can then form the basis for *goal fulfillment analysis* [24]. Looking to the future, goal-oriented BPM will help solve important classes of problems associated with knowledge-intensive processes and the design of intuitive process dashboards.

5.4   Process Intelligence

Process intelligence generically refers to the inference of insights from process executions in order to then optimize the process. There are several kinds of problem process intelligence can solve. The first is to get **visibility** into what happens in a process, at the **appropriate level of abstraction**. For example, in a customer support process, we ideally want to be able to understand the user journey, from the time they begin searching for a solution to their problem to when they interact with a chatbot or file a support request, supported by an agent. Analysts want to make sense of what is happening and to do so at the appropriate level of abstractions that is useful for their analysis goals. We can do so for example by correlating search data, chatbot logs, and customer support interaction logs, selecting or abstracting (aggregating) events of interest from these logs [20].

Once we have visibility into what happens, we can achieve more goals thanks to AI. A common need is that of **explaining outcomes**, especially undesirable ones. In our support process, sometimes the process lasts too long, or the customer gives a negative feedback. In these cases, process owners often want to understand why processes have undesirable outcomes. Common ways to do this are root cause analysis or clustering algorithm that identify process patterns correlated (or causally related) to such outcomes. A second need is that of **predicting outcomes**, to then take corrective actions. This is relatively simple, especially when it reduces to a classification problem, and the success of this depends on the predictive information available in the abstracted process as well as on the ability of the ML algorithm to capture and represent the signals in the process that are predictive of the outcome.

Finally, process intelligence deals with **recommending next-best actions**. This is often considered the ideal outcome from a process intelligence effort, but in many enterprise context this can be very hard to achieve reliably, especially in contexts where errors are costly. Very often a next-best action approach has to be use-case specific and consider the nuances of each use case. For example, recommending the next-best agent to route a support request to is very different than recommending how an agent is supposed to advise customers. Ideally, the underlying process representation can support predictability for both such cases and includes the proper abstractions, but very often this is hard to achieve.

For example, learning recommendations for how to solve a problem may be tricky as the proper solution may change over time (for example, a company may have changed the procedures to reset a password), or because learning to recommend requires the ability to read and comprehend a conversation and "distill" the notion of successful problem resolution, or because support requests may be often inherently ambiguous and require clarifications. An interesting point here is that if and when we are successful in automating a process via a next-best action approach, then we really do not have a process any more, or at least we do not have a *designed* process (or at least, not a process designed by humans), but just AI acting based on experience (and some constraints, often rule-based). We would "just" have an *observed* process, constantly evolving as AI continues to learn from user feedback, along with possibly the ability of AI to describe its learned logic in process form.

5.5 Document Intelligence

Data is central to the operation of enterprise business processes. While workflow systems and business process management (BPM) systems deal with structured data as primary form of data, there is a great amount of process data in unstructured and semi-structured forms. This data often lives outside of the BPM systems and at times is attached to the process records, e.g., in form of document attachments or human conversation payload over emails, chats or other channels in workflow and process management systems. Additionally, a large number of business processes are not supported by a workflow system or a BPM system. For these processes, their data is captured in a mix of structured information as database records as well as documents in file systems and information in data and knowledge lakes.

An important class of data for business processes are **business documents**, as a representative of semi-structured and unstructured process information. Business documents play many roles in business processes. Documents contain information exchanged in the context of process execution (e.g., purchase orders, proof of delivery, invoices that are produced and exchanged in the context of a procure-to-pay process), as well as they can be used to describe processes and process flows (e.g., business process definition documents and process guides that are produced by process owners detailing the tasks, actions and different conditions/rules for tasks triggering and execution). We define **Document Intelligence** as the ability to read, understand and interpret business documents using artificial intelligence. Document Intelligence goes hand-in-hand with Process Intelligence. Document Intelligence provides abstractions, methods and tools for extracting process-related information, understanding business documents in the context of a process and interpreting the document content w.r.t. processes execution and process stakeholders, therefore enabling Process Intelligence in tasks such as next best action and outcome prediction over all process information space.

While the interest in Document Intelligence topic from AI researchers in academia and industry[5] has grown substantially, the work in Document Intelligence for business processes is still at a very early stage. There are many open topics such process instance discovery over business documents, process discovery and mining from business documents, process-aware understanding and interpretation of business document content, and understanding process flow charts and other visuals in business documents, process-aware question answering over documents, and many more.

Given the textual nature of the content of business documents, the work on Document Intelligence require advanced natural language processing and natural language understanding. Business documents are also not all alike. There are fully textual (also referred as unstructured business documents), to semi-structured (those with form-like content and tabular-like data such as invoices and tax forms) to those that have structure but lack semantics such as spreadsheets. Due to the difference in content types, more specialized AI techniques have been devised in the literature to extract information and understand the content of each type. Academic contests such as Document Visual Question Answering[6] focus on encouraging and measuring our progress towards different aspects of Document Intelligence. The state of the art in semi-structure document understanding systems include transformer-based models (e.g., LayoutLM [28]) and earlier pioneer work in document reading and understanding (e.g., DICR [27]).

On the other front, for process understanding from textual and document content, there has been pioneer work for action identification from textual sources (e.g., [21]), and early work focused on automated process discovery from textual sources and documents [18,22]. Beyond these lines of work, process-level understanding of the content of business documents and unstructured information require a more holistic approach that would need to turn process data into process knowledge, which can be represented and become available, actionable and query-able in knowledge graphs and knowledge lakes.

---

[5] https://sites.google.com/view/di2019.

[6] https://rrc.cvc.uab.es/?ch=17.

6   Motivating Scenario: AI-enabled Police Investigation Processes

In this tutorial, we focus on the police investigation processes around Missing Persons. Between 2008 and 2015, over 305,000 people were reported missing in Australia (aic.gov.au/), an average of 38,159 reports each year. In the USA (nij.gov/), there are as many as 100,000 active missing person's cases on any given day. The first few hours following a person's disappearance are the most crucial. The sooner police can put together the sequence of events and actions right before the person's disappearance, the higher the chance of finding the person. This entails gathering information about the person, including physical appearance and activities on social media in the physical/social environments of the person, person's activity data such as phone calls and emails, and information on the person detected by sensors (e.g., CCTVs).

The investigation process is a data-driven, knowledge-intensive and collaborative process. The information associated with an investigation (case process) is usually complex, entailing to collect and present many different types of documents and records. It is also expected that separate investigations may impact other investigation processes, and the more evidence (knowledge and facts extracted from the data in the data lake [5]) collected, the better-related cases can be linked explicitly. Although law enforcement agencies use data analysis, crime prevention, surveillance, communication, and data sharing technologies to improve their operations and performance, many challenges remain in sophisticated and data-intensive cases such as missing persons. For example, fast and accurate information collection and analysis are vital in law enforcement applications [26]. From the policymakers' perspective, this trend calls for adopting innovations and technologically advanced business processes that can help law enforcers detect and prevent criminal acts. AI-enabled Processes in law enforcement processes have the potential to help investigators to understand a potential pool of data evidence. A set of intelligent services can be used to facilitate this process, prepare the big process data for analytics, summarize the big process data, construct narratives, and enable analysts to link narratives and easily dig for facts.

There are several tasks and activities, during and investigation process, that can benefit from an AI-enabled approach. For example, a cognitive assistant in a form of a smart phone application, can facilitate data collection from the scene, or extract important insight from the images taken from the crime scene. This in turn can assist investigators in decision making and choosing the best nest steps. In our previous work, we presented iCOP [26], a system composed of a framework and a set of techniques to assist knowledge workers (e.g., a criminal investigator) in knowledge-intensive processes (e.g., criminal investigation) to benefit from AI-enabled processes, collect large amounts of pieces of evidence and easily dig for the facts. The system will facilitate augmenting police officers with Internet-enabled smart devices (e.g., phones/watches) to assist them in the process of collecting evidence, access to location-based services to identify and locate resources (CCTVs, cameras on officers on duty, police cars, drones and more), organize all these islands of data in a Knowledge Lake and feed them into a scalable and extensible Process Data Analytics Pipeline. In this tutorial, We focus on a motivating scenario where a criminal investigator will be augmented by smart devices to collect data and to identify devices around the investigation location and communicate with them to understand and analyze evidences.

7 Conclusion and Future Directions

The continuous demand for business process improvement and excellence has prompted the need for intelligently understanding/analyzing the process-related data to facilitate decision-making. In this context, AI-enabled Processes are considered a new direction in business process management, especially as process-related data increasingly come to show all typical properties of big data. In this context, intelligent processes could be able to learn from domain experts' knowledge and experience. This tutorial focused on a motivating scenario in a police investigation, an example of data-centric and knowledge-driven processes, and introduced intelligent Data Lakes and Knowledge Lakes as the back-end for AI-enabled Processes. AI-enabled approaches may involve from process modeling to process execution and monitoring. Future directions in the field of AI-enabled Processes will focus on organizing, curating, analyzing, and visualizing processes-related data using AI-enabled approaches. This may include: (i) data-Centric AI-enabled techniques to systematically engineering the process-related data; and (ii) intelligent approaches to learn from knowledge worker's experience and use this knowledge to adapt to new situations and facilitate decision-making.

Acknowledgements. We acknowledge the AI-enabled Processes (AIP) Research Centre (https://aip-research-center.github.io/) for funding part of this research.

References

1. van der Aalst, W.M.P.: Process Mining - Data Science in Action, 2nd edn. Springer, Heidelberg (2016). https://doi.org/10.1007/978-3-662-49851-4
2. Alonso, G., Casati, F., Kuno, H.A., Machiraju, V.: Web Services - Concepts, Architectures and Applications. Data-Centric Systems and Applications, Springer, Heidelberg (2004). https://doi.org/10.1007/978-3-662-10876-5
3. Beheshti, A., Benatallah, B., Motahari-Nezhad, H.R.: ProcessAtlas: a scalable and extensible platform for business process analytics. Softw. Pract. Exp. **48**(4), 842–866 (2018)
4. Beheshti, A., Benatallah, B., Motahari-Nezhad, H.R., Ghodratnama, S., Amouzgar, F.: A query language for summarizing and analyzing business process data. CoRR abs/2105.10911 (2021). https://arxiv.org/abs/2105.10911
5. Beheshti, A., Benatallah, B., Nouri, R., Chhieng, V.M., Xiong, H., Zhao, X.: CoreDB: a data lake service. In: Proceedings of the 2017 ACM on Conference on Information and Knowledge Management, CIKM 2017, Singapore, 06–10 November 2017, pp. 2451–2454. ACM (2017)
6. Beheshti, A., Benatallah, B., Nouri, R., Tabebordbar, A.: CoreKG: a knowledge lake service. Proc. VLDB Endow. **11**(12), 1942–1945 (2018)
7. Beheshti, A., Benatallah, B., Sheng, Q.Z., Schiliro, F.: Intelligent knowledge lakes: the age of artificial intelligence and big data. In: Leong Hou, U., Yang, J., Cai, Y., Karlapalem, K., Liu, A., Huang, X. (eds.) WISE 2020. CCIS, vol. 1155, pp. 24–34. Springer, Singapore (2020). https://doi.org/10.1007/978-981-15-3281-8_3
8. Beheshti, A., Benatallah, B., Tabebordbar, A., Motahari-Nezhad, H.R., Barukh, M.C., Nouri, R.: DataSynapse: a social data curation foundry. Distrib. Parallel Databases **37**(3), 351–384 (2019). https://doi.org/10.1007/s10619-018-7245-1

9. Beheshti, A., et al.: iProcess: enabling IoT platforms in data-driven knowledge-intensive processes. In: Weske, M., Montali, M., Weber, I., vom Brocke, J. (eds.) BPM 2018. LNBIP, vol. 329, pp. 108–126. Springer, Cham (2018). https://doi.org/10.1007/978-3-319-98651-7_7

10. Beheshti, A., Yakhchi, S., Mousaeirad, S., Ghafari, S.M., Goluguri, S.R., Edrisi, M.A.: Towards cognitive recommender systems. Algorithms 13(8), 176 (2020)

11. Beheshti, S., Benatallah, B., Motahari-Nezhad, H.R.: Scalable graph-based OLAP analytics over process execution data. Distrib. Parallel Databases 34(3), 379–423 (2016). https://doi.org/10.1007/s10619-014-7171-9

12. Beheshti, S.-M.-R., Benatallah, B., Motahari-Nezhad, H.R.: Enabling the analysis of cross-cutting aspects in ad-hoc processes. In: Salinesi, C., Norrie, M.C., Pastor, Ó. (eds.) CAiSE 2013. LNCS, vol. 7908, pp. 51–67. Springer, Heidelberg (2013). https://doi.org/10.1007/978-3-642-38709-8_4

13. Beheshti, S.-M.-R., Benatallah, B., Motahari-Nezhad, H.R., Allahbakhsh, M.: A framework and a language for on-line analytical processing on graphs. In: Wang, X.S., Cruz, I., Delis, A., Huang, G. (eds.) WISE 2012. LNCS, vol. 7651, pp. 213–227. Springer, Heidelberg (2012). https://doi.org/10.1007/978-3-642-35063-4_16

14. Beheshti, S., et al.: Process Analytics - Concepts and Techniques for Querying and Analyzing Process Data. Springer, Cham (2016). https://doi.org/10.1007/978-3-319-25037-3

15. Beheshti, S., Nezhad, H.R.M., Benatallah, B.: Temporal provenance model (TPM): model and query language. CoRR abs/1211.5009 (2012). http://arxiv.org/abs/1211.5009

16. Beheshti, S., Tabebordbar, A., Benatallah, B., Nouri, R.: On automating basic data curation tasks. In: Barrett, R., Cummings, R., Agichtein, E., Gabrilovich, E. (eds.) Proceedings of the 26th International Conference on World Wide Web Companion, Perth, Australia, 3–7 April 2017, pp. 165–169. ACM (2017)

17. Bresciani, P., Perini, A., Giorgini, P., Giunchiglia, F., Mylopoulos, J.: Tropos: an agent-oriented software development methodology. Auton. Agent. Multi-Agent Syst. 8(3), 203–236 (2004)

18. Chambers, A.J., et al.: Automated business process discovery from unstructured natural-language documents. In: Del Río Ortega, A., Leopold, H., Santoro, F.M. (eds.) BPM 2020. LNBIP, vol. 397, pp. 232–243. Springer, Cham (2020). https://doi.org/10.1007/978-3-030-66498-5_18

19. Darimont, R., Delor, E., Massonet, P., van Lamsweerde, A.: GRAIL/KAOS: an environment for goal-driven requirements engineering. In: Proceedings of the 19th International Conference on Software Engineering, pp. 612–613 (1997)

20. Nezhad, H.R.M., Benatallah, B., Casati, F., Saint-Paul, R.: From business processes to process spaces. IEEE Internet Comput. 15(1), 22–30 (2011)

21. Nezhad, H.R.M., Gunaratna, K., Cappi, J.M.: eAssistant: cognitive assistance for identification and auto-triage of actionable conversations. In: Barrett, R., Cummings, R., Agichtein, E., Gabrilovich, E. (eds.) Proceedings of the 26th International Conference on World Wide Web Companion, Perth, Australia, 3–7 April 2017, pp. 89–98. ACM (2017). https://doi.org/10.1145/3041021.3054147

22. Park, H., Motahari-Nezhad, H.R.: Learning procedures from text: codifying how-to procedures in deep neural networks. In: Champin, P., Gandon, F., Lalmas, M., Ipeirotis, P.G. (eds.) Companion of the The Web Conference 2018 on The Web Conference 2018, WWW 2018, Lyon, France, 23–27 April 2018, pp. 351–358. ACM (2018). https://doi.org/10.1145/3184558.3186347

23. Ponnalagu, K., Ghose, A., Narendra, N.C., Dam, H.K.: Goal-aligned categorization of instance variants in knowledge-intensive processes. In: Motahari-Nezhad, H.R., Recker, J., Weidlich, M. (eds.) BPM 2015. LNCS, vol. 9253, pp. 350–364. Springer, Cham (2015). https://doi.org/10.1007/978-3-319-23063-4_24

24. Santipuri, M., Ghose, A., Dam, H.K., Roy, S.: Goal orchestrations: modelling and mining flexible business processes. In: Mayr, H.C., Guizzardi, G., Ma, H., Pastor, O. (eds.) ER 2017. LNCS, vol. 10650, pp. 373–387. Springer, Cham (2017). https://doi.org/10.1007/978-3-319-69904-2_29

25. Santiputri, M., Ghose, A.K., Dam, H.K.: Mining task post-conditions: automating the acquisition of process semantics. Data Knowl. Eng. **109**, 112–125 (2017)

26. Schiliro, F., et al.: iCOP: IoT-enabled policing processes. In: Liu, X., et al. (eds.) ICSOC 2018. LNCS, vol. 11434, pp. 447–452. Springer, Cham (2019). https://doi.org/10.1007/978-3-030-17642-6_42

27. Tecuci, D.G., et al.: DICR: AI assisted, adaptive platform for contract review. In: The Thirty-Fourth AAAI Conference on Artificial Intelligence, AAAI 2020, The Thirty-Second Innovative Applications of Artificial Intelligence Conference, IAAI 2020, The Tenth AAAI Symposium on Educational Advances in Artificial Intelligence, EAAI 2020, New York, NY, USA, 7–12 February 2020, pp. 13638–13639. AAAI Press (2020). https://aaai.org/ojs/index.php/AAAI/article/view/7106

28. Xu, Y., et al.: LayoutLMv2: multi-modal pre-training for visually-rich document understanding. In: Zong, C., Xia, F., Li, W., Navigli, R. (eds.) Proceedings of the 59th Annual Meeting of the Association for Computational Linguistics and the 11th International Joint Conference on Natural Language Processing, ACL/IJCNLP 2021 (Volume 1: Long Papers), Virtual Event, 1–6 August 2021, pp. 2579–2591. Association for Computational Linguistics (2021). https://doi.org/10.18653/v1/2021.acl-long.201

29. Yu, E.S.: Towards modelling and reasoning support for early-phase requirements engineering. In: Proceedings of ISRE 1997: 3rd IEEE International Symposium on Requirements Engineering, pp. 226–235. IEEE (1997)

Continuously Testing Distributed IoT Systems: An Overview of the State of the Art

Jossekin Beilharz[1]([✉]), Philipp Wiesner[2], Arne Boockmeyer[1], Lukas Pirl[1], Dirk Friedenberger[1], Florian Brokhausen[2], Ilja Behnke[2], Andreas Polze[1], and Lauritz Thamsen[2,3]

[1] Hasso Plattner Institute, University of Potsdam, Potsdam, Germany
{jossekin.beilharz,arne.boockmeyer,lukas.pirl,andreas.polze}@hpi.de,
dirk.friedenberger@guest.hpi.de
[2] Technische Universität Berlin, Berlin, Germany
{wiesner,florian.brokhausen,i.behnke}@tu-berlin.de
[3] Humboldt-Universität zu Berlin, Berlin, Germany
lauritz.thamsen@hu-berlin.de

Abstract. The continuous testing of small changes to systems has proven to be useful and is widely adopted in the development of software systems. For this, software is tested in environments that are as close as possible to the production environments. When testing IoT systems, this approach is met with unique challenges that stem from the typically large scale of the deployments, heterogeneity of nodes, challenging network characteristics, and tight integration with the environment among others. IoT test environments present a possible solution to these challenges by emulating the nodes, networks, and possibly domain environments in which IoT applications can be executed. This paper gives an overview of the state of the art in IoT testing. We derive desirable characteristics of IoT test environments, compare 18 tools that can be used in this respect, and give a research outlook of future trends in this area.

Keywords: Internet of Things · Cyber-physical systems · Fog computing · Edge computing · Testing · Iterative software development

1 Introduction

The Internet of Things (IoT) has the potential to transform our lives by connecting everyday objects to the Internet for smarter cities, factories, houses, and more. To realize this vision, distributed software systems will need to integrate IoT devices – usually equipped with sensors and actuators – allowing them to continuously monitor and interact with their environments. These distributed software systems of the IoT will span from devices to clouds and, in many cases, also include intermediate resources at the edge or fog level [7]. Examples of distributed IoT systems include those that control and manage traffic and transportation [19,36], those that enable telemedicine and remote patient

© Springer Nature Switzerland AG 2022
H. Hacid et al. (Eds.): ICSOC 2021 Workshops, LNCS 13236, pp. 336–350, 2022.
https://doi.org/10.1007/978-3-031-14135-5_30

monitoring [10,18], and those that detect and predict failures as well as optimize processes in urban infrastructures and manufacturing [9,13,22].

A major remaining challenge to practically developing and deploying distributed IoT systems is the difficulty of adequately testing them [15]. This is complicated due to a number of factors, including the large number of devices, the heterogeneity of devices, mobile nodes resulting in dynamic topologies, network disconnections and node failures, as well as a tight integration of systems with their respective environments. At the same time, properly testing IoT systems in application domains such as traffic and transportation management, patient monitoring, and factory processes is absolutely critical. Consequently, the need for adequate testing of distributed IoT systems has been widely recognized and many solutions have been put forward. Prominent examples include hardware testbeds like StarBED [21] and FIT-IoT [1], hybrid approaches such as Chameleon [14], as well as simulators like IoTSim [34] and iFogSim [11].

Hardware testbeds allow to execute actual application code in realistic settings, yet can be limited in terms of scalability and flexibility. Hybrid test environments address these limitations by incorporating both actual hardware and virtual nodes. Simulations on the other hand enable to flexibly assess the behavior of distributed applications over various scales and possible infrastructures. However, they usually lack the ability to evaluate the non-functional properties of actual application code.

All these approaches have in common that it is typically hard to test distributed IoT systems within their actual environment. Field testing regularly requires a large and coordinated effort, so distributed IoT systems cannot be tested continuously, while lab testing routinely resorts to merely replaying sensor data, so that the distributed IoT systems, despite being equipped to interact with environments, cannot actually influence their domains. This runs contrary to generally understood and widely adopted principles and best practices of iterative software development, where continuous testing of small changes to systems in environments that mirror production environments as closely as possible is a key mechanism for fast feedback and trust in changes. We, therefore, argue that there is a significant lack of approaches and tools for continuously testing IoT systems.

In this paper, we compare currently available IoT test environments to provide an overview over the current state of the art and expose the perceived research gap. For our comparison, we selected test environments that

1. focus on testing software systems on geo-distributed, heterogeneous computing infrastructures such as IoT and edge/fog architectures,
2. allow to run actual system code (i.e., not merely simulating communication),
3. and have the ability to include virtual nodes, allowing tests at large and various scales (i.e., no hardware-only testbeds).

We only discuss general-purpose test environments of which details have been published (i.e., no proprietary offers such as IoTIFY[1] or AWS IoT Device Simulator[2]).

---

[1] https://iotify.io/.

[2] https://aws.amazon.com/solutions/implementations/iot-device-simulator/.

We report the following aspects of IoT test environments in our comparison: how and with which capabilities the tools provide nodes, how the network between nodes is realized, whether domain environments are integrated, as well as general aspects such as project maturity and ongoing development.

The main contributions of this paper are:

- A description of key characteristics of IoT test environments, which can be used to distinguish proposed solutions.
- A point-by-point comparison of state-of-the-art IoT test environments that meet the outlined selection criteria.
- A discussion of current trends and considerable gaps in the state of the art of IoT testing.
- An outlook on future work to close these gaps and an overview of our work in this area.

The remainder of this paper is structured as follows: Sect. 2 describes central characteristics of test environments. These are used in Sect. 3 to evaluate and compare concrete test environments. Section 4 discusses the results of our comparison. Section 5 presents the research outlook. Section 6 covers related work. Lastly, Sect. 7 concludes this paper.

2   Characteristics of Test Environments

Continuously testing IoT systems and applications requires a test environment that reproduces reality as close as possible. To classify and compare existing test environments we derive several quantifiable characteristics from generally desirable properties of test environments.

To be able to continuously develop and test distributed IoT systems in iterative software development processes, we need to be able to deploy and run actual code in flexible, yet realistic environments. To facilitate large-scale deployments while also allowing the realistic testing of system behavior, we believe the support for both virtualized nodes as well as hardware nodes in test environments is crucial. The testing of large-scale deployments further requires the distributability of not only the nodes, but also the network representation and the simulation of the domain environment. Another important aspect relating to the three feature dimensions here—nodes, networks, and the domain environments—is the meaningful testing of fault tolerance of IoT systems by precisely injecting faults. Lastly, because IoT systems are inherently integrated tightly with their specific environment through sensors and actuators, we believe that the simulation of the domain environment is a key characteristic for test environments.

The remainder of this section discusses the characteristics that will be used in the following comparison of test environments in Sect. 3. We identified 16 different attributes, which are organized into four overarching categories, regarding general features, the nodes, the network and the domain environment.

2.1 General

First, we describe general attributes of test environments. We present the *initial year* of *publication* along with the information if the project is *actively maintained*, which is assessed based on whether there has been a new release or active collaboration (e.g., commits to the repository) in 2021.

As the *maturity* of a project is subjective, we try to formalize it as follows. An empty circle (○) denotes the lowest maturity, meaning that the specific test environment only exists as a concept in form of a publication, but no actual tool is available. We did not assess whether such concepts are actively maintained, as this cannot be sensibly judged. The second degree of maturity (◐) is reported if the tool is available only as a prototype without good documentation. If there is a full system available with detailed documentation, we denote it as the third degree of maturity (●).

Next, we classify if a test environment is *offered as a service*. This indicates whether there is a service where the test environment can be used without manually deploying and operating it.

Lastly, we asses the property *scriptable scenarios*, which is fulfilled if the execution of experiments can be controlled via a script. With the capability to predefine schemes to alter parameters and characteristics of a simulation at runtime, much more complex scenarios can be implemented. This is highly important when systematically approaching an investigation with a test environment.

These general information about a test environment can serve as an indicator for the applicability to current challenges, but they are also used to identify recent trends in test environments in Sect. 4.

2.2 Nodes

An essential aspect of test environments is which type of nodes can be used. The attributes investigated here determine if a scenario or application of interest can at all be properly implemented or analyzed with a given test environment.

The first attribute, *hardware integration*, classifies the test environments according to their capability to integrate physical hardware nodes. The availability of hardware integration enables the inclusion of embedded systems and facilitates testing of applications in realistic environments.

The *virtualization type* describes how virtual nodes are represented, namely via virtual machines (V), containerized nodes (C), or a combination of the two (VC). Depending on the application under test, the differentiation between containerized and virtualized nodes can be crucial. Virtual machines enable a more realistic execution environment for the application under test, while containerization is a more light-weight approach.

For the *energy consumption* characteristic, we investigate if a test environment facilitates modeling (or, in the case of hardware nodes, monitoring) the power consumption of nodes and network. In any use case where energy is a scarce resource, for example for battery-constrained IoT devices, this feature allows testing the effect of software changes to a node's power usage.

The *distributability* describes whether virtual nodes of the test environment can be spread across multiple physical host nodes, enabling large-scale scenarios.

Finally, we investigate the possibility of *fault injection*. For individual nodes, examples include the purposeful shutdown or internal failure of a given node at a given time. By simulating such faults, the robustness of an application or network setup towards faults can be tested. We analyze the characteristics of distributability and fault injection as well for the network and the domain environment categories.

2.3   Network

Regarding network, we first analyze the *network type*, namely how network is emulated within the test environment. Traffic shaping (TS) allows users to change network parameters, like delay or bandwidth. Examples of this are the Linux Traffic Control (*tc*) or the more advanced NetEm. Tools based on software-defined networks (SDN) use a virtualized network such as provided by Mininet or MaxiNet. Lastly, network simulators (NS) can be used to model the underlying network. In our understanding, network simulators can simulate different kinds of networks, also future ones, without having them physically available. Common network simulators are ns-3 or OMNet++ with INET.

Network *distributability* regards the possibility of the test environment to span the network across multiple physical hosts, leveraging more complex routing schemes in a physical network. For traffic shaping-based approaches this comes naturally if nodes are distributed, for network simulation-based approaches also the simulation has to run in a distributed manner.

Fault injection entails active support of the tool to purposefully alter network connections at runtime, e.g., the increase of latency or the loss of packets. Such capabilities are important when comparing fault tolerance of network setups and in general testing of network robustness.

2.4   Domain Environment

As IoT applications run in embedded, real-life settings like traffic control, water management or smart homes, simulating the domain environment is of high importance. First, we asses the general *domain environment support* of a tool, meaning whether there is an API for connecting domain-specific simulators that can interact with the test environment at runtime. Examples include a traffic simulation, such as SUMO, that can send the coordinates of mobile nodes to the test environment for it to adapt its networking parameterizations.

Similar to the node and network categories, we also asses the *distributability* of the domain environment to see if the execution of the environment can be spread across multiple hosts.

Last, we report the capability of *fault injection* inside the domain environment. We define this functionality to be present when the test environment supports to alter the domain environment during runtime in a way that is expected to introduce faults in the application running on the nodes.

Table 1. An overview of test environments for IoT systems. A gray background marks works where code is not openly available. All characteristics are described in Sect. 2.

	General					Nodes					Network			Domain Env.		
	Initial Publication	Actively Maintained	Maturity	Offered as a Service	Scriptable Scenarios	Hardware Integration	Virtualization Type	Energy Consumption	Distributability	Fault Injection	Network Type	Distributability	Fault Injection	Domain Env. Support	Distributability	Fault Injection
EMU-IoT [27]	2019	O	◐	O	●	O	C	O	●	O	O	-	-	O	-	-
ELIoT [23]	2017	O	◐	O	O	O	C	O	●	O	O	-	-	O	-	-
IOTier [24]	2021	-	O	O	●	O	C	O	●	●	TS	●	●	O	-	-
Fogify [30]	2020	●	●	O	●	O	C	●	●	●	TS	●	●	O	-	-
MockFog [12]	2019	●	◐	O	●	O	V	O	●	●	TS	●	●	O	-	-
Blockade [33]	2014	●	●	O	O	O	C	O	O	●	TS	O	●	O	-	-
Distem [28]	2013	O	●	O	●	O	C	O	●	●	TS	●	●	O	-	-
Fogbed [6]	2018	O	◐	O	O	O	C	O	●	O	SDN	●	O	O	-	-
EmuFog [20]	2017	O	◐	O	O	O	C	O	●	●	SDN	●	●	O	-	-
Dockemu [26]	2015	O	◐	O	●	O	C	O	●	O	NS	O	O	O	-	-
EmuEdge [35]	2019	O	◐	O	●	●	VC	O	O	●	TS	O	●	O	-	-
Héctor [2]	2019	O	◐	O	●	●	V	O	O	●	TS	O	●	O	-	-
Sendorek et al. [29]	2018	-	O	O	●	●	V	O	O	O	SDN	O	O	O	-	-
Chameleon [14]	2015	●	●	●	O	●	V	O	●	O	SDN	●	O	O	-	-
StarBED [21]	2002	●	●	●	●	●	V	O	●	O	SDN	●	O	O	-	-
UiTiOt [16]	2017	-	O	O	O	●	C	O	●	O	NS	●	●	O	-	-
WHYNET [38]	2006	-	O	O	●	●	V	●	●	O	NS	●	●	O	-	-
MobiNet [17]	2005	-	O	O	O	●	V	O	●	O	NS	O	O	O	-	-

3 Comparison of Test Environments

Based on the selection criteria defined in Sect. 1, we selected 18 environments for testing IoT applications and evaluated them on the characteristics described in Sect. 2. Table 1 provides an overview of all evaluated tools. We clustered the test environments (1) by their ability to integrate real IoT devices in their experiments and (2) by the type of network modeling approach they use.

3.1 Test Environments Without Hardware Integration

First, we cover test environments that emulate IoT environments without the possibility to integrate real IoT devices in experiments. These test environments are further categorized based on whether they use network simulation, SDN-based solutions, or simple traffic shaping to emulate realistic network traffic.

Using No Network Model. EMU-IoT [27] is a container-based test environment with a focus on defining, orchestrating and monitoring reproducible experiments. Although the authors describe the many challenges faced by developing IoT test environments, their implementation does not consider any kind of network emulation and has no mechanism for injecting faults into the system.

ELIoT [23] is based on Docker containers and supports the IoT protocols CoAP and LWM2M by using the open-source projects Leshan and coap-node. While ELIoT includes the interaction with the environment for the use case described in the paper, this interaction is only modeled within the nodes (i.e., they implemented a simple calculation of an illuminance sensor value based on the time of day). It does not integrate an environment emulation that would allow for two-way interaction between IoT systems and the environment.

Using Traffic Shaping. IOTier [24] is a virtual testbed for tiered IoT environments that is unfortunately not openly available. Nodes are represented via resource-constrained containers while networking is based on NetEm. A special focus is grouping emulated components into tiers with comparable capabilities, and enabling inter-tier as well as intra-tier communication. It features a testbed controller in which operators can define desired runtime states over time. However, there is no API for integrating simulators of domain environments. Its simulation engine uses fixed-increment time progression and can modify experiments via scheduled and conditional events.

Fogify [30] appears to be one of the most capable tools according to our criteria. It uses Infrastructure-as-Code descriptions for containerized deployments to define experiment settings (i.e., Docker Compose) and features the possibility to adapt configurations at runtime (e.g., for injecting faults). Fogify uses Virtual eXtensible LAN (VXLAN) for overlay networks and is distributable across multiple physical hosts. We classified this tool as being able to model energy consumption as this feature is described in the paper. However, this is currently not implemented in code. Although Fogify has an API for interacting with experiments during runtime, there is not yet a uniform way to integrate simulations of domain environments.

MockFog [12] is a tool for automated execution of fog application experiments. It consists of three modules: one for infrastructure setup, one for application management, and one for experiment orchestration, which enables the scripting of scenarios. The experiment infrastructures are set up automatically in public cloud environments via dockerized application containers. Hence, applications must support running inside Docker and must be available as container image. Experiment descriptions can be used to generate events, such as traffic scenarios and network or machine failures.

Blockade [33] is a test environment based on Docker containers and traffic shaping. The user creates a setup similar to a Docker Compose file and Blockade manages the set-up as well as tear-down processes. Each node is implemented as a separate Docker container. Blockade offers basic networking capabilities by using the Docker network and integrates the manipulation of network parameters via, e.g., NetEm settings.

Distem [28] is a virtual testbed using Linux Containers (LXC) that can be executed on multiple physical hosts. One focus of Distem is resource allocation and assignment to achieve realistic setups for special devices (like IoT devices). Network parameters can be adapted using NetEm. Distem can be used via the command line and allows scriptable scenarios via its Ruby library.

Using Software-Defined Networking. Fogbed [6], as described by the original paper, uses Mininet for networking and is hence bound to a single host. The latest prototype additionally extends MaxiNet, which enables emulating environments that span several physical machines. Fogbed furthermore enables the testing of third-party systems such as resource management, virtualization, and service orchestration through standard interfaces.

EmuFog [20] is a fog computing emulation framework based on the distributable network emulator MaxiNet. The framework does not resort to simulations but is able to span an emulated network of thousands of virtual devices over multiple physical machines. EmuFog focuses on the networking components of fog computing by embedding a network topology generator, enhancer, and node placement algorithm. Applications have to be deployed as Docker containers.

Using Network Simulation. Dockemu [26] is the only tool without hardware integration that uses network simulation. It utilizes the network simulator ns-3 to model the communication between nodes, which in turn are represented by Docker containers. The paper recognizes the importance of providing realistic conditions and environmental factors for the applications under test. The tool itself, however, is restricted to controlling properties of nodes and the network but does not include mechanisms to provide a domain environment in which the application operates.

3.2   Test Environments with Hardware Integration

Next, we describe hybrid tools that offer the possibility to integrate real IoT devices in otherwise emulated environments to make experiments more realistic.

Using Traffic Shaping. EmuEdge [35] is an openly available, hybrid simulator that can represent nodes using containers, virtual machines, and physical devices. It supports OS-level as well as system-level virtualization and can interface simulators and real testbeds. Networking is based on networking namespaces (*netns*) and can replay real-world network traces.

Héctor [2] is an IoT testing framework with the main goal of representing devices realistically. Devices are emulated with QEMU in system mode, allowing fine grained performance moderation of individual devices and testing on the target platform, including its corresponding microarchitecture. Specifically, this allows testing of devices that are not able to run Docker containers (e.g., micro-controllers). Physical as well as emulated devices can be part of the network, which itself can have emulated properties such as added delay and packet loss.

Using Software-Defined Networking. Sendorek et al. [29] describe an elaborated concept for a software-defined virtual test environment for IoT systems. Their system supports three so called "immersion levels" that range from fully virtualized environments for low-cost, scalable experiments to environments with real devices and sensors for testing under realistic conditions. The authors do not cover distributability or fault injection in their concept.

Chameleon [14] builds upon OpenStack to deliver a testbed that can be used like a cloud. Chameleon is both a concept with an open-source implementation and a platform service supported by hardware at University of Chicago and at the Texas Advanced Computing Center that includes different nodes and setups including GPUs, FPGAs as well as ARM and x86 cores. In addition to bare metal nodes, nodes virtualized with KVM can be used. Besides the concept of an OpenStack-based testbed, the Chameleon project has some insights regarding the operational side of such a testbed, like user management, fair resource allocation with leases and lease reapers, security attacks etc.

StarBED [21] is a large-scale general purpose network testbed based on co-located physical nodes which uses SpringOS to build experiment topologies and drives experiments. Its updated fourth version implements additional features, such as wireless network emulation and a background traffic generator. Although StarBED aims to enable Internet-scale experiments, it apparently lacks the possibility to emulate IoT characteristics (e.g., resource constraints, heterogeneous network capacities) and mainly acts as a resource management system.

Using Network Simulation. UiTiOt [16], meanwhile in its third version, is a test environment for large-scale wireless IoT applications. Instances of the application under test are executed using Docker Swarm on top of an OpenStack instance. The network connections between the application instance (e.g., IEEE 802.11a/b/g, ZigBee) are emulated using the wireless emulator QOMET. Apart from the virtual resources, UiTiOt can integrate physical nodes into the network. The authors also introduce a web interface for users of the testbed and a load-balanced database for receiving and storing logs from the application under test.

WHYNET [38] is a hybrid testbed that focuses on mobile communication and applications, using a combination of simulation, emulation, as well as physical nodes and connections. It simulates the network via the QualNet simulator and the sensor network simulation framework sQualNet, which is one of the few tools that model energy consumption. Using the TWINE framework [37], it emulates the network stack and the execution of applications to provide a scalable but realistic test environment. WHYNET includes a basic concept of mobility but does not allow the integration of domain-specific simulators for this purpose.

MobiNet [17] focuses on the evaluation of applications and network setup in ad hoc wireless networks. The tool allows the testing of different deployment schemes for applications and includes the simulation of movement of nodes. The core of MobiNet takes care of emulating the physical, data link, and network layers. Edge nodes can be distributed across machines and can host multiple virtual nodes for large-scale environments. Unfortunately, the code for MobiNet is not publicly available.

4 Discussion

We identified themes that emerged in our comparison of test environments in each of our categories of characteristics: general characteristics, and those that relate to representation of nodes, network and domain environment.

4.1 General

Testing of IoT systems is an active research area and many solutions try to help the developers of IoT systems in this regard. In our comparison, most systems were initially published within the last five years. These works include mature and widely adopted projects, but also ideas and research prototypes. Only two of the examined test environments are offered as a service.

4.2 Nodes

In our comparison we investigated the ability of test environments to use virtual and hardware nodes for the testing of IoT systems. For the virtual nodes, both containers and virtual machines are used, with recent works showing a tendency to use more lightweight container virtualization. This choice of virtualization type correlates with the integration of hardware nodes: Systems that include hardware nodes mostly use virtual machines, while the others mostly use containers. The ability to execute some nodes on actual hardware is missing from more than half of the test environments, even though this is especially important in many IoT use cases because often highly customized hardware is used. While energy consumption modeling is crucial to test the behavior of battery-constrained IoT devices, this is barely considered in the tools covered. A better integration of power models, for example using simulators built for this purpose [32], would be an important next step for virtual test environments.

4.3 Network

The environments included in our comparison contain a mix of different methods to model the network. This includes two systems that do not even include the ability to specify a network topology, seven systems that support traffic shaping (usually via tc and NetEm), as well as full network simulation (four systems) and software defined networking (five systems). The scalability to large networks that need to be realized on multiple execution nodes is possible in almost all test environments that can distribute nodes. Network distributability only seems to still be a challenge when network simulators are used. Fault injection is an important feature for IoT testing, but dedicated support for defining and executing specific failure scenarios is missing from many IoT test environments.

4.4 Domain Environment

Despite the tight integration of distributed IoT systems with their environment, support for the simulation of domain environments is missing from all testing tools included in our comparison. Accordingly, system developers have to resort to expensive and time-consuming field testing, when they want to test the interaction of IoT systems with their environment. While some environmental factors can be integrated in the testing by feeding applications recorded streams of sensor data, this integration is naturally limited and cannot model the manipulation of the environment by IoT systems.

5 Research Outlook

While many tools exist that tackle the problem of testing distributed IoT systems, there are still important open challenges.

Research Gap. Currently, there is limited support for assessing key system requirements such as high resilience and low energy consumption. However, the biggest gap in our view is the missing integration with domain environment simulations. This integration is particularly important for IoT systems, because the tight coupling and interaction with the environment is a fundamental property of the Internet of Things. The integration of domain environment simulations like traffic or infrastructure simulations would allow for meaningful and continuous testing of these interactions.

An Ideal IoT Test Environment. As we have derived the characteristics described in Sect. 2 from our understanding of the needs of a test environment for continuous testing, an ideal IoT test environment would fulfill all these characteristics. Specifically, an ideal test environment would:

- support testing on virtual and hardware nodes,
- model and monitor the energy consumption,
- include a network representation that allows complex network topologies and dynamic changes thereof,
- integrate domain environment simulations,
- enable the distribution of nodes, networks, and domain environments across multiple physical nodes to allow the testing of large-scale deployments,
- and also support fault injection on these three dimensions to evaluate the fault tolerance of the system under test.

The Marvis Testing Framework. We are working on a framework towards our vision of an ideal IoT test environment, called Marvis [3]. By combining virtual nodes (containers) with hardware nodes, Marvis offers capabilities for hybrid setups to combine the advantages of scalability and realism. Nodes can communicate via a simulated network realized by the network simulator ns-3.

A focus of our work is the integration of domain environment simulators to enable the continuous testing of the often intricate interactions between the IoT system and the environment. Currently, Marvis integrates the traffic simulator SUMO to demonstrate this, allowing the testing of interactions between the real software systems that run on the nodes and the movement of road users in the traffic simulation. This integration is bidirectional, meaning both, the change of the movement of road users by the applications under test, and the change of connectivity in the network simulation by the traffic simulation is possible.

Besides this, Marvis also offers fault injection in the three feature dimensions: It is possible to inject faults in the nodes (e.g., start and stop nodes, or execute commands), the network simulation (e.g., connect or disconnect nodes, change network parameters like delay), or the domain-specific simulation (e.g., changing speed of vehicles).

6 Related Work

Testing has been recognized as an important topic in IoT systems research since its beginning. Consequently, several related works provide an overview of testing research, environments and frameworks.

Tonneau et al. [31] presented an extensive work focusing on the question of choosing the right wireless sensor network testing platform for specific environment characteristics in 2015. It is the only related work in which all presented testbeds consist of devices carrying real sensors – no platforms were presented that only simulate or emulate devices under test. Tonneau et al. considered seven platform features: experimentation, scale, repeatability, mobility, virtualization, federation, and heterogeneity.

Dias et al. [8] identified the motivation and challenges of testing planetary-scale, heterogeneous IoT applications and devices. Surveyed testing tools were chosen with no specific properties in mind, making 16 IoT testing platforms that were available in 2018 part of the survey. Tools were compared based on ten properties, including supported IoT layers, test level, test method, supported platforms, and scope (market/academic). The authors conclude that further research and development in the area of IoT testing is necessary, given the criticality of many IoT systems and the challenges of testing them.

A journal article from the same year by Chernyshev et al. [5] discusses the state of IoT research, simulators and testbeds. They defined a set of relevant research topics, including eight goals for the IoT. Furthermore, they performed a comparative study of nine simulation tools, categorized by the scope of coverage of the IoT architecture layers, as well as a comparison of three large-scale IoT hardware testbeds. They identified three open challenges concerning IoT testing: A lack of support for common IoT communication standards, a lack of end-to-end service simulation across all IoT layers, and a large discrepancy between simulator and real-world test results.

Patel et al. [25] compared a total of 26 simulators, emulators, and physical testbeds for the IoT. The authors discussed these groups of test environments independently from each other on characteristics such as scope, scale, and security measures. While there is no specific survey system or selection method given, the comparison is followed by a short analysis of the usage of simulators, emulators, and physical testbeds in the different stages of software development.

Bures et al. [4] performed a systematic mapping study on interoperability and integration testing of IoT systems. Rather than comparing specific tools, they analyzed 115 out of 803 identified papers in the general area of IoT. The literature study was guided by seven research questions regarding research trends, researchers, publication media, topics, challenges, and limitations mentioned in the surveyed works. They conclude that there is a need for more specific testing methods for IoT systems.

7 Summary

This paper presented the current state of the art in continuous testing of distributed IoT systems. Specifically, we described desirable characteristics for test environments in this context and compared IoT test environments that allow to run system code on virtual nodes. Many solutions have been put forward, implementing various approaches to providing execution hosts and realizing network conditions. However, no currently available solution provides support for domain simulations, even though IoT systems form cyber-physical systems that make sense of and interact with their surroundings.

We believe that systems that monitor and affect the real world should be tested comprehensively, especially in critical application domains such as traffic management, patient monitoring, and manufacturing. Future work should therefore focus on providing comprehensive test environments, including simulation of domains and modeling of system characteristics such as energy consumption.

References

1. Adjih, C., et al.: FIT IoT-LAB: a large scale open experimental IoT testbed. In: 2015 IEEE 2nd World Forum on Internet of Things (WF-IoT). IEEE (2015)
2. Behnke, I., Thamsen, L., Kao, O.: HéCtor: a framework for testing IoT applications across heterogeneous DGE and cloud testbeds. In: 12th International Conference on Utility and Cloud Computing Companion. ACM (2019)
3. Beilharz, J., et al.: Towards a staging environment for the Internet of Things. In: 2021 IEEE International Conference on Pervasive Computing and Communications (PerCom Workshops). IEEE (2021)
4. Bures, M., et al.: Interoperability and integration testing methods for IoT systems: a systematic mapping study. In: de Boer, F., Cerone, A. (eds.) SEFM 2020. LNCS, vol. 12310, pp. 93–112. Springer, Cham (2020). https://doi.org/10.1007/978-3-030-58768-0_6
5. Chernyshev, M., Baig, Z., Bello, O., Zeadally, S.: Internet of things (IoT): research, simulators, and testbeds. IEEE Internet Things J. **5**, 1637–1647 (2017)

6. Coutinho, A., Greve, F., Prazeres, C., Cardoso, J.: Fogbed: a rapid-prototyping emulation environment for fog computing. In: 2018 IEEE International Conference on Communications (ICC). IEEE (2018)
7. Dastjerdi, A.V., Buyya, R.: Fog computing: helping the Internet of Things realize its potential. Computer **49**, 112–116 (2016)
8. Dias, J.P., Couto, F., Paiva, A.C., Ferreira, H.S.: A brief overview of existing tools for testing the Internet-of-Things. In: 2018 IEEE International Conference on Software Testing, Verification and Validation Workshops (ICSTW). IEEE (2018)
9. Geldenhuys, M.K., Will, J., Pfister, B., Haug, M., Scharmann, A., Thamsen, L.: Dependable IoT data stream processing for monitoring and control of urban infrastructures. In: IEEE International Conference on Cloud Engineering. IEEE (2021)
10. Gontarska, K., Wrazen, W., Beilharz, J., Schmid, R., Thamsen, L., Polze, A.: Predicting medical interventions from vital parameters: towards a decision support system for remote patient monitoring. In: Tucker, A., Henriques Abreu, P., Cardoso, J., Pereira Rodrigues, P., Riaño, D. (eds.) AIME 2021. LNCS (LNAI), vol. 12721, pp. 293–297. Springer, Cham (2021). https://doi.org/10.1007/978-3-030-77211-6_33
11. Gupta, H., Vahid Dastjerdi, A., Ghosh, S.K., Buyya, R.: iFogSim: a toolkit for modeling and simulation of resource management techniques in the Internet of Things, Edge and Fog computing environments. Pract. Exp. Softw. **47**, 1275–1296 (2017)
12. Hasenburg, J., Grambow, M., Bermbach, D.: Mockfog 2.0: automated execution of fog application experiments in the cloud. IEEE Trans. Cloud Comput. (2021)
13. Kang, H.S., et al.: Smart manufacturing: Past research, present findings, and future directions. Int. J. Precis. Eng. Manuf.-Green Technol. (2016)
14. Keahey, K., et al.: Lessons learned from the Chameleon testbed. In: 2020 USENIX Annual Technical Conference (USENIX ATC 2020) (2020)
15. Kim, H., Ahmad, A., Hwang, J., Baqa, H., Le Gall, F., Ortega, M., Song, J.: IoT-TaaS: towards a prospective IoT testing framework. IEEE Access (2018)
16. Ly-Trong, N., Dang-Le-Bao, C., Huynh-Van, D., Le-Trung, Q.: UiTiOt v3: a hybrid testbed for evaluation of large-scale IoT networks. In: 9th International Symposium on Information and Communication Technology. ACM (2018)
17. Mahadevan, P., Rodriguez, A., Becker, D., Vahdat, A.: MobiNet: a scalable emulation infrastructure for ad hoc and wireless networks. ACM SIGMOBILE Mobile Comput. Commun. Rev. **10**, 26–37 (2006)
18. Malasinghe, L.P., Ramzan, N., Dahal, K.: Remote patient monitoring: a comprehensive study. J. Amb. Intell. Human. Comput. **10**, 57–76 (2019)
19. Masek, P., et al.: A harmonized perspective on transportation management in smart cities: the novel IoT-driven environment for road traffic modeling. Sensors (2016)
20. Mayer, R., Graser, L., Gupta, H., Saurez, E., Ramachandran, U.: EmuFog: extensible and scalable emulation of large-scale fog computing infrastructures. In: 2017 IEEE Fog World Congress (FWC). IEEE (2017)
21. Miyachi, T., Chinen, K.i., Shinoda, Y.: StarBED and SpringOS: large-scale general purpose network testbed and supporting software. In: 1st International Conference on Performance Evaluation Methodolgies and Tools. ACM (2006)
22. Mohammadi, M., Al-Fuqaha, A.: Enabling cognitive smart cities using big data and machine learning: approaches and challenges. IEEE Commun. Mag. **56**, 94–101 (2018)

23. Mäkinen, A., Jiménez, J., Morabito, R.: ELIoT: design of an emulated IoT platform. In: 2017 IEEE 28th Annual International Symposium on Personal, Indoor, and Mobile Radio Communications (PIMRC). IEEE (2017)
24. Nikolaidis, F., Marazakis, M., Bilas, A.: IOTier: a virtual testbed to evaluate systems for IoT environments. In: 2021 IEEE/ACM 21st International Symposium on Cluster, Cloud and Internet Computing (CCGrid). IEEE (2021)
25. Patel, N.D., Mehtre, B.M., Wankar, R.: Simulators, emulators, and test-beds for internet of things: A comparison. In: 2019 Third International Conference on I-SMAC (IoT in Social, Mobile, Analytics and Cloud). IEEE (2019)
26. Petersen, E., Cotto, G., To, M.A.: Dockemu 2.0: evolution of a network emulation tool. In: 2019 IEEE 39th Central America and Panama Convention. IEEE (2019)
27. Ramprasad, B., Fokaefs, M., Mukherjee, J., Litoiu, M.: EMU-IoT - a virtual Internet of Things lab. In: 2019 IEEE International Conference on Autonomic Computing (ICAC). IEEE (2019)
28. Sarzyniec, L., Buchert, T., Jeanvoine, E., Nussbaum, L.: Design and evaluation of a virtual experimental environment for distributed systems. In: 21st Euromicro International Conference on Parallel, Distributed, and Network-Based Processing. IEEE (2013)
29. Sendorek, J., Szydlo, T., Brzoza-Woch, R.: Software-defined virtual testbed for IoT systems. Wireless Commun. Mobile Comput. **2018**, 1–11 (2018)
30. Symeonides, M., Georgiou, Z., Trihinas, D., Pallis, G., Dikaiakos, M.D.: Fogify: a fog computing emulation framework. In: 2020 IEEE/ACM Symposium on Edge Computing (SEC). IEEE (2020)
31. Tonneau, A.S., Mitton, N., Vandaele, J.: How to choose an experimentation platform for wireless sensor networks? A survey on static and mobile wireless sensor network experimentation facilities. Ad Hoc Networks (2015)
32. Wiesner, P., Thamsen, L.: LEAF: Simulating large energy-aware fog computing environments. In: 2021 IEEE 5th International Conference on Fog and Edge Computing (ICFEC). IEEE (2021)
33. Worstcase: Blockade (2021). https://github.com/worstcase/blockade
34. Zeng, X., Garg, S.K., Strazdins, P., Jayaraman, P.P., Georgakopoulos, D., Ranjan, R.: IOTSim: a simulator for analysing IoT applications. J. Syst. Architect. **72**, 93–107 (2017)
35. Zeng, Y., Chao, M., Stoleru, R.: EmuEdge: a hybrid emulator for reproducible and realistic edge computing experiments. In: 2019 IEEE International Conference on Fog Computing (ICFC). IEEE (2019)
36. Zhao, Y., et al.: Continuous monitoring of train parameters using IoT sensor and edge computing. IEEE Sens. J. **21**, 15458–15488 (2021)
37. Zhou, J., Ji, Z., Bagrodia, R.L.: TWINE: a hybrid emulation testbed for wireless networks and applications. In: INFOCOM, vol. 6. Citeseer (2006)
38. Zhou, J., Ji, Z., Varshney, M., Xu, Z., Yang, Y., Marina, M., Bagrodia, R.: WHYNET: a hybrid testbed for large-scale, heterogeneous and adaptive wireless networks. In: 1st International Workshop on Wireless Network Testbeds, Experimental Evaluation & Characterization. ACM (2006)

Towards Scalable Blockchains Using Service-Oriented Architectures

Ali Dorri[1($\boxtimes$)], Raja Jurdak[1], Amin Beheshti[2], and Alistair Barros[3]

[1] School of Computer Science, Queensland Univeristy of Technology,
Brisbane, Australia
{ali.dorri,r.jurdak}@qut.edu.au
[2] School of Computer Science, Macquarie University, Sydney, Australia
amin.beheshti@mq.edu.au
[3] School of Information Systems, Queensland Univeristy of Technology,
Brisbane, Australia
alistair.barros@qut.edu.au

Abstract. In recent years, blockchain applications beyond cryptocurrency has received tremendous attention due to its salient features which includes distributed management, security, anonymity, and immutability. However, conventional blockchains suffer from lack of scalability, high complexity, privacy, and governance. In this paper, we study the existing solutions introduced to address these limitations. We categorize these solutions into four groups which are: i) grouping nodes where the participating nodes are formed into smaller groups, ii) side channels where selected nodes form a child ledger attached to the main ledger to communicate privately, iii) optimized consensus algorithms that aim to reduce the overheads associated with committing new blocks, and iv) Blockchain-as-a-Service (BaaS) that employ service computing concepts and offload the blockchain management overheads to the cloud. A detailed discussion on BaaS is proposed along with a study of the existing cloud architectures. We elaborate on the advantages of employing blockchain to address challenges in service computing such as service recommendation. Finally, we discuss future research directions.

Keywords: Blockchain · Service computing · Scalability

1 Introduction

Blockchain technology has received tremendous attention due to its salient features which includes distributed management, trust, anonymity, and immutability. Blockchain can be perceived as a distributed tamper-resistant database that is jointly managed by all the participating nodes. The database stores the communications between the participating nodes that is known as transactions. The distributed nature of the blockchain allows any participating node to store data by forming a block. Such nodes are known as validators. To ensure consistency

© Springer Nature Switzerland AG 2022
H. Hacid et al. (Eds.): ICSOC 2021 Workshops, LNCS 13236, pp. 351–362, 2022.
https://doi.org/10.1007/978-3-031-14135-5_31

and increase the network security, the validators must follow a consensus algorithm. Consensus algorithms are distributed leader selection algorithms that randomly selects one validator as the leader whose block is committed in the blockchain. The security of the blockchain is largely tied to the randomness and unpredictability of the leader selection of the consensus algorithm.

Blockchain initially was introduced in Bitcoin [20] and since then has been widely applied in various other cryptocurrencies. In recent years, blockchain attracted attention beyond monetary applications such as the Internet of Things (IoT) [21]. IoT is the collection of millions of sensors/actuators that sense the environment, process the captured data, and act accordingly. The current IoT ecosystem employs a centralized brokered communication model where all communication, authentication, and authorization is happening through the central controllers which suffers from single-point-of-failure. IoT devices collect a huge volume of privacy-sensitive information which are stored and processed by central controllers which in turn risks the privacy of the users. IoT devices come from various manufacturers with low or no security configurations which raises security concerns. The distributed nature of the blockchain along with its anonymity and transparency features make it a plausible solution to address the outlined limitations in IoT [13].

Despite significant advantages applying blockchain in IoT is not straightforward and involves the following challenges: i) scale and overheads: in conventional blockchains transactions are broadcast to and verified by all participants which incurs significant bandwidth and processing overhead. Additionally, the conventional consensus algorithms require the validator to solve a computationally demanding puzzle. These features increase the blockchain management overhead and thus reduce its scalability, ii) complexity: in monetary applications blockchain is employed only for coin exchange, however, IoT demands various services, e.g., access control or smart contracts which increases the complexity in developing IoT-based blockchains, iii) privacy: blockchain is an append-only database where data is anonymized by using a Public Key (PK) as the identity of the transaction generator. In IoT-based blockchain, a huge volume of personalized information about the user is permanently stored in the blockchain. Applying machine learning algorithms to deanonymize a user will risk user privacy [16], and iv) Governance: blockchain aims to establish trust among participants without reliance on trusted third parties (TTPs) that makes governance challenging. However, IoT involves many applications that require the government oversight, e.g., supply chain.

In recent years multiple solutions have been introduced to address the aforementioned limitations which includes sharding, hierarchical blockchain, and optimized consensus algorithms (see Sect. 3 for more details). Cloud service providers also started offering Blockchain-as-a-solution (BaaS) services [9,23]. In BaaS the cloud provider manages the blockchain infrastructure where the end-users connect and utilize the service as in conventional services, such as Software-as-a-Service (SaaS) or Plateform-as-a-Service (PaaS).

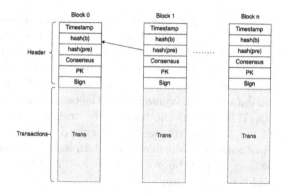

Fig. 1. A high-level view of blockchain structure.

This paper studies the existing solutions to address the blockchain limitations in IoT with a special focus on service computing-based solutions. We first explore the existing solutions introduced to address limitations of blockchain in IoT. Next, we discuss solutions that employ service computing and outline their limitations. Finally, we conclude the paper by outlining future research directions.

The rest of the paper is organized as follows: Sect. 2 provides a background discussion on blockchain. Section 3 outlines the existing solutions to increase the blockchain scalability in IoT. Section 4 explores the intersection of blockchain and service computing, while Sect. 5 outlines the future research directions. Finally, Sect. 6 concludes the paper and outlines the key findings.

2   Blockchain Technology: An Overview

This section outlines the fundamental concepts of the blockchain technology.

Blockchain is a distributed database that stores data in the form of blocks. Unlike conventional databases where a central node manages writes to the database, in blockchain all participating nodes jointly manage the database and decide on the data to be stored, i.e., committed on the ledger. Each node in the network, known as validator, may choose to commit new blocks into the blockchain that involves following a consensus algorithm. The consensus algorithm ensures randomness and unpredictability in selecting the validator of the blocks that protects against malicious activities and ensures consistency of the database among all participants.

Figure 1 shows the two core fields of a block in blockchain, namely the header and transactions, noting that blocks can include additional fields depending on implementation requirements. The header includes a timestamp which is the time when the current block was generated. *hash(b)* is the hash of the content of the current block. *hash(pre)* is the hash of the previous block that creates a chained ledger of blocks and ensures blockchain immutability. Modifying the content of a block alters its corresponding hash which will not match with the hash stored in the next block in the ledger. *consensus* field includes information about the consensus algorithm (detailed later in this section). Finally, *PK* and *Sign* are the PK and signature of the validator that generated the block.

In blockchain, communications between the participating nodes are known as transactions, which are stored in the *transactions* part. The transactions are essentially the main data that are committed in the blockchain. Each transaction includes *timestamp*, *T_ID* that is the hash of the transaction content, *P_T_ID* that is the hash of the previous transaction generated by the same node (or in the same ledger), *metadata* which is additional data included in the transaction, e.g., hash of the data exchanged between IoT devices, and *PK* and *Sign* are the PK and signature of the transaction generator.

Blockchain achieves distributed trust, transparency, and immutability. Blockchain establishes a trusted network over untrusted participants benefiting from the consensus algorithm where the nodes achieve agreement over the data on the ledger without relying on central controllers. The participating nodes have full visibility on the data stored on the ledger which is essential for verifying the transactions and thus ensuring blockchain security. This in turn introduces high level of transparency.

In 2014 Ethereum blockchain [24] introduced the concept of distributed applications (DApp) and smart contracts that is a program run distributively on blockchain. To compensate the computational resources spent by the blockchain nodes, the contract generator shall pay a fee, known as gas, that is paid for running the contract. The immutability of the blockchain makes it impossible to modify the content of the smart contract, thus the code written in a contract is the law which is enforced by blockchain. This features enables smart contracts to replace the third parties.

Blockchain is categorized in two main groups based on the read/write permissions of the underlying nodes: i) permissionless: where any node may join the blockchain and participate as a validator by verifying and committing new transactions and blocks, and ii) permissioned: where only authorized nodes may join the blockchain. Nodes may have different read/write permissions, thus only selected nodes are permitted to function as validator. In a permissioned blockchain, there is a degree of trust among the participating nodes given that authorized nodes join the blockchain.

Having discussed the fundamental concepts of the blockchain, we next study the limitations of blockchain in IoT.

3   Blockchain Limitations in Enterprise IoT

Blockchain has received tremendous attention in non-monetary applications including the Internet of Things (IoT) due to its salient features. In [6] the authors introduced a blockchain solution to secure communications in smart cities. The authors in [19] employed blockchain to address the trust and privacy limitations in supply chain and introduce traceability. In [15] the authors employed blockchain to facilitate trading energy among participating nodes in the smart grid. Utilizing blockchain, trusted third parties can be eliminated that increases the benefit of the end users.

Despite its significant advantages, applying blockchain in IoT is not straightforward and involves challenges which includes scale and overheads, complexity, privacy, and governance (see Sect. 1). Various works in the literature are introduced to optimized the blockchain for IoT. The existing methods that aim to address the outlined limitations can be categorized as solutions to reduce overhead by grouping the nodes, side channels, optimized consensus algorithm, and Blockchain-as-a-Service (BaaS) solutions which are discussed in greater detail in the rest of this section.

Grouping Nodes: Blockchain broadcasts all transactions which then needs to be verified by the participating nodes. However, this incurs significant bandwidth and computational overhead. To address this challenge, most of the existing works move from a purely distributed system to a decentralized blockchain where selected nodes only participate in managing the ledger. We categorize these solutions into three main groups and detailed below:

- Hierarchical: Most of the existing works propose to divide the blockchain in different layers, i.e. hierarchies. The nodes in each hierarchy manage a unique blockchain by broadcasting transactions and committing blocks. To connect blockchains in different hierarchies, hash of each blockchain is stored in the upper layer blockchain. This protects immutability of the blockchains in hierarchies. In [7] the authors introduced a hierarchical method to share and analyze data of smart vehicles. Federated learning is employed to analyze the data in the edge of the network which is then processed in the blockchain.
- Sharding: Sharding refers to dividing the network into smaller groups, i.e. shards, where the transactions of the participating nodes in each shard is managed within the shard. Intra-shard transactions are very limited and involve a long delay. Sharding increases the blockchain scalability linearly, i.e., if n shards are introduced, the scalability of the blockchain improves n times. In [10] the authors employed sharding to improve blockchain scalability. Optimized Byzantine consensus algorithm is introduced to improve the throughput in each shard. An efficient shard formation algorithm based on Intel SGX hardware is proposed to improve the performance while allocating a node to a particular shard.
- Clustering: The participating nodes are grouped into clusters and for each cluster a node is selected as Cluster Head (CH). The cluster members only communicate with the CH, while the CHs jointly manage the blockchain. In [14] the authors introduced a cluster-based optimization method for blockchain. CHs manage access control lists to manage access to the cluster members.

The reduced number of transactions broadcast among all participating nodes in hierarchical and sharding methods enhances the user privacy as fewer information about the user is publicly available to the nodes. Although the outlined methods improve scalability and privacy to some degree, complexity is not improved and even these methods may add extra complexity. Sharding methods make intra-shard communication complicated. In hierarchical methods, creating the hierarchies and connecting them creates further complication.

Side Chains: A side chain, also known as side channel, is a deviation from the main blockchain that is established between a group of small nodes, usually two nodes. Let us explain side chains using an example scenario. Assume node A is buying energy from node B in a peer-to-peer energy market. Node A wishes to pay the energy price to node B as they receive energy which involves micro-payments, i.e., paying small amount of money. In blockchain, the transaction generator has to pay a small fee, known as transaction fee, to the validator that commits their transaction to the blockchain. This makes micro-payments challenging in blockchain as the total transaction fee will be expensive. To address this challenge, A and B create a side chain. Initially, A locks a particular amount of money say x in the main blockchain and then generates the side chain that includes A and B. Inside the side channel, A transfers the small amounts of money to B. The total amount of payments shall not exceed x. At the end of the energy trading, a transaction is generated that reflects the total amount of money paid to B and the money left for A. Using the outlined steps, the volume of transactions in the blockchain is reduced. The privacy of the users is also increased to some degree as details of the transaction exchanges are hidden from the main blockchain nodes. However, side chains do not reduce the complexity of the blockchain.

Optimized Consensus Algorithm: Consensus algorithms are core to blockchain security. Proof of Work (PoW), that is the underlying consensus algorithm of Bitcoin, is the most widely applied consensus algorithm. In PoW, the validators must find a nonce value in a way that the hash of the nonce value along with the block content starts with a particular number of leading zeros. The latter is defined based on the blockchain difficulty and ensures that roughly one block is generated per 10 min. PoW demands significant computational power from the underlying nodes which in turn consumes significant energy. Additionally, PoW throughput, i.e., the total number of transactions that can be committed to the blockchain per second, is very limited (about 7 transactions) while IoT consists of millions of devices that generate huge volume of transactions [4,22].

In recent years, various optimized consensus algorithms have been introduced. Generally, the consensus algorithms for permissioned blockchains achieve less overhead and higher throughput as compared with the permissionless blockchain due to the level of trust among the participants. Proof of Elapsed Time (POET) is a consensus algorithm that introduces time based consensus algorithm [8]. All validators are assumed to be equipped with Intel CPUs with Trusted Execution Environment (TEE). Before committing a new block, each validator has to sleep for a random period of time defined by the TEE. Although POET significantly reduces the overhead associated with generating new blocks, it relies on Intel CPUs and TEE. In [14] the authors introduced a Distributed Time-based Consensus (DTC) algorithm which conceptually is similar to POET, yet is hardware-agnostic. DTC employs neighbor monitoring to ensure each node has waited for a random period of time before committing a new block. Additionally, DTC limits the number of blocks each validator can generate per a particular time-period to protect against malicious nodes flooding the network with new blocks.

Ethereum [24] introduced Proof of Stake (PoS) blockchain where the chance of a potential validator to be selected as the validator of the next block is proportional to the amount of assets they lock in the blockchain. Similarly, Proof of Authority (PoA) [11] uses the reputation of a validator as a factor to determine the chance of the node to be selected as the validator of the next block.

In [12] the authors introduced Tree-chain a novel consensus algorithm that bases the validator selection on hash function output. Tree-chain moves away from a single ledger by introducing multiple parallel ledgers where each validator is allocated and commits transactions to a single ledger. The transactions in each ledger have the same pattern, known as consensus code range (CCR), which is the most significant characters of the hash function output. Two level of randomization is introduced which are: i) the validators are randomly allocated to CCRs based on the hash of their PK, and ii) transactions are randomly allocated to a validator based on the hash of the transaction.

Despite various efforts, the existing consensus algorithms suffer from: i) lack of throughput management where the blockchain throughput can be adjusted based on the load in the network, ii) resource consumption, and iii) delay in committing transactions into the ledger.

Having discussed three fundamental methods to optimize blockchain in IoT, we next dig deeper into blockchain optimization using service computing and BaaS.

4   When Service Computing Meets Blockchain

Service computing, also referred to as cloud computing in the rest of this paper, has grown significantly in recent years. Traditionally, an organization must maintain physical servers on-premises to offer services that significantly increases cost and complexity. To address these challenges, cloud services emerged where the service provider maintains the physical servers that allows the organizations to focus only on the functions that is important to them [5]. Depending on the level of abstraction, cloud services are categorized in three groups as shown in Fig. 2: i) Infrastructure-as-a-service (IaaS): where the cloud provider manages the hardware, network communications between different nodes, and the virtualization of the Operating System (OS) while customer manages the rest of the tasks, ii) Platform-as-a-service (PaaS): where in addition to what is managed in IaaS, the cloud manages the OS and the runtimes, and iii) Software-as-a-Service (SaaS): where the cloud provider manages applications and data as well as OS and hardware.

In recent years convergence of blockchain and cloud computing has received significant attention. The existing works mainly focus on two aspects: blockchain for service computing and service computing for blockchain.

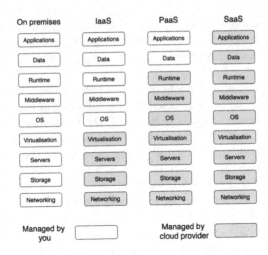

Fig. 2. A summary of cloud architectures.

4.1   Blockchain for Service Computing

Blockchain has been employed by some researchers as a tool to address the challenges in service computing. The main challenges are discussed below.

Security and Privacy: Security and privacy in cloud computing are critical given that huge volume of privacy-sensitive information is captured and monitored. Unauthorized access to the data compromises the privacy of the users. Blockchain anonymity and auditability facilitate addressing the outlined challenges. Blockchain-based access control mechanisms can be employed that allow decentralized management of data access. Any access to the data is recorded in the blockchain which in turn introduces higher security and auditability. Blockchain participants are known by changeable PKs which introduces a level of anonymity.

Service Discovery and Recommendation: Various service providers offer different services with various conditions which makes it challenging for the customers to identify a proper service. Blockchain can be employed as a distributed database where services offered by all service providers are recorded. This enables the clients to query only one place to find a proper service. Blockchain users can rate each service which will be permanently stored in the ledger thanks to the blockchain immutability and transparency. The recommendation given to a particular service can be employed as a metric by other users to decide on which service to opt.

4.2   Service Computing for Blockchain Scalability

By allowing a cloud service provider to manage the blockchain, service computing can reduce the cost and complexity in offering blockchain applications. This lead to the introduction of Blockchain-as-a-Service (BaaS). In [18] the authors defined four layers for BaaS architectures which are:

- Blockchain Infrastructure Layer: This layer facilities the communications with the computational resources that can be physical machines, virtual machines, or Docker containers. Peer-to-peer machines are established to facilitate communications.
- Blockchain Framework Layer: This layer is the core blockchain layer where one of the existing blockchain solutions, such as Ethereum, Hyperledger, Corda, etc., is employed.
- Middleware Layer: This layer consists of a variety of tools that facilitate the management and monitoring of the ledger that includes monitoring, resource scheduling, data analysis, and access control.
- Application Layer: This layer deploys applications on top of the existing blockchain infrastructure. Blockchain has received significant attention to be applied in various application domains including smart home, smart grid, health data sharing, etc.

BaaS builds on top of the outlined four-layered architecture to shift the complexity and overhead in managing the ledger from the end-users to the cloud service providers. This enables the end users to employ blockchain as a software or platform without being concerned about the management of the ledger. Various cloud service providers offer BaaS. We discuss some of the most well-known BaaS providers below.

Microsoft has released Azure Blockchain Workbench [2] that facilitates easy development and benchmarking of blockchain platforms. The blockchain workbench can be integrated with other Microsoft services that facilities management of the ledger, e.g., Active Directory can be employed to manage user accounts and logins. Azure blockchain enables businesses to integrate their business with blockchain using REST-based and message-based API for client development and system-to-system integration. IoT devices and data flow can be integrated with Azure using the provided APIs [3] which in turn facilities developing blockchain-based IoT.

The IBM blockchain platform [17] facilitates development of enterprise blockchain solutions. IBM manages the hyper ledger fabric blockchain which is part of the Hyperledger project that moves toward open-sourced blockchains. IBM blockchain supports multi-cloud environment that allows users from any third party cloud provider to join the blockchain. IBM employes Hedera Hashgraph consensus mechanism along with Hyperledger fabric to facilitate blockchain interoperability that allows multiple public and private blockchains to communicate and exchange information. IBM built a wide range of applications using their cloud service which includes supply chian, oil and gas, healthcare, retail and consumer goods, manufacturing, and media and entertainment.

Amazon Web Services (AWS) offers centralized and decentralized BaaS solutions [1]. In a centralized setting, known as Amazon Quantum Ledger Database (QLDB), a central read-only database is managed that offers immutability and cryptographically verifiable transactions. The decentralized setting allows untrusted participants to communicate in a peer-to-peer maner using conventional blockchains. AWS manages 25% of all Ethereum workloads and also supports integration with Hyperledger.

BaaS solutions address the limitations of the conventional blockchains which includes scale and overhead, governance, and complexity. The cloud service provider manages the blockchain with powerful devices which address the scalability problem. The cloud providers also enable authorized nodes to monitor the blockchain through various tools that introduces governance. Finally, the cloud providers manage the underlying blockchain network, while the end users only utilize the blockchain similar to conventional cloud services which in turn reduces complexity. Blockchain designers benefit from huge support and documentation provided by cloud providers. Cloud providers also offer learning resources that facilitates developing blockchain-based solutions on top of their BaaS.

Having discussed the benefits of BaaS, we next discuss the limitations of such solutions and outline future research directions.

5 Future Research Directions

BaaS solutions offer greater flexibility and facilitate adaption of blockchain technology in large scale networks. However, in BaaS the cloud provider remains a central point of trust, i.e., the participating nodes shall trust to the cloud provider. The cloud provider essentially has full control over the blockchain network. This deviates from the core blockchain concept, i.e., distributed management. A truly distributed blockchain shall run by different users without any assumption about the trust between participating nodes. In BaaS, on the other hand, a central cloud provider manages the blockchain nodes. Although nodes might be located in different geographical locations, they are managed and owned by the same entity. This also makes protecting the user privacy challenging as the cloud provider will have oversight over the blockchain.

As a future research direction, one can consider developing a modular blockchain architecture that allows various services to be plugged into the blockchain core tasks, i.e., transaction/block generation and verification. Decoupling application-specific tasks from blockchain and pushing those to services, reduces the number of transactions need to be committed in the main ledger and thus increases blockchain scalability. The modular blockchain management allows any node to define a service to be plugged into other applications, or define an application using the already existing services in the blockchain.

Another interesting future research direction is to employ service computing to address interoperability challenges in conventional blockchain. Blockchains in different applications or developed by different organizations have their own transaction format which in turn makes it challenging for the transactions from one blockchain to be transferred to the another ledger. Service computing can be employed to facilitate transferring transactions between ledgers.

Blockchain can also be employed to address service computing challenges. One interesting future research direction is to enable blockchain-based service marketplaces where service providers index services and users can query to locate the proper service. This also allows various services to be combined and offered in the form of a single service.

6 Conclusion

Applying blockchain for non-monetary applications has received tremendous attention due to its salient features which includes, security, privacy, auditability, and distributed management. However, adopting blockchain in large scale networks, such as the Internet of Things (IoT), is not straightforward and involves various challenges including scalability, complexity, privacy, and governance. In this paper, we studied the existing solutions to address these limitations which includes grouping nodes, side chains, optimized consensus algorithms, and Blockchain-as-a-service (BaaS). We focus on the integration of service computing with blockchain. We first discussed the advantages of using blockchain in service computing, e.g., for service recommendation. Then, we studied BaaS and the existing cloud providers.

References

1. Blockchain on AWS: AWS blockchain (2021). https://aws.amazon.com/blockchain/. Accessed 20 Nov 2021
2. Azure Blockchain: Azure blockchain workbench (2021). https://azure.microsoft.com/en-au/features/blockchain-workbench. Accessed 20 Nov 2021
3. Azure IoT and Blockchain: Azure blockchain workbench-IoT benchmark (2021). https://github.com/Azure-Samples/blockchain/blob/master/blockchain-workbench/iot-integration-samples/ConfigureIoTDemo.md
4. Beheshti, A., et al.: iProcess: enabling IoT platforms in data-driven knowledge-intensive processes. In: Weske, M., Montali, M., Weber, I., vom Brocke, J. (eds.) BPM 2018. LNBIP, vol. 329, pp. 108–126. Springer, Cham (2018). https://doi.org/10.1007/978-3-319-98651-7_7
5. Beheshti, S., et al.: Process Analytics - Concepts and Techniques for Querying and Analyzing Process Data. Springer, Cham (2016). https://doi.org/10.1007/978-3-319-25037-3
6. Biswas, K., Muthukkumarasamy, V.: Securing smart cities using blockchain technology. In: 2016 IEEE 18th International Conference on High Performance Computing and Communications; IEEE 14th International Conference on Smart City; IEEE 2nd International Conference on Data Science and Systems (HPCC/SmartCity/DSS), pp. 1392–1393. IEEE (2016)
7. Chai, H., Leng, S., Chen, Y., Zhang, K.: A hierarchical blockchain-enabled federated learning algorithm for knowledge sharing in internet of vehicles. IEEE Trans. Intell. Transp. Syst. **22**(7), 3975–3986 (2020)
8. Corso, A.: Performance analysis of proof-of-elapsed-time (PoET) consensus in the sawtooth blockchain framework. Ph.D. thesis, University of Oregon (2019)
9. Dabbagh, M., Choo, K.R., Beheshti, A., Tahir, M., Safa, N.S.: A survey of empirical performance evaluation of permissioned blockchain platforms: challenges and opportunities. Comput. Secur. **100**, 102078 (2021)

10. Dang, H., Dinh, T.T.A., Loghin, D., Chang, E.C., Lin, Q., Ooi, B.C.: Towards scaling blockchain systems via sharding. In: Proceedings of the 2019 International Conference on Management of Data, pp. 123–140 (2019)
11. De Angelis, S., Aniello, L., Baldoni, R., Lombardi, F., Margheri, A., Sassone, V.: PBFT vs proof-of-authority: applying the CAP theorem to permissioned blockchain (2018)
12. Dorri, A., Jurdak, R.: Tree-chain: a lightweight consensus algorithm for IoT-based blockchains. In: 2021 IEEE International Conference on Blockchain and Cryptocurrency (ICBC), pp. 1–9. IEEE (2021)
13. Dorri, A., Kanhere, S., Jurdak, R.: Blockchain for Cyberphysical Systems. Artech House, Boston (2020)
14. Dorri, A., Kanhere, S.S., Jurdak, R., Gauravaram, P.: LSB: a lightweight scalable blockchain for IoT security and anonymity. J. Parallel Distrib. Comput. **134**, 180–197 (2019)
15. Dorri, A., Luo, F., Kanhere, S.S., Jurdak, R., Dong, Z.Y.: SPB: a secure private blockchain-based solution for distributed energy trading. IEEE Commun. Mag. **57**(7), 120–126 (2019)
16. Dorri, A., Roulin, C., Jurdak, R., Kanhere, S.S.: On the activity privacy of blockchain for IoT. In: 2019 IEEE 44th Conference on Local Computer Networks (LCN), pp. 258–261. IEEE (2019)
17. IBM Blockchain: IBM blockchain solution (2021). https://www.ibm.com/au-en/blockchain. Accessed 20 Nov 2021
18. Li, X., Zheng, Z., Dai, H.N.: When services computing meets blockchain: challenges and opportunities. J. Parallel Distrib. Comput. **150**, 1–14 (2021)
19. Malik, S., Dedeoglu, V., Kanhere, S.S., Jurdak, R.: TrustChain: trust management in blockchain and IoT supported supply chains. In: 2019 IEEE International Conference on Blockchain (Blockchain), pp. 184–193. IEEE (2019)
20. Nakamoto, S.: Bitcoin: a peer-to-peer electronic cash system. Decentralized Business Review, p. 21260 (2008)
21. Panarello, A., Tapas, N., Merlino, G., Longo, F., Puliafito, A.: Blockchain and IoT integration: a systematic survey. Sensors **18**(8), 2575 (2018)
22. Schiliro, F., et al.: iCOP: IoT-enabled policing processes. In: Liu, X., et al. (eds.) ICSOC 2018. LNCS, vol. 11434, pp. 447–452. Springer, Cham (2019). https://doi.org/10.1007/978-3-030-17642-6_42
23. Singh, J., Michels, J.D.: Blockchain as a service (BaaS): providers and trust. In: 2018 IEEE European Symposium on Security and Privacy Workshops (EuroS&PW), pp. 67–74. IEEE (2018)
24. Wood, G., et al.: Ethereum: a secure decentralised generalised transaction ledger. Ethereum Project Yellow Paper **151**(2014), 1–32 (2014)

Snapshot of Research Issues in Service Robots

Patrick C. K. Hung[1](✉), Farkhund Iqbal[2], Saiqa Aleem[2], and Laura Rafferty[1]

[1] Faculty of Business and IT, Ontario Tech University, Oshawa, ON, Canada
{patrick.hung,laura.rafferty}@ontariotechu.ca
[2] College of Technological Innovation, Zayed University, Abu Dhabi, UAE
{farkhund.iqbal,saiqa.aleem}@zu.ac.ae

Abstract. A service (social) robot is defined as the Internet of Things (IoT) consisting of a physical robot body that connects to one or more Cloud services to facilitate human-machine interaction activities to enhance the functionality of a traditional robot. Many studies found that anthropomorphic designs in robots resulted in greater user engagement. Humanoid service robots usually behave like natural social interaction partners for human users, with emotional features such as speech, gestures, and eye-gaze, referring to the users' cultural and social background. During the COVID-19 pandemic, service robots play a much more critical role in helping to safeguard people in many countries nowadays. This paper gives an overview of the research issues from technical and social-technical perspectives, especially in Human-Robot Interaction (HRI), emotional expression, and cybersecurity issues, with a case study of gamification and service robots.

1 Introduction

A service (social) robot consists of a physical robot hardware component to interact with humans, connected through a network infrastructure as a cyber-physical system supported with Cloud services, such as Softbank Robotics Pepper and Amazon Astro Robot. In tandem with the increasing sophistication of Artificial Intelligence (AI), social robots behave in some ways like humans by using their sensors and actuators, with features such as speech, gestures, movements, and eye-gaze. For example, a robot could use gesture, motion, color, and sound to express emotion (e.g., happy, calm, sad, angry, etc.). The research found that users are more open to anthropomorphic design due to the Uncanny Valley theory (Mori et al. 2012). The Uncanny Valley theory describes the disturbing effect of imperfect human likenesses that have dominated human-robot social interaction (Mathur and Reichling 2009). For example, Herse et al. (2018) show that it is much easier for an embodied humanoid robot with emotional expression to gain users' trust to release personal information than a disembodied interactive kiosk. Human-Robot Interaction (HRI) is a research area of understanding, designing, and evaluating robots for use by or with humans, respectively. The HRI interactions can be modeled in bilateral or multilateral relations by a top-down design approach as follows (Kanda et al. 2004):

(1) Develop situated modules for various scenarios;
(2) Define the necessary execution order of the situated modules with episode rules for sequential transition;

© Springer Nature Switzerland AG 2022
H. Hacid et al. (Eds.): ICSOC 2021 Workshops, LNCS 13236, pp. 363–376, 2022.
https://doi.org/10.1007/978-3-031-14135-5_32

(3) Add episode rules for reactive changes; and
(4) Modify implemented episode rules, and specify episode rules of negation to suppress execution of the situated modules for a particular long-term context.

A developer can program social robots using a high-level programming language such as Python and Java, an emulator application with blocky coding, or a choreograph-based tool with drag-and-drop features (Miller et al. 2018). In general, there are three types of automation in social robots:

(1) Hard Automation: Do a specific, highly repetitive task like iRobot vacuum;
(2) Programmable Automation: Do a variety of tasks above hard automation; and
(3) Autonomous (Independent): Make decisions based on the use of sensors and recognize faults to take corrective actions by the robot itself.

Table 1. Sample social robots and their operating systems

Vendor	Social robot	Operating system	Sources
ASUS	Zenbo	Android 10.0	https://zenbo.asus.com
Softbank	Pepper	Android NAOqi	https://pepper.generationrobots.com
Ubtech	Lynx Robot	Android 5.0	https://www.ubtrobot.com
Blue Frog robotics	Buddy	Android	https://buddytherobot.com
Ubtech	Walker	Android, ROS	https://assets-new.ubtrobot.com
Sanbot	Sanbot	Robot Operating System (ROS)	http://en.sanbot.com
Softbank robotics	Nao	Linux NAOqi	https://www.softbankrobotics.com
Sony	Aibo	Linux	https://www.sony.com
Hanson robotics	Sofia	Ubuntu Linux	https://www.hansonrobotics.com
Mayfield robotics	Kuri	ROS	https://www.crunchbase.com/organization/mayfield-robotics
Mji robotics	Tapia	Android	https://mjirobotics.co.jp/tapia-en/
Aeolus	Aeolus	Aeolus ROS	https://robots.nu/en/robot/aeolus-robot
PAL robotics	Reem	Ubuntu Linux	https://pal-robotics.com
Sharp	RoBoHon	Android	https://robohon.com
Slamtec	Athena	Linux, ROS	http://www.slamtec.com
RobotElf technologies	Robelf	Android	https://www.robelf.com

(continued)

Table 1. (*continued*)

Vendor	Social robot	Operating system	Sources
LuxAI	QTrobot	Ubuntu, ROS	https://luxai.com/
Ingen dynamics	Aido Robot	Android, Linux	https://aidorobot.com
Robot Temi	Temi	Android	https://www.robotemi.com

Emotions are an essential component of human cognition and behavior caused by an identifiable source, such as an event or seeing emotions in other people. Emotions are necessary to either promote or threaten the survival of different situations that people encounter. They prepare the body for behavioral responses, help in decision-making, and facilitate interpersonal interaction. A social robot may mimic how people display emotions as an interaction strategy to recognize emotion in the human and then reflect the emotion in response using computer vision (Bartneck et al. 2020). Developers and designers often use different ways to convey emotions through facial expressions or the robot's body language, such as body movements and prosody. Some social robots may also have an avatar with animated faces as Graphical User Interface (GUI) on the tablet face (screen), such as Misty Robotics Misty II and ASUS Zenbo. For example, Zenbo represents a type of social robot in terms of mid-size (height, length, and depth), features (e.g., facial expression, emotion detection, voice recognition, color expression, etc.), movement (e.g., head degree of freedom, etc.), and sensors (e.g., touch sensor, passive infrared sensor, sonar, gyroscope, etc.). Zenbo can also be controlled or interacted with voice commands and a virtual keyboard. Extant research has programmed robotic emotion and behavior in a completely autonomous mode or a semi-autonomous mode, in which the robot follows a predefined script by a human behind (Homburg 2018).

Robot Operating System (ROS) is a set of software libraries and tools for developing robot applications (ROS 2021). In addition to ROS, Linux and Android are still the most common Operating Systems (OS) the social robots adopted in the market. Referring to Table 1, we surveyed 44 social robots in the current market. While public documentation on OS versions is limited for many social robots on the market, our survey found that 19 are based on Linux and Android. Thus, most identified robot apps rely on the user-based permissions model (discretionary access control) according to the Linux Security kernel.

This paper gives an overview of the research issues from technical and social-technical perspectives, especially in HRI, emotional expression, and cybersecurity issues with case studies. The paper is organized as follows: we discuss emotional expression in social robots in Sect. 2; the cybersecurity issues of a social robot in Sect. 3; a case study in gamification is discussed in Sect. 4, and we give the conclusions and future works in Sect. 5.

2 Literature Review of Emotional Expressions

Building on this, "Emotional design" studies customers' personal hidden needs into specific products. For example, Kansei engineering is a tool that translates customers' feelings into concrete product parameters in the robotic engineering field and supports future product design (Schütte et al. 2004). Emotions can motivate and modulate user behaviors in interaction as a necessary component of human cognition and behavior. Emotions are an essential component of human cognition and behavior caused by an identifiable source, such as an event or seeing emotions in other people. Emotions are necessary to either promote or threaten the survival of different situations that people encounter. They prepare the body for behavioral responses, help in decision-making, and facilitate interpersonal interaction. Anger, sadness, and happiness are always classified as a set of core emotions. Emotion-aware (affective) computing aims to recognize, interpret, process, and simulate the emotional states of humans and respond with appropriate reactions (Lee et al. 2014; Liu et al. 2017). Prior research found that people's decisions might be influenced differently according to the emotional expressions by computing systems. For example, people may form more positive impressions of avatars when compared to human agents. Affective computing is an interface of interpreting the emotional state of humans and behaving appropriately in response to those emotions (Bartneck et al. 2020).

For social robots, emotion is one of the most critical design features because it needs to understand and express feelings to make them friendly and companionable. Many social robots have an avatar with animated faces on the tablet face (screen), such as Misty Robotics and Zenbo. For example, Zenbo has 24 emotional facial expressions. A social robot may mimic how people display emotions as an interaction strategy to recognize emotion in the human and then reflect the emotion in response by computer vision (Bartneck et al. 2020). Social robots are often designed to interpret and express human emotions driving their behavior. Developers and designers usually use different ways to convey emotions through facial expressions or the robot's body language, such as body movements and prosody. Altering the appearance, tone, and movement of a robot and user makes it easy to express emotions because it is simple to tell code what to do. Emotion models capture the user's emotional state, represent the robot's emotional state, and drive the robot's behavior. For example, the Ortony, Clore, and Collins (OCC) model specify 22 emotion categories based on balanced reactions to situations. Russel's two-dimensional (2D) space of arousal and valence captures a wide range of emotions on a 2D plane that still has sufficient expressive power for HRI. A three-dimensional (3D) continuous space consisting of pleasure, arousal, and dominance has been used on many social robots to model the emotional state, including Kismet (Bartneck et al. 2020).

Researchers have been actively working on a robot's emotional state to interact with a human. For example, Park et al. (2007) present an emotion expression system to support five emotion expression types (face, emoticon, sound, text, and gestures). A plurality of different sensors senses information about internal/external stimuli. Next, Kwon et al. (2007) present an emotion interaction system composed of the emotion recognition, generation, and expression components by facial expression, voice, language, gesture, and physiological signals. Read and Belpaeme (2012) explore non-linguistic utterances' potential and interpolate between 9 prototypical facial expressions with the dimensions that represent pleasure, arousal, and dominance. Then, Cohen et al. (2011) use a "facial robot" called iCat to show emotional expressions with dynamic body postures. Shih et al.

(2017) design a robot that ex-tracts the face and computes the facial image feature using Support Vector Machine (SVM) to classify the facial expression into different emotional states.

Further, Chew and Chua (2020) present a robot using its emotion recognition and body language automated to demonstrate the Chinese words, to increase learners' understanding and enhance their memory of the terms learned. Kita and Mita (2015) use Kinect for face tracking and the color of face changes representing the temperature environment to detect the emotion in a sympathetic, parasympathetic nervous system. Thus, Tielman et al. (2014) present a study of the role of the adaptive expression of gestures and emotion in robot-child interaction through voice, posture, whole-body poses, eye color, and gestures. Arora and Chaspari (2018) present two hybrid architectures for privacy-preserving emotion recognition from speech based on a Siamese neural network to retain emotion-dependent content and sup-press information related to a speaker's identity based on a publicly available Interactive Emotional Dyadic Motion Capture (IEMOCAP) dataset. Finally, Latif et al. (2020) discuss federated learning for speech emotion recognition using a publicly available dataset considering privacy concerns by involving multiple participants to learn a shared model without revealing their local data collaboratively. Due to the limited computing resources in a social robot, an emotion-aware social robot often transmits complex computation tasks to Cloud services with sensitive information.

3 Cybersecurity Issues in Robots

Cyber-physical systems integrate cyber and physical components that introduce new security threats beyond what a regular computing system may tackle. Thus, traditional access control policies and mechanisms are inadequate for cyber-physical systems. Most of the cybersecurity problems in robotics are due to the lack of awareness among software developers for robots (Apa and Cerrudo 2017). The software controlling robots need to be secured, which means that the methodologies, tools, and development frameworks used must be secured. For example, Vilches et al. (2018) present a Robot Security Framework (RSF) to perform systematic security assessments in robots from four main layers: physical, network, firmware, and application, but the RSF does not consider the cyber-physical spaces. First, the interaction of subjects and objects in the physical- and cyber spaces should be coordinated, constrained, and secured simultaneously. Secondly, subjects and objects may roam among different domains and types in the physical- and cyber-spaces. Thus, a convergent access control model is needed to react to the state changes of the cyber-physical spaces (Akhuseyinoglu and Joshi 2020). For example, Security-Enhanced Linux (SELinux) is an example of a convergent access control model for Linux built by the United States National Security Agency (NSA).

The International Organization for Standardization (ISO) specifies requirements and guidelines for the inherently safe design and protective measures for personal care robots. However, it only covers human-robot types of physical contact applications, but not on the data security and privacy perspective (ISO 2014). However, while social robots attract much new research, security and privacy issues are still thoroughly investigated. In this regard, Lee et al. (2011) reveal the importance of privacy-sensitive designs to foster better adoption of service robots, stressing the new privacy risks they bring to users. Next, Krupp et al. (2017) adopt an experimental approach to identify the different privacy categories and propose respective privacy enhancements. On the other hand, Portugal et al. (2017) analyze the potential security issues in the ROS concerning the different layers pertinent to the robotic network architecture and propose new hierarchical mechanisms for improved security. Lastly, Yousef et al. (2017) further present a security risk assessment and analysis of the specific PeopleBot mobile robot platform.

Security vulnerabilities in robots thus raise significant concerns, not only for manufacturers and programmers but also for those who interact with them. These social robots interact with their surroundings and acquire large amounts of data. This increases concerns for data security and privacy issues due to the wealth of data collected, stored, and processed. Obtaining sensitive information from social robots comes in many forms, such as personalized, location-specific, or user-centric information (MacDermott et al. 2020). Surveillance enabled by social robots enables greater observation and profiling of individuals via a collection of personal information or feelings through social bonding (Calo 2012). If this data is intercepted or redirected to a malicious user/system, this data could be used in a myriad of malicious means. Calo (2012) posits the question: '*privacy-friendly robots, an ethical responsibility of engineers?*' They suggest taking a middle ground, where engineers and regulators come together, and their rationalities are reconciled will help with future deployments. For example, Secure by Design, Default, and in Deployment (Doddson et al. 2020) is envisioned to formalize cyber security design, automate security controls, and streamline auditing to have more apparent transparency on data collection and usage, Robotic issues are not limited to one but to many aspects that could exploit any vulnerability/security gap to target robotic systems and applications (Yaacoub et al. 2021). Social robots are connected to a local network or the Internet through Wi-Fi or Ethernet to provide remote access and control. Security threats may result in completely preventing the access and control of the robot (Yousef et al. 2017; Fosch-Villaronga et al. 2021). Lack of secure networking renders the communication between robots/machines and humans insecure and prone to various attacks, such as Man-in-the-middle (MITM) attacks, eavesdropping, sniffing, and replay. The more functions are performed across interconnected systems and devices, the more opportunities for weaknesses in those systems to arise, and the higher the risk of system failures or malicious attacks (Michels and Walden 2018). Attackers can compromise the control of robots; such an attack on a social robot may affect the safety and well-being of people.

When worrying over the interception of data, it is important to consider and map all the potential avenues of data traversal. At the lowest level, we would consider the data on the device itself, and then we would map if it interacted with any gateway devices, Cloud-based storage, or Enterprise access. These security issues, coupled with the continued growth of the Internet of Things (IoT), present a much larger attack surface for attackers to exploit in their attempts to disrupt or gain unauthorized access to networks, systems, and data. Potential threats to the data include data interruption (deletion), exposure in the network, modification of data at rest and in transit, all of which would cause a privacy breach. A further issue is that data may be transmitted without encryption.

Further, Yousef et al. (2018) analyzes vulnerabilities discovered in social robots and the software used to interact with these platforms. Notably, with many social robots, if the robot server requires login credentials, the username information is sent in plain text without being encrypted over the network to the server. Thus, if there are any connection issues or delays, this connection would be vulnerable to a Man-in-the-Middle (MITM) attack. For example, suppose an attacker interrupts the connection between the server (robot) and the client for some time. In this case, it might look like a bad network connection without anything noticed or any action taken from the server and the client.

One additional security issue for robots is that they are often created and hardcoded with default user credentials. By default, the remote server is not configured to require the user (or client) to specify a username or password to connect to it. As a result, an attacker could potentially log in and connect to the robot when accessing the robot network and providing the robot's Internet Protocol (IP) address. In addition, most of the information sent from the server-side to the client is sent in plain text in many social robots. This would cause a lack of integrity due to the use of weak message authentication protocols that can be easily compromised, leading to the alteration of sensitive robotic data stored or in transit (Yaacoub et al. 2021).

Since robots rely on running software programs and applications to perform tasks, these programs are vulnerable to application attacks, such as Denial of Service (DoS) and Distributed Denial of Service (DDoS) attacks, along with the code execution and rootkit attacks. Also, physically, attackers could gain access by physically stealing the robot or accessing the data if there are insufficient authentication measures on the physical device. For example, they access the screen lock via pin code, pattern, or biometrics.

Humanoid service robots can be a rich source of sensitive data about individuals and environments. This data may assist in digital investigations, delivering additional information during a crime scene investigation. In digital forensics, digital evidence is collected from a digital crime scene and preserved for further analysis and examination by standards allowing it to be accepted and presented in law courts. While digital forensics consists of different areas, including mobile forensics, network forensics, cloud forensics, memory forensics, etc., we suspect that robot forensic analysis will require a robust and multifaceted approach due to the advanced capabilities of such devices.

Worryingly, the capabilities of digital forensics tools are increasingly becoming obsolete. A dramatic improvement in the efficiency of both tools and research processes are mandatory to cope with the newer technologies and diverse OS. Robots comprise different technological ecosystems, including hardware, firmware, and OS. Robot data can reside on multiple platforms and often across different locations. Recovering a whole data trail requires piecing elements from various devices and locations. The limited time for which forensically important data is available is also an issue with cloud-based systems. Because said systems are continuously running data, it can be overwritten at any time. Time of acquisition has also proved a challenging task concerning cloud forensics.

The growing development of humanoid robotics and human interaction has allowed us to develop applications that can adapt to the diverse demands of modern society. Using non-invasive sensors, social robots allow collecting and using the extracted information to generate a wealth of useful data. Examples of data include specific robot information and artifacts, including call logs, calendar entries, friends and family listings, voice commands, and different modes of operation (these differ depending upon the social robot role). Other data of interest include the OS version and the last security patch installed on the device. Knowing the robot's running time and the amount of time spent on the processing can provide valuable information to forensic cases in terms of confirming certain events and providing timelines. In addition, log files (such as system event logs, diagnostic logs, kernel logs, server debugging logs) can identify actions performed by the robot and what specific time they occurred. This can help draw an accurate timeline of events during a particular period for post-event analysis.

New technologies bring in a new dimension beyond the personal and individual sphere: the possibility of making large-scale attacks at no cost in real-time (Fosch-Villaronga 2021). Cybersecurity for social robots requires a holistic approach that addresses technology, people, skills and processes, and governance. Robots are data-driven technologies, and a cyber-attack may compromise the adequacy of the robot's operation and the users' safety.

4   Case Study: Gamification and Service Robots

Service robots need to be efficient in achieving the task at hand, and at the same time, maintain user attention and engagement on a long-term basis, beyond the novelty period. Many researchers highlight the main challenge to maintaining user engagement beyond the novelty period in service robots (Pu et al. 2018). To address the challenge of lack of long-term interaction, the application of gamification principles can be an option. The gamification principles can be utilized to achieve long-term interaction and user engagement in service robots. Gamification means the "application of game mechanics into non-game context" (Deterding et al. 2011). It has been explored and implemented in various contexts, such as education, marketing campaigns, and health care programs (Baptista and Oliveria 2019). It also seems promising regarding service robots' user' engagement and motivation for adopting specific behaviour for the long term. The gamification principles in service robots can be used as tools for behaviour driving and learning for its user.

Gamification principles include the Mechanics, Dynamics and Emotions (MDE) framework (Robson et al. 2015). The first principle of mechanics is related to defining outcomes and goals in a gamified scenario.

- The first step in determining mechanics is understanding the target audience and the context of a service robot's user's situation. This step aligns with the principles of relevance to the learner's environment theory proposed by Knowles (1996), which is considered an important factor for the adult learning process.
- The second step in mechanics is defining the learning objectives, which will target the expected outcomes in a specific context. The second principle of gamification is dynamics that examine the user's involvement during the gamification process and its impact on their strategies. This principle lets the user actively engage and solve the problem independently for rewards implanted into the process. It will also help drive the knowledge transfer using engagement and motivations (Sogunro 2015). The second principle of dynamically formed steps three and four of gamification design (Huang and Soman 2013).

- The third step involves structuring experience by breaking it down into stages to achieve a specific target for learning. It also ensures that learners achieve the target with a more manageable scope. This step allows a learner to select its path to reach the target or complete the task and eliminate the perception of a bad experience (Marache-Fransciso and Brangire 2013).
- The fourth step includes identifying resources that involve reviewing all stages that can be gamified. The stages have clear rules and feedback. In this step, the user has all components of the scenario, including the goals that need to be achieved. The third principle of gamification Emotions invokes the user's emotional state during the gamification process. It is based not only on the pragmatism but also on the fun factor that the gamification process tries to achieve. It is a product of the first two principles, i.e., how a user utilizes game mechanics and creates dynamics. This principle helps define the final step of the gamification design process (Huang and Soman 2013).
- The fifth step is about applying gamification elements. Different gaming elements can be used in different contexts (Dicheva et al. 2015; Mekler et al. 2017). For example, the game elements in the service robot's context can be points, badges, levels, feedback, the progress of activity, reward points, avatars, and leaderboards.

The MDE framework helps the designer and the user define the gamified experience in service robots. The service robot designers can use gamification principles to select the appropriate mechanics and dynamics to retain control of the user's engagement. The user's emotions will then follow it. For long-term interaction, the principle of emotions is considered very important. In an optimized long-term experience, the user's responses and dynamics during service robots help define the mechanics. Thus, understanding the MDE principles and their relation to each other is important for successfully implementing gamification in service robots (Robson et al. 2015). Several studies reported that gamification results in more user engagement (Looyestyn et al. 2017). The main impact of gamification can motivate the user to use the social robot. It also highlights that goals and context should be considered important during the gamification process for service robots.

However, very little work is reported in the literature about using gamification principles in social robots to increase users' motivation, interaction, and engagement. Donnermann et al. (2021) perform the empirical investigation for social robots and gamification for technology-supported learning. They reported no increase in motivation and engagement in the learning process. Therefore, they are not sure that it may be due to gamification or social robots in learning environments. Some studies reported no effect (Schroeder and Adesope 2014) or no impact of service robot presence in the learning environment context (Li 2015). Another study reported by Fiengold-Polak et al. (2021) use the gamified system for long-term stroke rehabilitation using a socially assistive robot. Their studies showed that the level of acceptance of social robots among patients was high.

One of the main challenges of using gamification in service robots is its poor implementation. The lack of planning and strategy for the gamification process can result in its poor implementation. In service robots, the definition of specific behaviour that encourages users to achieve the target goal is very important. The game mechanic must be designed by keeping the targeted audience in mind. The difference in users' personalities should be considered while designing gamification. The bad process and gamification design will result in poor engagement of the user. The unrealistic expectations can also be the reason for poor engagement in gamified service robots. The importance of gamification in social robots for picking processes can enable users to avoid poor performance symptoms, such as cognitive disengagement, boredom, haste-induced error and fatigue. Thus, more attention and investigation are required from the research community to facilitate the more extended interaction with service robots beyond the novelty period by utilizing the gamification design process.

5   Future Research Works

Recently AI technologies have been applied to robotic and toy computing. Robotic computing is one branch of AI technologies, and their synergistic interactions enable and are enabled by robots. Social robots can now easily capture a user's physical activity state (e.g., walking, standing, running, etc.) and store personalized information (e.g., face, voice, location, activity pattern, etc.) through the camera, microphone, and sensors AI technologies. Social robots comprise a physical humanoid robot component that connects through a network infrastructure to Cloud services that enhance traditional robot functionalities. Humanoid robots often behave like natural partners for social interaction for human users, with features such as speech, gestures, and eye-gaze when referring to users' data and social context. In addition, social robots can interact with humans by performing tasks that adhere to specific social cues and rules. Emotional expressions are one of the most critical components in HRI and the research challenges are clearly explained in the literature. The foundations of this article will set the baseline for understanding how HRI is likely to influence and change our new practices and lifestyle.

To our best knowledge, there is not much focus on a convergent access control model for social robots with emotional expression in a cyber-physical environment in the literature. For example, Cao et al. (2019) present a topology-aware cyber-physical access control (TA-CPAC) model based on the topology configuration and related attributes. The TA-CPAC model can adjust permissions to subjects dynamically with role hierarchy, role mapping, and separation of duty. Shah and Nagaraja (2019) present an information flow model derived from lattice-based access control that includes multiple levels of confidentiality and integrity and the need for compartmentation arising from conflicts of interest. Further, Zhang et al. (2019) present an access control policy in the emotion-aware interactive robot environment in the edge Cloud environment. Gupta et al. (2020) present an Attribute-and Role-Centric Google Cloud Platform Access Control (GCPAC) model based on the dynamic roles approaches that consider User Attributes (UA) to determine the roles of a user. Lastly, Akhusey-inoglu and Joshi (2020) present an attribute-based Cyber-Physical Access Control model (CPAC) and a Generalized Action Generation Model (GAGM) with cyber-physical components and cyber-physical interactions. As the next generation of social robots will become more complex and autonomous, making decisions independently, humans will become less aware of the robots' intent and internal processes (Nesset et al. 2021). Thus, it is essential to develop a robust access control management for social robots in this context (Zhang et al. 2019).

Based on the research works of cyber-security systems by Akhuseyinoglu and Joshi (2020), here are the recommendations for a future convergent access control model for social robots shown below.

1. Inclusion of Context Information: An access control model for social robots should include the social-environmental factors and other types of context information related to users and objects in cyber and physical spaces.
2. Dynamicity: An access control framework should consider the dynamic characters of social robots in cyber and physical spaces.
3. Mixed-Criticality: An access control framework for social robots should simultaneously handle emergency and non-emergency cases and facilitate the transition between them in cyber and physical spaces.
4. Proactivity and Adaptability: An access control mechanism for social robots should handle exceptions that occur in cyber and physical spaces.
5. Strong Coupling: An access control framework for social robots should capture both cyber elements and physical processes and strong coupling or interactions between them.

   The role of emotions in decision-making is another important perspective (Hieida et al. 2018). The social robot's decision-making process is performed computationally during HRI (Unhelkar et al. 2020). We also elaborate on how emotional expressions may affect the decision-making in access control for social robots shown below.
6. Emotional Expression: The decision-making process in access control for social robots should not be influenced by the emotional state of the robot. The decision-making process should be transparent for autonomy for both functional and ethical reasons (Nesset et al. 2021).

References

Apa, L., Cerrudo, C.: Hacking Robots Before Skynet, pp. 1–17. IOActive Inc., Seattle (2017)

Arora, P., Chaspari, T.: Exploring Siamese neural network architectures for pre-serving speaker identity in speech emotion classification, In: Proceedings of the 4th Workshop on Multimodal Analyses Enabling Artificial Agents in Human-Machine Interaction, (MA3HMI 2018) - In conjunction with the 20th ACM International Conference on Multimodal Interaction (ICMI 2018), pp. 15–18. ACM (2018)

Akhuseyinoglu, N.B., Joshi, J.: A constraint and risk-aware approach to attribute-based access control for cyber-physical systems, Comput. Secur. **96**, 101802, 18 (2020)

Bartneck, C., Belpaeme, T., Eyssel, F., Kanda, T., Keijsers, M., Sabanovic, S.: Human-Robot Interaction – An Introduction. Cambridge University Press, Cambridge (2020)

Baptista, G., Oliveira, T.: Gamification and serious games: a literature meta-analysis and integrative model. Comput. Human Behav. **92**, 306–315 (2019)

Calo, R.: Robots and privacy. In: Lin, P., Bekey, G., Abney, K. (eds.) Robot Ethics: The Ethical and Social Implications of Robotics, pp. 187–202. MIT Press, Cambridge (2012)

Cao, Y., Huang, Z., Ke, C., Xie, J., Wang, J.: A topology-aware access control model for collaborative cyber-physical spaces: specification and verification, Comput. Secur. **87**, 101478, 17 (2019)

Chew, E., Chua, X.N.: Robotic Chinese language tutor: personalising progress assessment and feedback or taking over your job? Horizon **28**(3), 113–124 (2020)

Cohen, I., Looije, R., Neerincx, M.A.: Child's recognition of emotions in robot's face and body. In: Proceedings of the 6th ACM/IEEE International Conference on Human-Robot Interaction (HRI 2011), pp. 123–124. ACM (2011)

Deterding, S., Dixon, D., Khaled, R., Nacke, L.: From game design elements to gamefulness: defining "gamification". In: Proceedings of the 15th International Academic MindTrek Conference: Envisioning Future Media Environments (MindTrek 2011), pp. 9–15. ACM

Dicheva, D., Dichev, C., Agre, G., Angelova, G.: Gamification in education: a systematic mapping study. J. Educ. Technol. Soc. **18**(3) (2015). Int. Forum Educ. Technol. Soc. 75–88

Dodson, D., Souppaya, M.P., and Scarfone, K.: Mitigating the risk of software vulnerabilities by adopting a Secure Software Development Framework (SSDF). In: White Paper, the National Institute of Standards and Technology (NIST), USA, p. 27 (2020)

Donnermann, M., et al.: Social Robots and gamification for technology-supported learning: an empirical study on engagement and motivation. Comput. Hum. Behav. Comput. Hum. Behav. **121**(4), 106792, 9 (2021)

Fosch-Villaronga, E., Mahler, T.: Cybersecurity, safety and robots: strengthening the link between cybersecurity and safety in the context of care robots. Comput. Law Secur. Rev. **41**, 05528, 13 (2021)

Gupta, D., Bhatt, S., Gupta, M., Kayode, O., Tosun, A.S.: Access control model for Google Cloud IoT. In: Proceedings of the IEEE 6th International Conference on Big Data Security on Cloud (BigDataSecurity), IEEE Intl Conference on High Performance and Smart Computing, (HPSC) and IEEE International Conference on Intelligent Data and Security (IDS), pp. 198–208. IEEE (2020)

Hieida, C., Horii, T., Nagai, T.: Decision-making in emotion model. In: Proceedings of Companion of the 2018 ACM/IEEE International Conference on Human-Robot Interaction (HRI 2018), pp. 127–128. Association for Computing Machinery, ACM/IEEE (2018)

Herse, S., et al.: Do you trust me, blindly? Factors influencing trust towards a robot recommender system. In: Proceedings of the 27th IEEE International Symposium on Robot and Human Interactive Communication (RO-MAN), pp. 7–14. IEEE (2018)

Homburg, N.M.: Designing HRI experiments with humanoid robots: a multistep approach. In: Proceedings of the 51st Hawaii International Conference on System Sciences, AIS, pp. 4423–4432 (2018)

Huang, W.H.-Y., Soman, D.: Gamification of Education, Research Report Series: Behavioural Economics in Action, Rotman School of Management, University of Toronto, p. 29 (2013)

ISO: Robots and robotic devices—Safety requirements for personal care robots, The International Organization for Standardization (ISO) 13482:2014 (2014)

Kanda, T., Ishiguro, H., Imai, M., Ono, T.: Development and evaluation of inter-active humanoid robots. In: Proceedings of the IEEE, vol. 92, issue number 11, pp. 1839–1850. IEEE (2004)

Kita, S., Mita, A.: Emotion identification method using RGB information of human face. In: Proceedings of SPIE - The International Society for Optical Engineering, vol. 9435, p. 6. The SPIE Digital Library (2015)

Krupp, M., Rueben, M., Grimm, C., Smart, W.: A focus group study of privacy concerns about telepresence robots. In: Proceedings of the 26th IEEE International Symposium on Robot and Human Interactive Communication (RO-MAN), pp. 1451–1458. IEEE (2017)

Kwon, D.S., et al.: Emotion interaction system for a service robot. In: Proceedings of the 16th IEEE International Symposium on Robot and Human Interactive Communication (RO-MAN 2007), pp. 351–356. IEEE (2007)

Knowles, M.S.: Adult learning. In: Craig, R.L. (ed.) The ASTD Training and Development Handbook: A Guide to Human Resource Development, 4th edn. McGraw-Hill, New York (1996)

Latif, S., Khalifa, S., Rana, R., Jurdak, R.: Federated learning for speech emotion recognition applications. In: Proceedings of the 19th ACM/IEEE International Conference on Information Processing in Sensor Networks (IPSN 2020), Article No. 9111050, pp. 341–342. ACM/IEEE (2020)

Lee, E., Kim, G.W., Kim, B.S., Kang, M.A.: A design platform for emotion-aware user interfaces. In: Proceedings of the 2014 workshop on Emotion Representation and Modelling in Human-Computer-Interaction-Systems (ERM4HCI 2014), pp. 19–24. ACM (2014)

Lee, M., Tang, K., Forlizzi, J., Kiesler, S.: Understanding users! Perception of privacy in human-robot interaction. In: Proceedings of the 6th ACM/IEEE International Conference on Human-Robot Interaction (HRI 2021), pp. 181–182. ACM/IEEE (2011)

Li, J.: The benefit of being physically present: a survey of experimental works comparing copresent robots, telepresent robots and virtual agents. Int. J. Hum. Comput. Stud. 77, 23–37 (2015)

Liu, B., He, J., Geng, Y., Huang, L., Li, S.: Toward emotion-aware computing: a loop selection approach based on machine learning for speculative multithreading. IEEE Access 5, 3675–3686 (2017)

Looyestyn, J., Kernot, J., Boshoff, K., Ryan, J., Edney, S., Maher. C: Does gamification increase engagement with online programs? A systematic review. PloS one 12(3), e0173403, 19 (2017)

MacDermott, Á., Carr, J., Shi, Q., Baharon, M.R., Lee, G.M.: Privacy preserving issues in the dynamic Internet of Things (IoT). In: Proceedings of the 2020 International Symposium on Networks, Computers and Communications (ISNCC 2022), pp. 1–6. IEEE (2020)

Marache-Francisco, C., Brangier, E.: Process for gamification: from the decision of gamification to its practical implementation. In: Proceedings of CENTRIC 2013: The Sixth International Conference on Advances in Human-oriented and Personalized Mechanisms, Technologies, and Services, IARIA, pp. 126–131 (2013)

Mathur, M.B., Reichling, D.B.: An uncanny game of trust: Social trustworthiness of robots inferred from subtle anthropomorphic facial cues. In: Proceedings of the 4th ACM/IEEE International Conference on Human-Robot Interaction (HRI 2009), pp. 313–314. ACM/IEEE (2009)

Mekler, E.D., Brühlmann, F., Tuch, A.N., Opwis, K.: Towards understanding the effects of individual gamification elements on intrinsic motivation and performance. Comput. Hum. Behav. 71, 525–534 (2017)

Miller, J., Williams, A.B., Perouli, D.: A case study on the cybersecurity of social robots. In: Proceedings of Companion of the 2018 ACM/IEEE International Conference on Human-Robot Interaction (HRI 2018), pp. 195–196. ACM/IEEE (2018)

Michels, J.D., Walden I.: How Safe is Safe Enough? Improving Cybersecurity in Europe's Critical Infrastructure Under the NIS Directive, Queen Mary School of Law Legal Studies Research Paper, p. 291 (2018)

Mori, M., MacDorman, K.F., Kageki, N.: The uncanny valley. Robot. Autom. Mag. 19(2), 98–100 (2012)

Nesset, B., Robb, D.A., Lopes, J., Hastie, H.: Transparency in HRI: trust and decision making in the face of robot errors. In: Proceedings of Companion of the 2021 ACM/IEEE International Conference on Human-Robot Interaction (HRI 2021 Companion), pp. 313–317. ACM (2021)

Park, C., Ryu, J., Sohn, J., Cho, H.: An emotion expression system for the emotional robot. In: Proceedings of the 2007 IEEE International Symposium on Consumer Electronics, pp. 1–6. IEEE (2007)

Portugal, D., Pereira, S., Couceiro, M.S.: The role of security in human-robot shared environments: a case study in ROS-based surveillance robots. In: Proceedings of the 26th IEEE International Symposium on Robot and Human Interactive Communication (RO-MAN 2017), pp. 981–986 (2017). IEEE

Pu, L., Moyle, W., Jones, C., Todorovic, M.: The effectiveness of social robots for older adults: a systematic review and meta-analysis of randomized controlled studies. Gerontologist 59(1), e37–e51 (2018)

Read, R., Belpaeme, T.: How to use non-linguistic utterances to convey emotion in child-robot interaction. In: Proceedings of the 7th ACM/IEEE International Conference on Human-Robot Interaction (HRI 2012), pp. 219–220 (2012). ACM/IEEE

Robson, K., Plangger, K., Kietzmann, J.H., McCarthy, I.: Is it all a game? Understanding the principles of gamification. Bus. Horiz. **58**(4), 411–420 (2015)

ROS: The Open Robot Operating System (ROS) Project (2021). https://www.ros.org/

Schroeder, N.L., Adesope, O.O.: A systematic review of pedagogical agents' persona, motivation, and cognitive load implications for learners. J. Res. Technol. Educ. **46**(3), 229–251 (2014)

Schütte, S.T.W., Eklund, J., Axelsson, J.R.C., Nagamachi, M.: Concepts, methods and tools in Kansei engineering. Theor. Issues Ergon. Sci. **5**(3), 214–231 (2004)

Shah, R., Nagaraja, S.: An access control model for robot calibration. In: Proceedings of the 24th ACM Symposium on Access Control Models and Technologies (SACMAT 2019), pp. 11 (2019). ACM

Shih, W., Naruse, K., Wu, S.: Implement human-robot interaction via robot's emotion model. In: Proceedings of the IEEE 8th International Conference on Awareness Science and Technology (iCAST), pp. 580–585. IEEE (2017)

Sogunro, O.A.: Motivating factors for adult learners in higher education. Int. J. High. Educ. **4**, 22–37 (2015)

Tielman, M., Neerincx, M., Meyer, J.J., Looije, R.: Adaptive emotional expression in robot-child interaction. In: Proceedings of the ACM/IEEE International Conference on Human-Robot Interaction (HRI 2014), pp. 407–414. ACM/IEEE (2014)

Unhelkar, V.V., Li, S., Shah, J.A.: Decision-making for bidirectional communication in sequential human-robot collaborative tasks. In: Proceedings of the 2020 ACM/IEEE International Conference on Human-Robot Interaction (HRI 2020), pp. 329–341. ACM/IEEE (2020)

Vilches, V.M., et al.: introducing the Robot Security Framework (RSF), a standardized methodology to perform security assessments in robotics. In: Proceedings of the Symposium on Blockchain for Robotic Systems 2018 - MIT Media Lab, p. 20 (2018)

Yaacoub, J.-P., Noura, H.N., Salman, O., Chehab, A.: Robotics cyber security: vulnerabilities, attacks, countermeasures, and recommendations. Int. J. Inf. Secur. 1–44 (2021). https://doi.org/10.1007/s10207-021-00545-8

Yousef, K., AlMajali, A., Hasan, R., Dweik, W., Mohd, B.: Security risk assessment of the PeopleBot mobile robot research platform. In: Proceedings of the International Conference on Electrical and Computing Technologies and Applications (ICECTA 2017), p. 5. IEEE (2017)

Yousef, K., AlMajali, A., Ghalyon, S.A., Dweik, W., Mohd, B.J.: Analyzing cyber-physical threats on robotic platforms. Sensors, Sensors **18**(5), 1643, 22 (2018)

Zhang, Y., Qian, Y., Wu, D., Hossain, M.S., Ghoneim, A., Chen, M.: Emotion-aware multimedia systems security. IEEE Trans. Multimedia **21**(3), 617–624 (2019)

Author Index

Printed in the United States
by Baker & Taylor Publisher Services